THE DINOSAUR HERESIES

THE DINOSAUR HERESIES

New Theories Unlocking The Mystery of the Dinosaurs and Their Extinction

ROBERT T. BAKKER, Ph.D.

ZEBRA BOOKS
KENSINGTON PUBLISHING CORP.

ZEBRA BOOKS

are published by

Kensington Publishing Corp.
475 Park Avenue South
New York, NY 10016

First Zebra Books printing: October, 1988

Printed in the United States of America

To a dear friend, Professor Bernhard Kummel of Harvard University. Bernie grabbed me by the lapels back in 1974 and said, "Kid, you can't go on being an *enfant terrible* forever. You gotta write a book." So Bernie, here's your book.

PREFACE

I t all started very suddenly, in the spring of 1955. I was reading magazines in my grandfather's house in New Jersey, and I found that magical *Life* cover story—"Dinosaurs." Fold-out, full-color pictures of heroic creatures. *Allosaurus, Brontosaurus, Stegosaurus, Tyrannosaurus rex.* I discovered an entire world, far, far away in time, that I could visit, whenever I wanted, via the creative labors of the paleontologists. And I made up my mind then and there that I would devote my life to the dinosaurs. Since I was in the fourth grade, my parents weren't alarmed at my vow. Surely, they thought, it's just a phase that he'll grow out of. Lots of kids my age got hooked on dinosaurs for a while—it was a childhood disease, like mumps or chicken pox, and if left alone, most kids recovered and then had a lifetime immunity to dinosaurmania. But I was that rare exception, a terminal, chronic case. And my mother was patient enough to take me twice a year over the George Washington Bridge to the American Museum of Natural History in New York, where the best dinosaur show in the world played every day of the week on the fourth floor. My family valued scholarship, even if they couldn't quite understand the reverence I had for the study of fossils.

I owe a great deal to a few fine friends at Harvard. Bernie Kummel always encouraged me, even though we seemed to represent opposite extremes of college society—he a member of the Old Guard, I one of the sixties radicals. But we both loved fossils. Bryan Patterson taught me about rodents and giant ground sloths and elephants, and, most especially, how wonderfully complex the fossil history of life was. Steve Gould was always stimulating, and challenging, and fiercely protective when occasion demanded. I would not have survived Harvard without these three.

I must tip my well-worn cowboy hat to Ms. Kate Francis, of the Johns Hopkins University. Kate was a loyal friend, invaluable critic of my prose, and superb manuscript manipulator all through the first three drafts of this book. Maxine Mote was a soul mate at Hopkins, too, and helped with some key chapters. Many a time I sat for hours in the hallway at Hopkins discussing dinosaurs and evolution with my old friend from Yale, Steve Stanley, now a professor of paleontology at Hopkins. Steve is a clam-paleontologist at heart, but his mind roves far afield, wherever the fossil record of life leads to neat discoveries about how evolution works. Thank God for the WATS line—we can still have these long rambling talks long distance.

And a fond thank-you to all the Hopkins pre-meds who helped to dig at Como—Jan Koppelman, Robert Beck, Conrad Foley, Sue Reiss, and especially Julius Goepp. They're all doctors now, or almost.

To my editor, Maria Guarnaschelli, I owe an enormous debt for her patience, encouragement, and extraordinary creative energy. She is passionate about making good books, and she succeeds.

Constance Areson Clark loves dogs, old books, and the Badlands as much as I do, and most of my ideas about how evolution works have been explored during our walks in the rain in Baltimore or lingering breakfasts in Greybull, Wyoming. Constance, Wyoming, and I were destined to come together, and stay together.

And, finally, I must acknowledge my debt to hundreds of people I have never met. The fieldmen who dug dinosaurs in the 1880's. The skilled preparators who chiseled bones out of the rock in countless basement laboratories. The exhibit craftsmen who bent the ironwork to mount the skeletons. All the people who have kept the great museums going for the last century. I love museums more than any other institution the human race has invented. Museum people are always overworked and underpaid, and they all deserve sainthood, every one.

CONTENTS

PART 1

THE CONQUERING COLD-BLOODS: A CONUNDRUM

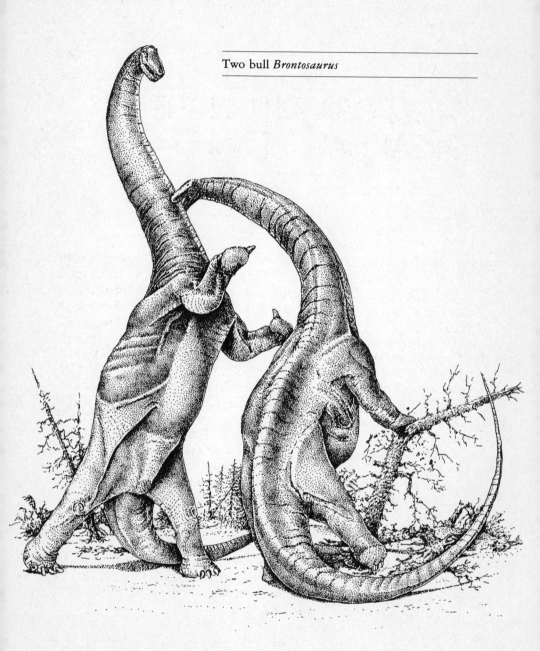

Two bull *Brontosaurus*

1

BRONTOSAURUS IN THE GREAT HALL AT YALE

I remember the first time the thought struck me! "There's something very wrong with our dinosaurs." I was standing in the great Hall of Yale's Peabody Museum, at the foot of the *Brontosaurus* skeleton. It was 3:00 A.M., the hall was dark, no one else was in the building. "There's something very wrong with our dinosaurs." The entire Great Hall seemed to say that. I had grown up with the dinosaurian orthodoxy about dinosaur ways—how they were swamp-bound monsters of sluggish disposition, plodding with somnolent strides through the sodden terrain of the Mesozoic Era. Dimwitted and unresponsive to change, the dinosaurs had ruled by bulk. Bizarre and exotic shapes ornamented their heads and bodies like the decadent opulence of a Byzantine palace. Books and museum labels solemnly preached the same message—the dinosaurs were failures in the evolutionary test of time. Stories of their mode of life were replete with negatives: *Brontosaurus* couldn't walk on land because its body was too heavy. *Diplodocus* couldn't feed on anything but soft water plants because its head was too small. Duckbill dinosaurs couldn't run quickly because their limb joints were too imperfect. *Pteranodon* couldn't flap its wings because they were too weak. Dinosaurs couldn't be warm-blooded because their brains were too small. And the final, ultimate failure of their character—dinosaurs couldn't cope with competition from the smaller, smarter, livelier mammals.

All of these "couldn'ts" together built up the orthodox view of dinosaurs as a dynasty of flawed creatures. And it was this orthodoxy that suddenly seemed so wrong as I stood looking at the *Brontosaurus* in the Great Hall. The public image of dinosaurs is tainted by extinction. It's hard to accept dinosaurs as a success when they are all dead. But the fact of ultimate extinction should not make us overlook the absolutely unsurpassed role dinosaurs played in the history of life.

Creatures with four legs first crawled slowly out of the ancient swamps 400 million years ago. Dinosaurs were not in this first evolutionary wave, nor in the second or third. Contrary to the cartoonists' view of geological history, dinosaurs aren't the most ancient of life forms, not even close. Dinosaurs as a clearly defined group don't make their grand entrance until 200 million years after the first four-legged beasts emerged from the primordial swamps. By the time they appear in the land ecosystem, the woodlands and waterways were already full of creatures, large and small, flesh-eater and vegetarian. For a brief twilight zone—five million years, short by geological standards—these earliest dinosaurs shared the terrestrial realm with a host of older clans. But then the dinosaurs seized power. They took over all the large roles in the land ecosystem. They filled the offices of mega-predator and mega-herbivore. Their control of the land ecosystem was complete. No nondinosaur larger than a modern turkey walked the land during the Age of Dinosaurs.

If we measured success by longevity, then dinosaurs must rank as the number one success story in the history of land life. Not only did dinosaurs exercise an airtight monopoly as large land animals, they kept their commanding position for an extraordinary span of time—130 million years. Our own human species is no more than a hundred thousand years old. And our own zoological class, the Mammalia, the clan of warm-blooded furry creatures, has ruled the land ecosystem for only seventy million years. True, the dinosaurs are extinct, but we ought to be careful in judging them inferior to our own kind. Who can say that the human system will last another thousand years, let alone a hundred million? Who can predict that our Class Mammalia will rule for another hundred thousand millennia?

If we measure success of a zoological dynasty by the defense

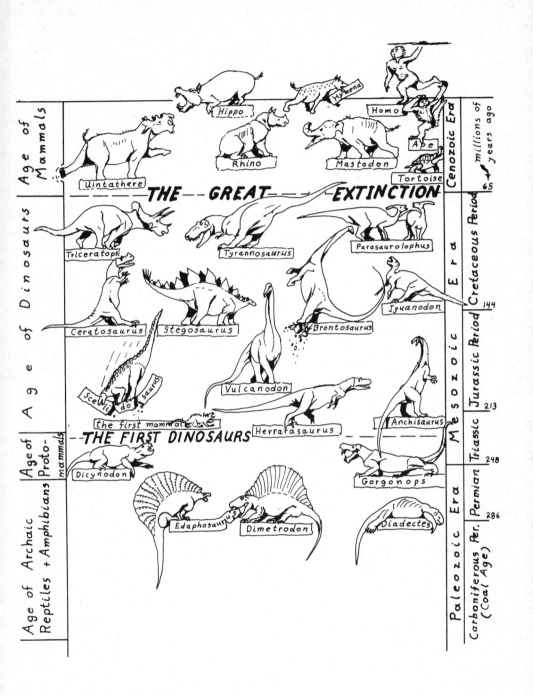

THE — GREAT — EXTINCTION

THE FIRST DINOSAURS

Age of Mammals

Age of Dinosaurs

Age of Archaic Reptiles + Amphibians

Proto-mammals

Age of

A g e   o f   D i n o s a u r s

Cenozoic Era

M e s o z o i c   E r a

Cretaceous Period

Jurassic Period

Triassic Period

Permian Per.

Carboniferous Per. (Coal Age)

P a l e o z o i c   E r a

millions of years ago

65

144

213

248

286

Uintathere

Rhino

Hippo

Hyaena

Homo

Ape

Mastodon

Tortoise

Triceratops

Tyrannosaurus

Parasaurolophus

Ceratosaurus

Stegosaurus

Brontosaurus

Iguanodon

Scelidosaurus

Vulcanodon

the first mammal

Herrarasaurus

Anchisaurus

Dicynodon

Gorgonops

Edaphosaurus

Dimetrodon

Diadectes

of its borders, then the Dinosauria must rank as the most robust of ruling clans. Dinosaurs were not unopposed in the world-game of competition and predation. As the early Dinosauria spread their species into every role open to large land creatures, the dinosaurs were driving out the last remnants of very advanced and very specialized clans, zoological tribes which had been evolving and perfecting their adaptive equipment for tens of millions of years. Like the Mongol hordes sweeping across the old cities of eastern Europe, dinosaurs wasted little time in expelling these well-established kingdoms. And during their long reign, the dinosaurs faced potential threats from dozens of new clans that evolved even higher grades of teeth and claws, bodies and brains. Despite the evolutionary vigor of the potential opposition, dinosaurs kept their ecological frontiers intact; no other clan succeeded in evolving to a large size as long as the Dinosauria existed.

Humans are proud of themselves. The guiding principle of the modern age is "Man is the measure of all things." And our bodies have excited physiologists and philosophers to a profound awe of the basic mammalian design. But the history of the dinosaurs should teach us some humility. The basic equipment of our mammal class—warm bodies clothed in fur, milk-producing breasts to nourish our young—is quite ancient. These mammalian hallmarks are as old as the dinosaurs themselves. Indeed, the Class Mammalia emerged, fully defined, in the world ecosystems just as the Dinosauria began their spectacular expansion. If our fundamental mammalian mode of adaptation was superior to the dinosaurs', then history should record the meteoric rise of the mammals and the eclipse of the dinosaurs. Our own Class Mammalia did not seize the dominant position in life on land. Instead, the mammal clan was but one of many separate evolutionary families that succeeded as species only by taking refuge in small body size during the Age of Dinosaurs. As long as there were dinosaurs, a full 130 million years, remember, the warm-blooded league of furry mammals produced no species bigger than a cat. When the first dinosaur quarry was opened in 1822 at Stonesfield, England, quarry men found the one-ton *Megalosaurus* and a tiny mammal.

So the popular image of dinosaurs as unprogressive behemoths is wrong. Political cartoonists use *Brontosaurus* as the ultimate symbol of ignorant lethargy and obsolete organization. In fact,

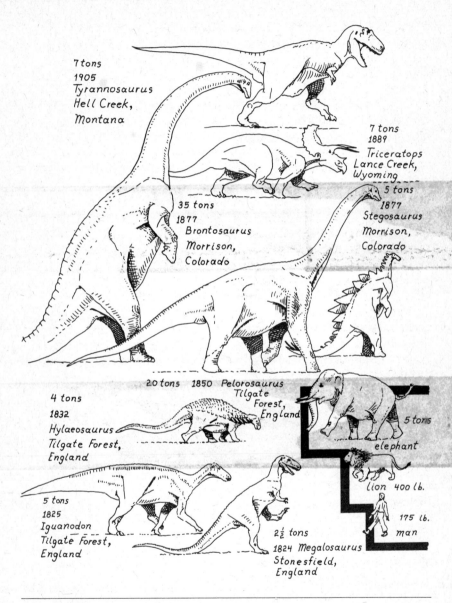

7 tons
1905
Tyrannosaurus
Hell Creek,
Montana

7 tons
1889
Triceratops
Lance Creek,
Wyoming

35 tons
1877
Brontosaurus
Morrison,
Colorado

5 tons
1877
Stegosaurus
Morrison,
Colorado

20 tons 1850 Pelorosaurus
Tilgate
Forest,
England

4 tons
1832
Hylaeosaurus
Tilgate Forest,
England

5 tons
elephant

lion 400 lb.

5 tons
1825
Iguanodon
Tilgate Forest,
England

2½ tons
1824 Megalosaurus
Stonesfield,
England

175 lb.
man

The nineteenth century discovers the age of Dinosaurian Giants. Scholars probing the fossil relics dug up in gravel quarries and clay pits early in the 1800s were astounded by the unprecedented size of dinosaurs. By 1889, samples from all the major dinosaur clans had been found.

dinosaurs evolved quickly, changed repeatedly, and turned out wave after wave of new species with new adaptations all through their long reign. Sir Richard Owen, best and brightest of Victorian anatomists, coined the name "Dinosauria," from Greek roots meaning "terrible lizard," in 1842. When Owen first penned the word "dinosaurs," paleontology was still a brand-new science. Baron Cuvier had invented the scholarly art of reconstructing form and function in fossil creatures only forty years earlier. Though careful study of the earth's crusts had gone on for only one human generation, the naturalists of 1840 already knew that an Age of Reptiles had preceded our own Age of Mammals. And the many skeletons already dug up showed that this Age of Reptiles was a time when fishlike reptilian forms swam in the seas and batlike reptilian species flew through the air. Owen invented the term "Dinosauria" to describe the huge *land* animals of this age. And his original definition is still good.

When the first dinosaur skeletons were hewn out of gravel quarries in England during the 1820s and '30s, the gentleman-naturalist immediately recognized their strange combination of characteristics. These great fossils combined traits found in lizards, in birds, in mammals, and in crocodiles as well. Owen was especially impressed by the advanced, birdlike shape of dinosaurs' hip bones, and he used their characteristics to set dinosaurs apart from all other animals with backbones. And so can we. A very good anatomical definition for the Dinosauria is "a vertebrate group close to crocodiles but with at least some important birdlike features in the hind leg."

Sir Richard Owen's astute observations are too often ignored. Twentieth-century paleontologists have fallen into the bad habit of reconstructing the dinosaurs' life functions by using crocodiles as a living model. But the earliest researchers of the nineteenth century proved beyond a doubt that the dinosaurs' powerful hind legs must have operated like the limbs of gigantic birds. And further birdlike characteristics turned up in the dinosaurs' backbone. Many species of dinosaur had hollow chambers in their vertebrae. In life, these bony caverns were filled with air sacs connecting to the lung, just as in many birds today. Later nineteenth-century discoveries made the dinosaur-bird connection very intimate. *Archaeopteryx,* the oldest fossil bird, was discovered in 1861 and made headlines because it looked so much like a small dinosaur with

One of the first dinosaurs discovered—the thirty-foot predator *Megalosaurus*, dug up in the 1820s in England. Here the great megalosaur is attacking a sea crocodile, *Teleosaurus*.

In the 1860s Thomas Henry Huxley argued that birds descended from
dinosaurs, and that view is now being revived—the six-feet-long predator
Deinonychus, shown here attacking an ostrich dinosaur, is a near-perfect
missing link between dinosaurs and modern birds.

feathers. The great Darwinian orator and advocate Thomas Henry
Huxley pounded the pulpit of evolutionary theory by pointing to
Archaeopteryx as the missing link between dinosaurs and modern
birds.

It's important to be clear about the reverse definition as well:
what dinosaurs are not. Dinosaurs are not lizards, and vice versa.
Lizards are scaly reptiles of an ancient bloodline. The oldest liz-

ards antedate the earliest dinosaurs by a full thirty million years. A few large lizards, such as the man-eating Komodo dragon, have been called "relicts of the dinosaur age," but this phrase is historically incorrect. No lizard ever evolved the birdlike characteristics peculiar to each and every dinosaur. A big lizard never resembled a small dinosaur except for a few inconsequential details of the teeth. Lizards never walk with the erect, long-striding gait that distinguishes the dinosaurlike ground birds today or the birdlike dinosaurs of the Mesozoic.

Snakes are lizard nieces—descendants of a close relative of lizards. Some lizards have lost their limbs and slither like snakes, but true snakes have specialized eyes and jaws. Snakes, of course, are not at all close to dinosaurs.

Crocodiles and their next of kin, alligators, are unquestionably dinosaur uncles, relatives of dinosaur ancestors. Baron Cuvier, Sir Richard Owen, and other early dinosaurologists discerned many important anatomical characteristics shared by dinosaurs and crocodiles. For example, dinosaur teeth are set in sockets—so are the teeth of crocodiles—whereas lizard and snake teeth are fused to the inside of the jawbone without sockets. Dinosaurs have a deep socket in the hip bones for the thigh, and so do crocodiles, but lizards do not. Crocodiles even show the beginnings of birdlike development in hip and thigh. Crocodiles first enter the chronicle of the rocks long after lizards but a few million years before dinosaurs.

Frogs and their short-legged relatives the salamanders are amphibians, not reptiles. Amphibians lay water-breathing eggs, and usually the newly hatched young breathe via gills for a while before becoming air-breathers. Like the reptiles, the amphibians have "cold blood." ("Cold blood" means that metabolism is so low that body temperature falls to air-ground temperature unless the animal can heat up by basking in the sun.) Amphibians have only a very distant kinship to dinosaurs.

Turtles are marvelous organic creations and very worthy objects for contemplation, but turtles aren't dinosaurs. Turtles have a scaly skin and a leathery or porcelaneous egg, points of resemblance to both lizards and crocodiles. But the body architecture of the turtle is so thoroughly unique that after nearly two centuries of research, turtle relationships are murky at best.

Are dinosaurs true members of the reptile class? Good ques-

Bird, crocodile, and mammal adaptations combined—the three-ton herbivorous dinosaur *Iguanodon*, first found in the 1820s in England. The spike of bone on the thumb must have been a dangerous weapon.

tion. Hard to answer—that's what this book is all about. The late nineteenth-century naturalists defined Reptilia by blood, skin, and sex. If an animal had "cold blood," skin covered with scales, and laid eggs on land, then it was a true reptile. Despite the obvious similarities of design between crocodiles and birds, therefore, the scaly, naked hide of crocodiles and their "cold blood" have persuaded most naturalists to separate them from the birds. Birds have their own class, the Aves. But crocodiles are left in the Reptilia with their more distant relatives, lizards, snakes, and turtles.

Birds and mammals differ from each other in extraordinarily numerous ways, in nearly all details of their joints, muscles, and other organs. But birds and mammals do share two key adaptations which color their entire evolutionary style: both have insulation for the skin (feathers for birds, hair for mammals) and both are "warm-blooded" (they have such a high metabolic rate that their bodies are generally heated from the inside). Mammals have their own zoological class. Although the "warm-bloodedness" of birds and mammals is very similar in physiological detail, it is quite clear that the "warm-blooded" condition evolved separately, once in birds, once in mammals.

Now, nineteenth-century science was self-consciously preoccupied with "progress." The Industrial Revolution had wrought such rapid advancement in machines, small and great, that mid-Victorian scientists could see no end to the upward perfection of technology. And Darwinism, in its vulgar "survival of the fittest" version, seemed to preach that there was a natural law guiding the continuous perfection of life forms through all geological time. Which was most perfect? *Homo sapiens,* of course—especially a male, English, Protestant *Homo sapiens.* And so our class, the Mammalia, had to be the highest zoological grouping. Birds were close because, like mammals, they had insulation and metabolic control of their body temperature.

Progress also meant freeing oneself from the uncomfortable whims of the environment—the sudden changes in heat and light. The poor reptile could bask happily on a rock in the sun but slipped back into a chilled torpor when clouds blotted out the warming rays. Not so the bird or mammal whose body furnaces burned so fast and so continually that blood and flesh remained warm. And Victorian biochemistry had progressed far enough to discover that most vital processes function best when the body temperature is nearly constant. English homes—upper-class ones, at least—were enjoying the dependable warmth of coal-fed furnaces, devices that finally made the damp winter climate cozy and comfortable. Clearly the highest zoological classes were the ones that had evolved an analogous metabolic adaptation.

The zoologists of the last century knew well that there was a case for a crocodile–bird relationship and an even better case for a dinosaur–bird kinship. But the scientists of the time nonetheless slipped into the habit of calling dinosaurs "reptiles."—cold-blooded, scaly creatures that laid eggs.

Nineteenth-century naturalists used their warm blood/cold blood dichotomy to classify all vertebrates into two grand divi-

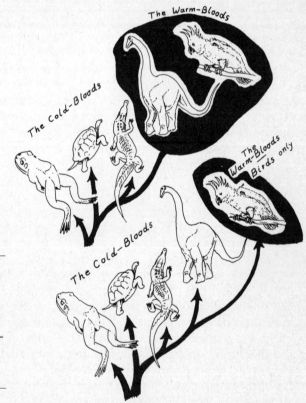

Two ways dinosaurs could be classified—as *cold-bloods* because dinosaurs share adaptations with crocodiles, or as *warm-bloods* because dinosaurs have many birdlike features.

sions, one above the other. At the bottom were the "lower vertebrates," the classes without metabolic control of their body heat. Here were lumped all the fishes, the Amphibia, and of course the Reptilia. At the top were the "higher vertebrates," the two classes—Aves and Mammalia. Dinosaurs were thrown into the Reptilia and so into the "lower vertebrates" by early naturalists, but an equally good case could have been made to classify dinosaurs as primitive birds. No one, either in the nineteenth century or the twentieth, has ever built a persuasive case proving that dinosaurs as a whole were more like reptilian crocodiles than warm-blooded birds. No one has done this because it can't be done.

Generally speaking badges are harmful in science. If a scientist pins one labeled "Reptile" on some extinct species, anyone who sees it will automatically think, "Reptile, hmmm . . . that means cold-blooded, a lower vertebrate, sluggish when the weather is dark and cool." There are never enough naturalists around, in any age; so most scientific orthodoxy goes unchallenged. There are just not enough skeptical minds to stare at each badge and ask the embarrassing question, "How do you know the label is right?"

Be kind to colleagues, ruthless with theories, is a good rule. A scientific theory isn't merely idle speculation, it's a verbal picture of how things might work, how a system in nature might organize things—atoms and molecules, species and ecosystems. But old paleontological theories too often aren't treated roughly enough. Old theories—like the reptilian nature of dinosaurs—are accepted like old friends of the family. You don't yell at old Aunt Cecilia. So hundred-year-old dinosaur theories live on without being questioned, and too often they are assumed to be totally correct. Even when such theory is caught in an error, it's likely to be excused.

Traditional dinosaur theory is full of short circuits. Like the antiquated wiring in an overaged house, the details sputter and burn out when specific parts are tested. I have enormous respect for dinosaur paleontologists past and present. But on average, for the last fifty years, the field hasn't tested dinosaur orthodoxy severely enough.

I'd be disappointed if this book didn't make some people angry. A lot of modern scientists—even some paleontologists—insist on saying that fossils are misleading. "Dead bones don't metabolize so how can physiology in dinosaurs be discussed?" Ecologists who study the Serengeti Plain or the rain forests of Burma are impressed by the complex ways animal species interact with each other and with their habitats: "How can a few spare bones capture all the organic subtlety of long-extinct systems?" Many people dismiss the record of the rocks as an incomplete and nearly unreadable document. Darwin himself did that; he didn't trust fossils to indicate the entire truth. But these views are wrongheaded. The Book of Job—oldest in the Bible—admonishes, "Speak to the Earth and it will teach thee." If we look and listen carefully, the record of the rocks can unlock the richly textured story of the dinosaurs and their ways.

The Stonesfield specter. From the very first discoveries of dinosaurs in the 1820s there was proof positive that our mammalian order had existed under the shadow of the gigantic dinosaurian monsters. Earliest of the dinosaur quarries was the road-gravel pit in the Stonesfield Slate, where giant jaws of *Megalosaurus* could be found with teeth so large that a single tooth was longer than the entire jaw of the mammals found in the same strata. Shown here, natural size, is *Phascolotherium* standing next to the lower jaw of *Megalosaurus*.

2
WYOMING REVERIE: MEDITATION ON THE GEOLOGICAL TEXT

From my Field Book, 1981

July 3, 6:35 A.M.

Como Bluff, Wyoming. 7,020 feet above sea level. No human being or human structure visible. Air clear, dry, cool. A pair of mule deer browsing along Rock Creek. Put the coffee water on the Coleman stove to heat up. No one else is awake in camp yet, but the smell of bacon will entice them out of their tents.

I have been in the business for twenty years—digging up fossil bones—but I'm still excited by the first dinosaur of the summer. I sit here on the crest of a little sandstone hogback, remnant of a stream that flowed a hundred million years ago, and look down on my crew's work of the last four days. It's becoming a sizeable hole, a proper dinosaur dig, twenty-five feet across, dug by pickax, army-surplus trenching shovel, icepick, and fingernails.

I saw my first dinosaur in that splendid Mecca for Mesozoic relics, the American Museum of Natural History in New York, at the age of nine. But those skeletons seemed tamed by civilization, mounted as they were on steel and plaster, posed for the benefit of countless parades of schoolchildren and tourists. A dinosaur in the rock is different. This one before me is huge, and its six-foot-long thigh bone, which would dwarf any elephant's, lies half exposed to the Wyoming sunrise. Its coal black form is clearly etched

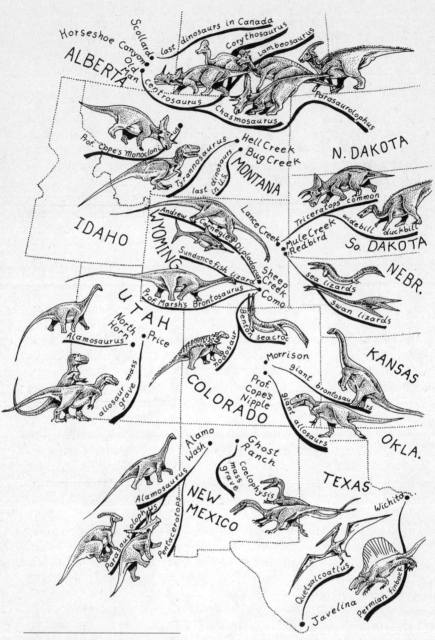

Horseshoe Canyon
Scollard
Old Man
last dinosaurs in Canada
Corythosaurus
Lambeosaurus
ALBERTA
Centrosaurus
Chasmosaurus
Parasaurolophus

Prof. Cope's monocloni...
Hell Creek
Bug Creek
N. DAKOTA
Tyrannosaurus
last dinosaurs in U.S.
MONTANA
Triceratops common
widebill
duckbill
IDAHO
Andrew
Carnegie's Diplodocus
Lance Creek
Mule Creek
Redbird
So. DAKOTA
WYOMING
Sundance fish lizard
Sheep Creek
Como
sea lizards
NEBR.
Prof. Marsh's Brontosaurus
Benton seacroc
swan lizards
UTAH
North Horn
Price
Morrison
giant brontosaurs
KANSAS
Alamosaurus
nodosaur
Prof. Cope's Nipple
COLORADO
giant allosaurs
allosaur mass grave
OKLA.
Alamo Wash
Ghost Ranch
Coelophysis mass grave
TEXAS
Alamosaurus
NEW MEXICO
Wichita
Parasaurolophus
Pentaceratops
Quetzalcoatlus
Permian finback
Javelina

The great dinosaur graveyards
of the American West

against the surrounding pale rock by thousands of careful chisel marks. This bone is a holy relic for me, as beautiful in its roughly hewn outline as Michelangelo's bound slaves struggling to free themselves from the enveloping marble.

From where I sit on the quarry's rim I can see the dinosaur's great trochanters, the attachment site of the immense hip muscles, and the bone surface pitted and rough where tendons and ligaments were anchored to the femur. A hundred thousand millennia ago, those tendons and muscles were full of dinosaur blood coursing through capillary beds, bringing oxygen to the cells that powered the stride of this ten-ton giant. Muscles pulsed in cycles of contraction and release, and the hind limb, fully twelve feet long from hip to toenails, swung through its stroke covering six feet with every pace.

Broken chips of bone lie under my boots, wretched fragments from now unidentifiable bones which had eroded long before we found the site. Even though I know I can't identify the bits of bone, I pick one up anyway because there is something special about the feel of dinosaur bone very early in the morning. Some of the broken bits are incredibly delicate bubbles of bone, a frothy texture of holes and vesicles that housed the living substance of the animal's cells. These bits crumble into shards if I rub them too hard, but in life the brittle bone crystals were embedded in a fabric of tough connective tissue, collagen fibers whose great tensile strength combined with the hardness of the bone crystals to produce a living bony architecture capable of resisting enormous loads of both compression and tension. Collagen has long since rotted away, along with all the muscle fibers and blood vessels. But the fossilized bone faithfully preserves the canals left by every capillary that made its passage through it to serve the dinosaur in life. Those living cells, now gone, left one other signature on this carcass. A black powder rubs into my gloves as I finger the bone chips. This powder is carbon dust mixed with granulated bone, the dried and distilled residue of all the cell membranes, cell fluids, and organelles whose work within the bone was ended when the dinosaur died.

Reverie is normal in Wyoming at sunrise. I suppose a nononsense laboratory scientist, clad in his white lab coat and steely-eyed objectivity, might think I was wasting my time communing

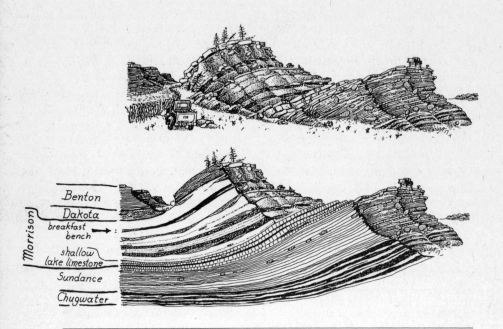

Benton
Dakota
breakfast
bench →
shallow
lake limestone
Sundance
Chugwater

Morrison

Our camp at Como Bluff and how the rock layers would look if cut through vertically

with the spirit of the fossil beast. But scientists need reverie. We need long walks and quiet times at the quarry to let the whole pattern of fossil history sink into our consciousness.

As I walk back to camp from the quarry, I climb through the ledges of hard rock, benches of limestone, each an irregular mosaic of ovoid nodules, each extremely hard and long-lasting in this dry climate and each a timekeeper and recorder of past climate. Taken altogether, this irregular staircase of rock is a chronicle of the dinosaurs' success throughout a great age in the history of life. The nodules grew from tiny mineral seeds in the well-drained soil of the floodplain where dinosaurs browsed the leaves of conifers, and birds with teeth glided from one tree crown to the other. In the rainy season, floods covered the landscape with chocolate-colored water full of mud and grit so that each flood added yet another layer of sediment to the gradually accruing stratigraphic pile. The seasonal flux of the water table—up near the surface during the rainy season, down during the dry season—stained the layers of sediment in blotches of green and mauve, blotches known to soil scientists by the Welsh word "gley."

Here and there the low places of the floodplain filled with black, stagnant water and putrefying leaves and branches, leaving a record of dark, carbon-rich mudstone and shale. Torrential spring rains cut channels into the plain and filled them with gravel, cross-bedded sand and mud. For five million years the floodplain here served as the arena for all these environmental agents, each performing its function in shaping the quantities of soil and leaving its own unique imprint on the sediments. And everywhere, in every habitat, there were dinosaurs—huge multi-ton brontosaurs swinging their long necks from treetop to treetop; predatory allosaurs, running on their enlarged hind legs, like some nightmarish bird; armored stegosaurs, ornamented with bony triangles along their back, brandishing a formidable set of spikes at the end of their muscular tail; little theropods, some no bulkier than a turkey, darting through the meadows and gallery forests along the stream courses, catching small prey.

The record of the rocks speaks eloquently here, without hesitation or ambiguity—this was an Age of Dinosaurs, a time when all the large ecological roles on the terrestrial stage were played by dinosaurs of one family or another. The domination of the dinosaurs extended across all the categories of large flesh-eater, large leaf-eater, and large omnivore. And dinosaurs spread their ecological hegemony across a worldwide empire, devoid of geographical limits. Dinosaurs are the unchallenged majority in all the fossil samples of large vertebrates on every continent from Australia to Siberia, New Jersey to Calcutta during this time. Dinosaurs like these lying in my Wyoming pit are being excavated by Iberian paleontologists in Portugal, by Chinese geologists in Yunnan Province, and by Zimbabwan naturalists along the banks of the Zambesi River in East Africa. No corner of the Mesozoic world withstood colonization by dinosaurs.

How much grander in scope the dinosaurs' history is than the cartoonists' view of prehistory, which consigns all extinct creatures to one Antediluvian Age. The day of the dinosaur was not merely one geological instant, played out by a single cast of species. Neither was it one dynasty of evolving dinosaur species. The dinosaurs' history was an extraordinary series of dynasties, one age followed by another and another, each filled with a complete cast of dinosaurs, and the entire dynastic series running through 130 million years.

No single spot on earth preserves this history in its entirety. But southeastern Wyoming comes close. When I walk north from the quarry to visit the staff at Rock Creek Fish Hatchery, I pass through the first half of the history of the dinosaurs. It is recorded in a thousand-foot-thick layer cake of sandstone, shale, and lime. In the rock strata near the fish hatchery's holding ponds, the dinosaurs make their debut. The sedimentary record here is a bizarre sandwich of thin-bedded maroon, pink, and brick red sandstone and mudstone, a formation that enjoys the delightful label Chugwater, named after a tiny stream where it was first discovered. The gaudily colored beds began as saline lakes, like those of Death Valley today, fetid bodies of soda-choked water too salty for fish to survive. Meandering rivers spread layers of sand on top of the mineral-rich muds accumulated on the lifeless lake bottoms. Dinosaurs were there. Small hunters from chicken to ostrich size prowled along the stream edges, hunting for their prey, leaving their unmistakable three-toed footprints and, very rarely, leaving behind their bony carcasses to be buried by Chugwater sand. In those days, the dinosaurs' empire was in its infancy. The evolutionary pioneers of the Dinosauria had to share this terrestrial realm with a host of short-legged and ugly reptiles, the beaked rhynchosaurs, the dog-faced cynodonts, and the dinosaurs' own ancestral stock, the big-headed thecodonts. This was the Triassic Period, the first of the three great Mesozoic ages.

The next was the Jurassic, the Golden Age of Giants, when the dinosaur clans burst out of their Triassic limitations. Wave after wave of ever-larger species filled the land habitats—long-necked brontosaurs, grotesquely armored stegosaurs, and a complete array of bird-limbed predators from ten pounds to five tons in live weight.

At Como Bluff, the land was covered with warm ocean water, the Sundance Sea, alive with stout-shelled squid and porpoiselike fish lizards for most of the late Jurassic's twenty million years. Above the Chugwater layers, the somber green-gray sandstones of the Sundance Formation enter this marine epoch in the stratigraphic

Sea croc of the Jurassic—the fish-tailed *Metriorhynchus,* about ten feet long

chronicle. Butch Cassidy's sidekick, the Sundance Kid, got his name from the same tiny Wyoming town that gave this Jurassic formation its formal geological nomenclature.

The waves of the Sundance Sea finally beat their last cadence 140 million years ago. But even now the rhythmic sea sound seems fresh, preserved in the ripple-marked sandstone surface and in the wave-winnowed piles of clam shells and fossil oysters, growing on top of each other, and the squid-pens known to the locals as "stone sea-gars." Dinosaurs were not here. Their fossils rarely appear in ocean beds. But rocks the same age as Sundance in India and Australia provide skeletons, proof that the land ecosystem was firmly under the control of the dinosaurs all through Jurassic times.

Sundance fossils and all the other rich Mesozoic marine beds underscore the one geographic limitation of the dinosaurs' world. Their empire was firmly landlocked. Unchallenged though they were on land, dinosaurs rarely went to sea, and so seemed to suffer that abhorrence of salt water which has limited many a human empire from Alexander to Napoleon. Reptilian leviathans are found in the Sundance outcrop—fish lizards, sea crocodiles, and the serpent-necked plesiosaurs. But these sea monsters are all from groups only distantly related to the Dinosauria proper.

If we proceed from the Fish Hatchery back to camp, climbing through the last Sundance sandstone ledge, the rock changes color from green to blotches of red and maroon, signalling the shift in the ancient habitats from shallow tropical sea to floodplain and river. This next layer of rock is the most famous dinosaur graveyard in the world: the Morrison Formation, named for a tiny Colorado town south of Boulder. It was the outcrops of the Morrison Formation here at Como Bluff that made *Brontosaurus* a household word in the 1880s. Union Pacific Railway station managers found huge bones along their right-of-way and cabled this news to Othniel Charles Marsh, stuffy but sagacious Yale paleontologist. Marsh hired the railway men to excavate the bones, crate them, and send them by boxcar to New Haven. News of the spectacularly complete Como dinosaurs galvanized the international community of scholars, who had been frustrated by the poor fragments of Jurassic dinosaurs available from French and English quarries.

American geology had been viewed as a scholarly backwater by most European scholars, whose tradition of analytical earth science was fully a century older. Como changed all that. For the first time Europe had to look to America for the lead in the paleontology of a major geological period. Woodcuts and lithographs of Marsh's *Brontosaurus* from Como appeared in European textbooks, travel guides, and popular nature studies. "Brontosaurus" was even transliterated into Russian, Chinese, and Japanese.

The famous brontosaur quarry lies in the midsection of the Morrison, in a zone full of the grey-blotched floodplain mudstone, about a hundred feet higher in the sediment-layer sequence than the topmost sandstones of the Sundance. Dinosaurs were everywhere here. In one afternoon's walk I counted seven immense carcasses eroding out of the mudstone along a four-mile transect. And not only bones. There are also trackways of the living giants, pressed into the limy mud of shallow lakes, and now hardened into creamy-gray calcareous mudstone.

For me, trackways and ripple marks have a special intimacy. Both can be so fresh-looking that they seem to hold the sounds made by the Jurassic world, the sucking noise of viscous mud being pulled by the cushionlike foot pads of brontosaurs as they stepped through the Jurassic muck. The size of their footprints almost defies the imagination. The largest are over three feet long and two feet across, and deep enough to hold sixty gallons of water—more than enough to bathe a three-year-old child or serve as a full-immersion baptism for the diminutive first dinosaurs of the Triassic Period. Ten miles north of Como, at Sheep Creek, a freshwater lake bed in the Morrison exposes an entire field of brontosaurs' footprints, dozens of tracks churned into each other, rendering the whole limestone slab a twisted craterland. A veritable symphony of noises must have filled the air as herds of brontosaurs executed their ponderous choreography.

We know that brontosaurs traveled in herds, sometimes. A rare glimpse into their social structure is provided down at Davenport Ranch, Texas. There the limestone records the passage of two dozen brontosaurs in a compact mass, the very largest prints at the front periphery, the very smallest in the middle of the group. So brontosaur bulls—or maybe senior cows—must have guarded their young against the attacks of the allosaurs. The footprints at Davenport Ranch contain a Mesozoic recording of just such a drama of attack and defense, for the three-toed trackways of a great allosaur reveal that it was prowling along the strandline near where the brontosaur had passed.

No rock formation provides a richer repertoire of dinosaur stories than the Morrison. Its quarries have been dug from southern Montana to the Cimarron River in Oklahoma, yielding hundreds of skeletons from every level and every fossil habitat.

The sudden extinction of dinosaurs is one of the most popularized topics in paleontology. Why, after all, did the last dynasties finally end in total extinction? In reality, however, the dinosaurs' history contains the drama of much more than a single death. They suffered three or four major catastrophes during their long predominance, each one thinning the ranks of the entire clan. And after each such fall, they recouped their evolutionary fortunes, rising again to fill the terrestrial system with yet another wave of new species and families of species. The final complete extermination did not come until sixty-five million years ago, at what geologists label the "Time of Great Dying," the grandest evolutionary disaster of all time.

At Como I can walk right through one of the earlier extinctions, a time when the Jurassic families, which seemed so secure after fifty million years of success, suffered sudden extinction. There's nothing dramatic about the spot marking the event—merely a one-foot-thick bed composed of gray mud laid down in a stream and green mudballs that the rainy season's floods had torn from the banks and deposited in the creek bed downstream. Beds like this are everywhere, scattered all through the sedimentary layers. This particular one records a sudden jolt in the fortunes of the dinosaurs at the end of the Jurassic Period. Below the level of this bed—it's called the Breakfast Bench Sandstone because it makes a convenient shelf for the Coleman stove in the morning—the record of stegosaurs, *Diplodocus, Brontosaurus,* and *Allosaurus* can be followed up through the Morrison Formation for three hundred feet, equivalent to five or ten million years.

But at Breakfast Bench these Jurassic threads are broken; the familiar stars of the Morrison disappear, and in their place a new cast enters to play the dominant roles. This introduces the Cretaceous Period, the third and last age that made up the Mesozoic. Instead of *Stegosaurus,* with its flamboyant triangular spikes, a different kind of dinosaur, an armor-clad herbivore, the nodosaur, is found. It was far less spectacular, but thoroughly protected by its armor coat of big and little plates that formed a mosaic over its entire back and neck. Instead of *Brontosaurus* and its close kin *Diplodocus,* there appeared the teeth and vertebrae of brachiosaurids. This family of long-necked giants had been rare in the Jurassic but seem to have taken advantage of the catastrophe that struck most Jurassic families by moving in to take their place at the opening of the Cretaceous. Opportunism such as this is a commonplace during times of extinction. As the preexisting dynasty loses its hold, families of animals that had previously been mere bit players on the ecological stage seize the leading roles.

The knife-toothed predators suffered too as the curtain fell on the Jurassic stage. Gone were the *Allosaurus* and horn-toting *Ceratosaurus,* replaced by those most famous of all dinosaurian hunters, the tyrannosaurs. Smaller roles changed hands also. From the mudstone of Breakfast Bench, one of the crew excavated a magnificently preserved turtle skull whose boxy shape and adaptive equipment were totally different from any of the long, low heads carried by the water-loving Jurassic turtles. This turtle head, like the nodosaur's armor and the brachiosaur's tails, strike the anatomist as a jarring discontinuity in the flow of adaptive forms through time. Anyone who cherishes notions that evolution is always slow and continuous will be shaken out of his beliefs by Breakfast Bench and the other geological markers of cataclysm. Our view of evolution must take into account the profoundly disorienting blows struck by the environment during these worldwide extinctions.

The white sandstone blocks studded with the rounded beach gravel of Pine Ridge, the sedimentary sign left by the Mid Cretaceous ocean, look west out over the eroded blister of the anticline. From this vantage point, the entire sequence of strata, from the red Chugwater in the center to the thin dark line made by the outcrop at Breakfast Bench, is visible. Pine Ridge is composed of Dakota sandstone, named for the Dakota Territory in the 1870s, long before the Dakota Sioux had given their name to the two Western states. East of Rock Creek, these Dakota outcrops are covered by a black mass of carbon-rich shale, the Benton Formation, created by an ancient sea and named after old Fort Benton, built in the 1860s as the Union Pacific spread into Wyoming. The

Benton tells its own story of revolution and overthrow in the organic world. In sandstone layers laid down in this mid-Cretaceous sea were entombed reptilian sea serpents of a distinctly Cretaceous cast—the long-bodied teleorhinids, sea crocodiles with heads that resemble frying pans. The long snout looks like the handle, the squarish cranium the pan. These *Meerkrokodiliers,* as our German colleagues call them, are not the lineal descendants of the sea crocodiles of the Jurassic, but are a new oceangoing group descended from Jurassic freshwater crocodiles.

Opportunism again. When the Jurassic sea crocodiles were exterminated by the Early Cretaceous disturbance, an ecological opportunity presented itself to any reptile that could swim and catch fish and that could adapt to fully oceanic conditions. And so the rivers gave to the sea a new player for the reorganized Cretaceous marine systems. All through vertebrate evolution the flow has been mainly in this direction—from the fresh waters to the ocean shore to the high seas. Just so did the rivers give us the first whales some fifty million years ago, descendants of some river-haunting predatory mammal, one of the many mammalian lines that were rushing in to fill the empty niches left by the final extinction of the dinosaurs and sea reptiles.

The Benton Sea supported a wonderful menagerie of Cretaceous oceanic reptiles. In Colorado, Benton-age shales produced a nearly complete elasmosaur skeleton, a fast-cruising type of plesiosaur that slipped through the tropical Late Mesozoic water with the propulsive power of four narrow, tapered flippers, snatching prey with its snakelike neck.

At Como, Othniel Charles Marsh's men found an armor-plated nodosaur lying on its back embedded in the now hardened deposits left by the mud on the sea floor. Finds like this were exceptions to the rule that dinosaurs did not go to sea. Was the nodosaur swimming in the Cretaceous shallows before it met its end? Or is it the remains of some terrestrial individual that died a death on dry land and then, in the form of a dried-up carcass, was washed out to a final oceanic resting place by flood-swollen river waters? The problem of oceangoing nodosaurs is especially perplexing because the Como carcass, upside down at the bottom of the Benton Sea, is not an isolated instance. Nodosaur carcasses lying on their backs cropped up in marine beds in Kansas in 1909 and several times since in similar sedimentary circumstances.

There are no duckbill dinosaurs at Como, because the Benton Formation is too old. The duckbill dynasty began past the midway point of the Cretaceous. A short trip east through the Laramie Mountains to Red Bird places us in duckbill country, the Lance Formation of the later Cretaceous, a mass of pale brown river sandstones with interbedded chocolate-colored mudstones, sometimes faintly discolored by the pink of oxidized iron. Nearby is Lance Creek, supposedly named for the cavalry lance carried by mounted Sioux warriors.

Lance outcrops give their name in turn to the final terrestrial epoch of the Cretaceous world, the Lancian Faunal Age—a time that witnessed the adaptive deployment of the most exotic and bizarre skulls of all the panoply of dinosaurs. *Triceratops* was here, the scientific etymology "three-horned face" being, in this case, an excellent shorthand description of this formidably armed herbivore. Over each reinforced eye socket grew a horn of such size as to threaten even the largest *Tyrannosaurus rex*. In life these weapons were long, sharp, and deadly because the underlying bone was covered with a horny sheath like that surrounding the cores of cattle and buffalo horns today. Out on the snout was a third, midline horn, and below it a toothless beak, deep and powerful like that of a multi-ton snapping turtle. This too was clothed in life by a shiny hornlike substance, giving the beak an ever-growing, self-sharpening edge. Plant-eater though it was, *Triceratops* could turn the branch-cutting apparatus of its beak into a defensive set of nippers strong enough to inflict wounds on even the largest antagonist.

Truculence, nippers, and horns seem to go together. Today, the great Indian one-horn rhino can turn into the terror of the mahouts as it charges domestic elephants. The largest *Triceratops* weighed nearly ten tons, bearing horns that, fully sheathed, were four feet long. No species that has ever evolved on land could withstand the full charge of such an animal.

Duckbill dinosaurs did not display the deadly cranial armature worn by *Triceratops*. Nonetheless the duckbill group enjoyed an extraordinary evolutionary flourish of head ornaments and adaptations in the final days of the Cretaceous. The term "duckbill" is a biomechanical misnomer. True, the duckbill dinosaurs did have wide, flattened beaks, which at a distance vaguely resembled that of a mallard. However, the edges of their beak were turned down into a sharp, cookie-cutter edge, sheathed in life by a self-sharpening horn. The entire apparatus was a leaf-cropping adaptation for slicing off mouthfuls of tough fodder in a single bite. Duckbill teeth were one of the true marvels of mastication, cited everywhere in texts on dental evolution. Instead of one single row of teeth along each jawline, the duckbill had multiple rows, which combined to make a leaf-shredding surface equivalent in function to an ever-sharp carrot grater. No evolutionary device has ever evolved to masticate tough plant fiber more effectively than the dental shredder of the duckbills.

Although the feeding devices of the duckbills have provoked no end of wonder among paleontologists since the first duckbill was excavated in the phosphate fertilizer mines of New Jersey in the 1850s, it is the array of duckbill head ornamentation that stirs up the most puzzlement and debate. The common Lance Creek duckbill, *Edmontosaurus,* seems built to a no-nonsense, practical design. Its skull houses the beak, teeth, jaw muscles, and sense organs. But close relatives from Alberta and New Mexico show no such restraint in their headgear: *Parasaurolophus* carried a double-hollow bony tube like a trombone slide on the back of its skull; *Saurolophus* had a solid bony spike in the same position; *Hypacrosaurus* sported a thin-shelled bony crest rising high above the full length of its forehead and skull table.

This cranial exuberance at first glance reminds one of all the head appendages some families of birds employ to show off in premating rituals, such as the combs of roosters, the domed foreheads of some species of geese, the crests of cassowaries. And perhaps here the first impression is the correct one. Dinosaurs had to have sex, although one would never guess so from the scrubbed Sunday school versions of dinosaur biology presented in the children's books. Sex and pre-mating ritual are parts of the basic evolutionary game: genes that produce adaptations which succeed in increasing their representation in the next generation are the genes that survive. The genes of the dinosaurs must have played by the same statistical rule. If a garish head crest and some accompanying behavior, such as a strutting head-bobbing walk, made the male *Parasaurolophus* more attractive to the female and more intimidating to his rivals, then eventually the genes responsible for this equipment and its use would be fixed in the species. For most of this century, American paleontologists avoided sexual interpretations of dinosaur structures.

The European contemporaries of American scientists weren't so prudish. The Swedish paleontologist Carl Wiman hired an American dinosaur hunter, Charles Sternberg, to quarry and crate duckbills from New Mexico and horned dinosaurs for the Swedish Museum at Uppsala. Sternberg sent a magnificent *Parasaurolophus* to Wiman, who noticed that the double-hollow tube of the crest was simply a U-shaped elaboration of the air tract from its nostril to its windpipe. Wiman was a broadly educated naturalist, well aware of the multitudinous ways in which modern species of bird, frog, and mammal make love by making noise—hooting, gurgling, chirping, and bellowing. So what was Wiman to think of the U-tube in the duckbill's air passages? It looked like a trombone, it *was* a trombone! If the duckbill inhaled or exhaled with force, the U-tube would be a resonating chamber, enriching the tone and amplifying the noise. Hollow crests in other duckbills also

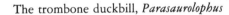

The trombone duckbill, *Parasaurolophus*

connected the throat to nostrils, and the variety of crest shapes from species to species would certainly produce a variety of hoots, wheezes, and amplified snorts specific to that species.

Even crestless duckbills like the Lancian *Edmontosaurus* had highly arched palates, and the vaulted roofs of their mouths could be used to modulate tones and increase decibel levels. And the crestless duckbill probably had additional sound equipment in its nasal compartment. The bone around the outer surface of the edmontosaur must have housed nasal diverticula, pouches of skin opening into the main nostril channel. Horses have similar diverticula, though of modest size compared to the edmontosaur's. Watch a stallion snort: The diverticulum shudders with pulses of forced air from the lungs, the sound controlled by sphincter muscles in lip and nose. The Late Cretaceous evenings in southeastern Wyoming must have been punctuated by reverberating snorts as the duckbills, driven by their genes, strove to impress each other.

The final hours of the Cretaceous are not to be found at Como or Lance Creek. This most profound of land extinctions may be witnessed if we go north, through Wyoming to northern Montana, to Hell Creek. Here, better than anywhere else in the world, the stratigraphic pile records in detail the events surrounding the extinction of the ultimate Great Dying.

Any attempt to analyze the events of the extinction of the dinosaurs runs into the fundamental difficulties that hinder the investigation of any of these mass murders of species. Most fossil bones owe their preservation to quick burial by sediment right after the death of their owner. But generally most spots in the terrestrial biosphere suffer erosion, not deposition. Only in slowly sinking basins, pieces of real estate hundreds of miles across, can we hope to see a long interval of time recorded by the preservation of fossils. If a broad, basin-like valley was near sea level, its rivers and estuaries could blanket the landscape with layers of mud and sand every flood season. The very weight of these blankets of mud and sand tended to push the land surface as if the basin itself were a sagging rubber bowl. If the sinking of the valley's surface kept up with the rate of buildup in the blankets' thickness, then the pile of sediment grew thicker and thicker, even though the average height of the land above sea level remained the same. The result, after ten or twenty million years, was a thick sandwich of sediment that might reach a vertical height of five miles.

Sinking basins don't sink forever. If they did, it would be possible to read the entire fossil record of life from bottom to top in one mine shaft sunk into a single valley. Instead, to understand the changing habitats of the end of the Cretaceous, it is necessary to hop from state to state, basin to basin, in order to piece together the disjointed narrative in the sediment, much as silent-movie buffs might try to reconstruct an entire lost feature by splicing fragments of film found in a dozen different studio storage vaults.

The fragment of the story recorded at Lance Creek carries us late into the Cretaceous, but not to the very end. In Hell Creek, Montana, and nearby Bug Creek, however, there is a sedimentary section, rich in fossils, that passes right through the last moments of the Cretaceous and continues into the next epoch, the Paleocene. Even at Bug Creek the strata do not record a year-by-year surveillance of the scene of the crime that would allow us to catch the perpetrator in the very act of extinction. In the best of basins, fossils weren't preserved every year, or even every hundred years. Big bones, such as those of dinosaurs, required big floods of mud to cover them, and these events didn't happen except at long intervals, perhaps hundreds or thousands of years apart. Even when buried, bones weren't necessarily safe. Acid groundwater might percolate through the sand, dissolve the bone mineral, and leave nothing behind but a gross, misshapen carbon stain where a duck-bill's skull once lay. Or a sudden shift in a river's course could erode part of the sedimentary layer it had deposited years before, and all the entombed bones would go tumbling down the new channel, breaking into irretrievable fragments. Paleontologists are grateful to streams for their blanketing of bones, but most streams also cannibalize. In one century they lay down deposits over the valley floor, in the next they might chew through their own sedimentary handiwork, churning and cracking buried bones and erasing the very fossil record they have previously preserved.

The movie-film analogy allows us to visualize the frustrating process of investigating the Cretaceous. Instead of a continuous film, one frame a year for each of the last million years of the Mesozoic, only short bursts of film remain intact, each a few dozen frames together, separated by hundreds of feet of totally missing footage. If something important, like the final extinction of dinosaurs, happened suddenly, within a few years, we wouldn't have a prayer of catching the deed in the film clips of sediment.

The best detective stories are those that command our rapt attention to every scrap of clue, so that we can solve the crime in the final chapter, just before the sleuth announces the identity of the murderer. Dinosaur extinction attracts the best of paleontological detectives. Up to Bug Creek and Hell Creek they go, digging quarries, running sediment through fine sieves to sift for the tiniest of bones and teeth. But much of the mystery remains. Only a few facts are clear. The final dying was sudden, compared to the immense length of the history of the dinosaurs: It took no more

The Mammalia take over. Four million years after the dinosaurs died, the mammalian hordes evolved into big tuber-digging herbivores, like *Psittacotherium* (at left), and big predators, like *Ancalagon* (at right).

than two million years—maybe much less—to exterminate all the Cretaceous dynasties. And there were opportunists waiting around for the dinosaurs to die: small, furry, insect-eating, berry-chewing mammals scurrying around the underbrush, fidgeting about, grooming their whiskers. As the dinosaurian clans were thinned out, with the extinction rate exceeding the production rate of new species, these Late Cretaceous furballs expanded their ecological sphere of influence. The fossils show new types of small, mammalian plant-eaters and insectivores blossoming in Montana at the very time the evolutionary fortune of the dinosaurs was sinking into its final, irrevocable decline. Passing upward through the sedimentary pile in Montana, exposed now in dry gulches, we can see the shifting census of evolutionary success. The mammals were diversifying rapidly near the very end of the Cretaceous, and dinosaurs dwindled until a level is reached in the layer of mud and sand through which no species of dinosaur passed. This layer marks the end of the Lancian Epoch, the end of the Cretaceous, the end of the Mesozoic. This time the dinosaurs would not recover.

3

MESOZOIC CLASS WARFARE: COLD-BLOODS VERSUS THE FABULOUS FURBALLS

Whenever I read Kipling's "Rikki-Tikki-Tavi," I root for the snake. There's something very irritating about the story's hero, that brave, ever-so-clever furry little mongoose who fearlessly confronts the Indian king cobra and its mate, defeating the slow-witted serpent by craft and nimbleness and thus saving the verandaful of upper-crust English colonialists. I like mongooses, but I don't like Kipling's fictitious beast. For one thing, real mongooses aren't so ingratiating or so stupid as to go down a cobra's burrow when it's occupied by its owner.

But the main reason I'm anti-Kipling is that his stories epitomize an all-pervasive bias in our popular and scientific culture against the Big Reptile. Kipling's cobra is a metaphor of size and strength without brains or honor. So the mongoose by comparison emerges as a noble and intelligent mammalian furball in contrast to its despicable reptilian foe. Snakes suffer such a terrible public image, being forced to serve as the very agent of evil in the Garden of Eden and as the synonym for deceit and ambush in popular slang. Crocodiles don't fare much better—the one in *Peter Pan* enjoys the dubious distinction of being only slightly less mean-spirited than the character it devours, Captain Hook. Big crocodilians, like big cobras, are dangerous, aggressive predators. A brackish-water crocodile grabbed the eminent Harvard entomologist Philip Darlington by the leg in 1944 on a South Sea island while that gentle scholar was studying mosquitoes for the Navy

Spectacular lizard bluff from Australia—*Chlamydosaurus*. The frilled lizard flips its huge scaly collar skin to frighten potential enemies. Growing up to three feet long or more, frilled lizards can sprint away at fast speeds on their hind legs.

Department. Darlington kicked his way free after being whirled around under water a couple of times, but not a few explorers have suffered the complete process crocodiles perform on their prey. Cobras and other venom-equipped snakes kill hundreds of village people, farmers, and migrants all through the tropics, a yearly toll far exceeding that of all the man-eating tigers, lions, and leopards together.

So there is some cause for the human species to be alarmed when confronted by a big reptile. However, in our culture, we react to these reptilian potential man-killers only with revulsion, not with respect. What a difference from the role reserved for mammalian man-eaters—the lion is so admired for strength and cunning that nearly every royal European household placed the tawny beast on its coat of arms, and both the Messiah of the Old Testament and the Emperor of Ethiopia were hailed as the Lion of Judah. I know admittedly little of heraldry, but rest certain that not even the shortest-lived Balkan principality adorned its royal crest with a Nile crocodile.

Some of my best friends are mammals. But like most other dinosaur paleontologists, I have very mixed emotions about the Mammalia as a class in vertebrate history. According to widely accepted theories, the Late Cretaceous mammals were among the chief ecological conspirators that manipulated the habitats until the Dinosauria were extinct. Most vertebrate paleontologists aren't dinosaur specialists but concentrate on the fossils of mammals instead. Any naturalist tends to identify emotionally with the objects of his research. Consequently, most mammal paleontologists view the Cretaceous extinctions not as a sad finale but as a grand opening, the dawn of the Age of Mammals.

Geologists generally have a fondness for dynamic terminology to label earth processes they study. A pulse of mountain-building activity is thus known as a revolution, and the Laramie Mountains, folded and raised in Late Cretaceous times east of Como, are described as the products of the Laramide Revolution. Tacked onto the bulletin board of the student offices in the University of Wyoming, where Late Cretaceous mammals are a specialty, is a poster in the best 1919 Bolshevik style. The earth explodes upward, *Triceratops* tumble over backward stunned into extinction, as a giant furry fist thrusts through the land surface clutching the banner "Join the Laramide Revolution." And to hear the mammal paleontologists talk, after a few pitchers of beer in the cowboy bars, it happened that way. With the geological equivalent of the "Hallelujah Chorus," the irresistible new wave of mammals swept aside the old order, replacing the sluggish brawn of the dinosaurs with the energetic intelligence of the Mammalia. Such talk is annoying. But we few dinosaur specialists huddle together at the dark end of the bar, muttering in our beers about the insults—insults not just against the Dinosauria, but impugning the honor of every turtle, crocodile, snake, frog, and salamander as well.

In European culture, the anti-reptile bias began centuries before the stratigraphic sequence was discovered. The very word "reptile" has a pejorative etymology. Derived from the Latin adjective *reptilis,* "creeping," the term originally was applied indiscriminately to anything low-down and loathsome—scorpions, centipedes, snakes, and lizards. Ever since classical antiquity the Reptilia meant roughly the equivalent of "creepy-crawlies." Aristotle, the ancient Greek naturalist, and the Christian philosophers who revised and edited his texts, put lizards and snakes low down in the scale of animate creation, far below cats, dogs, birds, and mongooses. The idea that all of nature could be arranged in an ascending scale of complexity and perfection was extraordinarily popular among medieval scholars. The principal criterion by which any species would be assigned its place on the *Scala Naturae* was how close it came to the unchallenged holder of the top rung, Man Himself. According to this view, the Creator, in His wisdom, put His best blueprint into production with the human race; the other mammals were close, but the scaly, crawling things were far from His Own Image. Even when Darwinism cleared away most of the creationist mythology from Western science, an evolutionary *Scala Naturae* was easily substituted for the theological one.

If all the bad-mouthing of reptiles came from bar-hopping mammal paleontologists emboldened by one too many beers, or the mystic musings left over from the Middle Ages, I wouldn't be much disturbed. But when the dean of reptile paleontologists, the late Al Romer of Harvard, wrote about the superiority of mammalness over the reptile condition, it made me shudder. Alfred Sherwood Romer did some superb and innovative research on the hind-limb muscles of dinosaurs in the late 1920s, showing that the thighs of duckbills operated more like those of giant birds than those of giant crocodiles. But dinosaurs were a minor diversion in Romer's long and distinguished career. He spent most of his field seasons, first in Texas, later in Brazil and Argentina, digging up mammal-like reptiles, a diverse lot of vertebrates that bridge the structural gap between a primitive lizardlike reptile and a genuine furry, milk-sucking mammal.

Romer inherited from his mentor, the magisterial anatomist William King Gregory, a passion for reconstructing, step by step, the evolutionary pathway that led from the first sprawling reptile of the Coal Age, 300 million years ago, to the first bona fide mammal, something that would look like a tiny 'possum, which emerged at the end of the Triassic. For Romer and Gregory, clearly the proper study for man was Mankind. Therefore the most important life thread in geological history was our own, leading backward through the Age of Mammals to the tiny, scurrying Mesozoic mammals and thence through all the successive stages of mammal-like reptiles. Gregory wrote a delightful evolutionary essay for the lay person, "Our Face from Fish to Man," which expressed elegantly the preoccupation with the single evolutionary trackway leading upward through the strata to the Mammalia and to *Homo sapiens*. Both Romer and Gregory did study what were perceived as evolutionary sideshows—Romer wrote about Coal Age amphibians with flattened, shovel-like heads, and Gregory executed definitive treatises about sailfish—but both scholars were true to their own class, the Mammalia, when it came to allocating the bulk of their labors.

Romer earned the everlasting gratitude and respect of all reptilian paleontologists with his *Osteology of the Reptiles,* a bountifully illustrated guide to skulls, limbs, and vertebrae of all the Reptilia, including dinosaurs and mammal-like reptiles. Romer's classification of reptiles, which places nearly every known fossil and living species in a formal hierarchy, is one of the most widely used among herpetologists and paleontologists. When I was a graduate student at Harvard's Museum of Comparative Zoology, I was fortunate to have a study carrel around the corner from Romer's office. He was always ready to talk about his first love, the evolution of mammal-like reptiles, at coffee break when he and the other senior paleontologist, Bryan Patterson, a pioneer in the analysis of Mid Cretaceous mammal teeth, sat on the basement stoop with the students and staff. For his affability, scholarship, and generous support of students, Al Romer is justly remembered as a prince among the reptile specialists.

But Romer did one thing I disagree with, vigorously. He wrote that after the close of the Cretaceous, the entire Reptilia became second-class, an overaged, unprogressive group that decayed steadily in biological importance down to the present time, the evolutionary equivalent of the senile Ottoman Empire gradually losing its grip over the eastern Mediterranean after its apogee in the fifteenth century.

Far from declining senile groups, the Reptilia and their cold-blooded cousins, the Amphibia, are today full of vigor, full of species, and full of ecological importance. To prove this, one need only stroll through any tropical rain forest in today's world. These habitats, the richest in vertebrate species, are quite literally crawling with frogs, snakes, and lizards—a hopping, slithering, scampering horde of highly specialized species whose numbers overwhelm those of the supposedly "higher" mammals.

To evaluate the fortunes of the classes, we need some simple system of scorekeeping. One of the best ways to score evolutionary success is to count species. The species is a self-perpetuating unit of interbreeding individuals, and two closely related species can be proved to be distinct only if they fail to interbreed freely in the wild. Most of the time closely related species have slightly different ways of making a living. For example, the high plains wolf hunted in packs across the prairie, cutting down elk and straggling buffalo calves. The coyote, a close relative of the wolf, usually traveled in smaller groups, snatching small prey and sneaking in to dine on wolf kills after the bigger predators had gone. In zoos, coyotes and wolves will mate and give healthy hybrids, but in the wild the two usually keep their genes to themselves. Thus, wolf and coyote are scored as separate species. When we count the total number of species in the Reptilia or Mammalia, we are scoring the number of different ecological roles filled by that class. We should send those barroom detractors who believe that reptiles are a moribund class into the tropical forests, armed with checklists, nets, and binoculars. Let the mammal chauvinists count species; the results will sober them up. (A genus is a group of closely related species; *Canis* is a genus name, the dog genus, and *Canis latrans* is the coyote species.)

If they do their work well, species census takers should score fifty nonflying mammals in a thousand-acre plot of the Congo Basin or the Burmese lowlands—squirrels, shrews, monkeys, mongooses, palm civets, antelopes, and elephants. This is a rich fauna of furballs by Temperate Zone standards; a New England woodland would score only two dozen species. Now set the census takers on the task of scoring every Congolese frog, serpent, and lizard. In that same thousand-acre plot the total score for Reptilia and Amphibia will be about 180, *three times* the mammal score. In Burma or Thailand the scores will be similar—the "cold-blooded" classes win two or three to one. So where is the proof that Mammalia are the best adapted class?

Herpetology is the study of amphibians and reptiles, the name being derived from the Greek for snake, *herpes,* and for learning, *logos.* Biologists, with their fondness for professional nicknames, lump all members of these two classes as "herps." An entire reptile–amphibian census is called a herpetofauna. (Herpes virus, that current scare in venereal epidemiology, gets its name from the alleged similarity between how the fever blisters spread and the crawling of snakes.) Not only is the herpetofauna of today's tropics much richer in species than the mammalian fauna, but the tropical herps display a veritable riot of adaptive diversity. The marine toad of South America is a five-pound warty predator that gulps down mice and rats, an ecological function that induces farmers to transport the toad all through the tropics to keep down rodent populations on plantations. Marine toads are toothless hunters but make up for their lack of dental armament by their poisonous saliva, which numbs their prey into submission. The big-mouthed marine toad sits in ambush along a small mammal trail and makes a short lunge to snap up the unwary rat; a moment or two inside the toad's mouth is all that is necessary to anesthetize the mammal.

Among herps, poison is a popular adaptation for defense as well as for offense. Arrow-poison frogs of the American tropics produce a potent toxin in their skin glands. Some species are dangerous to handle without gloves—my lab instructor in an undergraduate course at Yale fondled a pretty frog, just uncrated from Surinam, and was sick for two days. Amazonian hunters dry and concentrate arrow-poison frog toxins to smear on the cutting edges of their blowgun darts and arrows. One good dose from a dart and a thirty-pound monkey falls paralyzed from its treetop refuge. Poison even guards New England toads and salamanders—some species have enough skin toxins to make a hound dog retch. Wise old retrievers learn not to put toads into their mouths.

The amphibious branch of the herp kingdom is not limited to chemical adaptations. Kermit the Muppet frog has made the fly-catching tongue famous, but the ability to snap insects by tongue-flipping has also evolved in salamanders, the short-legged amphibians that look like scaleless lizards. Although most amphibian tooth equipment is modest by mammalian standards, there is a saber-toothed toad: the horned *Ceratophrys* of the New World tropics.

The sharp-edged upper fangs of a big, adult *Ceratophrys* can cut up the hand of an unwary herpetologist. *Ceratophrys* also claims the distinction of being an armor-plated toad. Embedded below the moist outer skin are wide bony plates protecting the shoulders and neck. If you grew up in northern New Jersey, as I did, you get the impression that frogs are a swamp-bound clan, because watery haunts offered the best frog-hunting ground. But most species of frog are tropical, and in the tropics fully half of the frog species can be land-living as adults, and many climb trees. In the flood-plain of the Congo River, three different families of species of tree frog clamber about the bushes and forests, snaring insects from leaves and bark.

Amphibians score significant subterranean successes, too. The New England mole salamander pushes its way through damp soil using its strong snout and thickly muscled torso. In the tropics sal-amanders are scarce, but the soil is churned up by hundreds of species of legless amphibians, the Apoda (Greek for "legless ones"). Several families of frog are well-equipped excavators, digging with pointed snout or spadelike feet. In the Malay Archipelago, herps reach the summit of their locomotor evolution. Here is found the flying frog. Spreading the thin membranes between its long fin-gers and toes, the flying frog launches off a forest perch and glides effortlessly to another tree a hundred yards away. It's not true

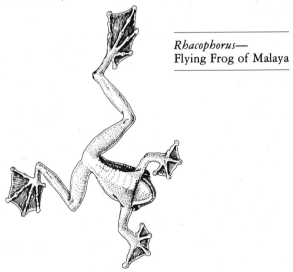

Rhacophorus—
Flying Frog of Malaya

powered flight, like the flapping progression of birds and bats, but the frog's powerless glide has a certain herp elegance.

All told, in habitats from all climates, the Class Amphibia scores three-thousand species, just as many as the total number of non-flying mammals. We're not living in the Age of Mammals, we're living in the Age of Frogs.

Mammal chauvinists also underestimate turtles. Compared to frogs and their amphibian kin, turtles don't score high in the species competition—only two hundred and thirty species fill the modern turtle clans—but turtle limbs, necks, backs, and skulls are true marvels of joint architecture. Most turtles can fold up everything that sticks out from the shell—neck, limbs, tail—and tuck it into the armored box, leaving little exposed. Several turtle species go even further in safeguarding their soft parts. Hinge lines have evolved in the top shell (the carapace) or in the bottom (the plastron) so that after it has pulled in its appendages, the turtle can close up the neckline and limb apertures, scaling its entire body into a nearly impregnable strongbox. The basic turtle shell itself is a most unusual structure that has evolved only through a bizarre bit of embryological hocus-pocus. Turtle shells have three layers: (1) the outer *horny* plates, a tough covering sheathing the bone beneath, like the horn-core sheaths of bison; (2) the outer *bony* plates, which lie just under the horny sheath and grow within the lower skin layer; and (3) the ribs, which arch around the body and fuse to the underside of the outer bony plates. Turtle hips and shoulder bones lie more deeply inside the body, *beneath* the ribs— a startling arrangement because in humans and all other vertebrates the ribs lie beneath the shoulder (reach around and feel your own ribcage beneath the edge of your shoulder blade if you don't believe this statement). In order to get the ribs up and on top of the shoulder, the turtle embryo inside its egg must grow the ribs much more rapidly than usual, pushing the developing ribcage up and between the shoulder and hip, and attaching the rib edges to the underside of the shell bones. No other vertebrate—not even the tanklike nodosaurian dinosaurs—has ever evolved a mobile body armor so complete and effective as the turtles'.

Turtle heads also command the respect of bioengineers. Turtle jaw and joints guide the chewing stroke into a long backward slide of the lower jaw against the upper. Toothless turtle beaks and palates are armed with horny cutting edges and multi-toothed shredding-crushing platforms. This basic masticatory apparatus is marvelously adaptable. Land tortoises shred tough grass. Giant sea turtles crush clams. Snapping turtles can slice up a dead trout or a drowned cow into chunks small enough to swallow. Right now, in the 1980s, my science is enjoying a turtle renaissance. Paleontologists in Utah are investigating the muscular-electrical phenomena of turtle chewing by using sophisticated electromyographs, high-tech gadgets that chart each muscle's physiological activity. Field paleontologists at Berkeley are mapping the historical details of turtle evolution through Cretaceous and Paleocene strata. An eminent New York anatomist is completely revising the turtle family tree. In scientific meetings all over the country the Turtle Renaissance is shaking old-time zoology out of its complacency with the message: Turtles are complex, turtles are successful, turtles are worthy objects of research.

The total turtle count—two hundred and thirty species— doesn't seem like an irresistible horde compared to the several thousand mammals in today's global ecosystem. However, turtles have scored quite an impressive ecological triumph in one very important role, that of freshwater predator-omnivore. Gavin Maxwell's *Ring of Bright Water* is an absolutely charming otters' tale, the story of these sleek-furred aquatic mammals that frolic in the Scottish streams, catching salmon and crayfish and stirring up warm bemusement in human onlookers. All through the Temperate Zone, otters delight the naturalist and the lay public. But how many other freshwater, semi-aquatic mammal predators can you name? Mink, of course. Relatives of otters on one hand, land weasels on the other, mink do hunt in streams. How many others? If you caught the excellent BBC series "Life on Earth," you saw footage of the swimming shrew, the Desman of the Pyrenees, a molelike furball that dives for aquatic worms and other freshwater small fry. Our own New England star-nosed mole goes hunting in water, using its starburst-shaped snout tip to feel out wriggling prey. Andean streams flowing through Peru are host to the fish-spearing mouse, *Ichthyomys,* that impales prey on its projecting front teeth. But if we go to a tropical lake or sluggish river, is it full of otters, mink, and paddling shrews? No, it is full of turtles. The mass of tropical turtledom far exceeds the Mammalia in numbers of species in the aquatic-predator category. A few tropical otters do exist, and Uganda can boast of the giant otter shrew (a full two pounds in weight), but a single Congolese river system can display a dozen and a half specialized turtles, swimming after prey, eating fallen fruit and leaves, walking along the river bottom, scavenging pieces of hippo carcass.

The entire subject of aquatic predation should embarrass the mammal chauvinist into silence. The lion may be king of the beasts on land, the top link in the terrestrial food chain. But in the Nile waters and in the great Rift Valley lakes of East Africa, the lion must fear for its crown. Here the king is the Nile crocodile. Contrary to the popular view, crocodiles are neither sluggish, nor stupid, nor lacking in maternal affection. Crocodile mothers guard their nest with aggressive vigilance for the three-month incubation necessary for hatching. When the hatchlings chirp as they struggle to wrest free of the shell, the mother will gently help her newborn, lifting them in her jaws from nest to water's edge. For months after hatching, the young crocodiles stay close to mother in the shallows, where she can drive away any potential threat. Field zoologists in Georgia and Florida tell the same story of maternal care of our Mississippi alligator. (Alligators and crocodilians differ in shape only in minor features, the broader, flatter snout of the 'gator being the most obvious; the term "crocodilian" encompasses all 'gators and crocodiles and their fish-eating kin, the frying-pan-headed gavial.)

The Harvard professor's close call with the South Sea crocodile is a warning that even the *numero uno* on nature's scale must be careful around crocodilians. Crocodiles are good hunters. An adult male will stick to one hunting territory for years, learning all the ins and outs of the watery passages among the reeds, gradually developing an ambush style calculated on the seasonal flux of fish, snails, turtles, and land mammals that come to the water's edge to drink. Adult crocodilians watch the shoreline, their heads submerged except for the bulging eyes and nostrils. If an antelope ventures close enough, the croc glides smoothly through the water, propelled by its deep, sculling tail. Five or six feet from the antelope may be close enough, then a quick lunge and the great reptilian jaws clamp shut on furry snout or leg. The thrashing victim is dragged under the water and stunned as the croc whirls around, rolling over and over. Sometimes a mammal victim escapes after one of its legs has been wrenched out of its socket. Lions, cheetahs, and baby elephants have died this way. Not only Man but his domestic servants can be croc prey. A giant South Seas croc snagged a horse from an Australian farmyard and dragged it back to the billibog.

Although such furry big game are key elements of many big-croc diets, most of the crocodilian clan subsist on less dramatic fare—fish of all kinds, aquatic turtles, swimming snakes, freshwater mollusks. When just out of the egg, young 'gators and crocs hunt aquatic insects, frogs, and other humble game. Everywhere in tropical waterways the crocodilian ensemble—two dozen species—are by far the most important large semi-aquatic predators. All crocodilians are large by modern reptilian standards. None are as small as a Scottish otter, but adult size does vary from species to species. Giants among living species are the slender-snouted gavial of the Ganges and the estuarine crocodile of the Pacific shores. In both, a big male can exceed twenty feet in length and half a ton in weight. Tiniest are the heavily armored West African dwarf crocodiles and dwarf caimans of South America. In one dwarf species, females probably breed at the tender young length of two and a half feet. But these dwarfs make up for their size with armor. All crocs have bony plates, sheathed in horn, embedded in the deep skin layer, and in dwarfs the plates make an especially tight-fitting mosaic, a flexible cuirass for chest and back. What's this protection for? Tigers and jaguars do pounce on careless little crocodilians caught basking on the shore, but the chief hunters of any given crocodilian species are other crocodilians.

Ecological science, for reasons not clear to me, lacks the lyric eloquence of geology. Ecological terms rarely have the color or dynamism of such geologisms as "rift," "thrust fault," "mountain-building revolution," "hogback." Ecology tends rather toward the gray-flannel-suit metaphors of marketplace and commerce: "resource partitioning," "energy budgets," and "investment." But modern ecological theory has given us one quite lovely term—the "guild." The clockmakers' guild in sixteenth-century Basle protected the interests of all the makers of timepieces in that Swiss Protestant city. Guild councils enforced quality control and regulated the entry of new artisans into the urban market. An ecological guild is all the species that follow a particular way of making a living in local habitats. Hence the top-predator guild of the Serengeti is filled today by lion and leopard, cheetah and spotted hyena. Working the same landscape is the small-predator-scavenging guild, the golden jackal, black-backed jackal, Egyptian vulture, and griffon vulture. Reviewing our ecological census figures, we would be compelled to conclude that most tropical guilds are dominated by the "cold blooded" clans. In the small semi-aquatic guild of predators, turtles are masters. The large-predator aquatic guild is firmly in the hands of a crocodilian cartel.

I would hope that by now in our census through the vertebrate guilds the delusions of mammal superiority would be shaken. But we are not even half done. Remember, all the world's non-flying mammals add up to 3,000 species. There are now, by conservative estimate, 3,000 lizard species and 2,700 of snakes.

European culture and its American offspring are more ignorant about lizards than about any other great divisions in the "cold-blooded" clans. Lizards don't abound in the cities of the Temperate Zone that served as cradles for Western science. Heidelberg, Paris, London, New Haven—all are great university towns, but all languish in a state of lizard impoverishment. Go to school in one of the Ivy League Colleges, take a field ecology course, and you will count yourself lucky to catch a glimpse of a little brown skink, speeding along a sunlit pathway. Turtles, frogs, salamanders, and snakes all outnumber lizard species in upper New York State or Massachusetts. Too bad, because the natural economy in the species-rich tropical world supports a dazzling lacertilian display.

"Lacertilian," the standard label for all lizards, comes directly from the Latin *lacerta,* the Roman name for the common Mediterranean wall lizard, a hefty four-pounder that has hunted big insects around human habitation since the Parthenon was built, and before. (Spanish conquistadors called the broad-snouted crocodilian of the Mississippi *el lacerto*—a label that quickly degenerated into "alligator.") Travel south from the Mediterranean wall lizard country, past the sandy barrier of the Sahara, and you will reach a lacertilian evolutionary epicenter. Patrolling along the shorefront of the Nile are six-foot monitor lizards, long-tailed hunters that can swim, dig up croc nests with their strong foreclaws, and race off to escape predators by climbing a tree or descending a burrow.

Locomotor triple threats, nearly all the dozens of monitor lizards can make quick progress in all modes, arboreal, terrestrial, and aquatic. Farther from water are snub-nosed savannah monitors, the lacertilian equivalent of badgers, truculent, stout-shouldered diggers that can exploit the buried resources of eggs, rodents, and fossorial reptiles.

Whenever I see a stocky golden-skinned savannah monitor, I get a lump in my throat and misty eyes—memories of a wonderful sweet-tempered pet I had my second year as a graduate student. Part of the task for my thesis was to measure metabolism during walking in lizards, so we could calculate an ecological energy budget for dinosaurs. After three frustrating months of failed experiments with skinks, race runners, and desert iguanas, I reluctantly concluded that even though it was small-brained, the average lizard was smart enough and mean enough not to consent to walking on a treadmill for an hour while wearing a lizard-size gas mask with two hoses attached to a great big Beckman oxygen analyzer with a battery of blinking lights emitting clicking instrument noises. But then I acquired G. Hawn, the gold-colored monitor. She ran beautifully, and we got excellent data. And she was a quick learner. We had to run the lizards when their stomachs were empty, because digestive metabolism complicates the measurements of exercise. G. Hawn learned that she would be fed two plump white mice, alive and fresh, and a raw egg at the end of a successful run. She soon was giving us two good long runs each week. Alas, she succumbed suddenly to a respiratory infection, untreatable because lacertilian veterinary medicine is still very crude. No mammal chauvinist can tell me that lizards can't be clever. I can't claim that my monitor responded to me with affection, but her vivacious character certainly elicited that response from her owner.

Both Nile monitor and savannah monitor are snail-boppers, a prey especially attractive to lizards. Adult lizards from the two species go about with swollen acorn-shaped teeth, pestles that can crack even the most resistant-shelled mollusks. Such evolutionary enthusiasm for shellfish might seem surprising, but in fact tropical habitats of all climates offer a tempting menu of multi-species escargot, because snails are among the most diverse of the land and aquatic animals. Although not many lacertilians are specialized swimmers—unlike their distant scaly relatives, the turtles and croc-

Dragon lizards, past and present. Today Komodo Dragons reach nine feet long and two hundred pounds, big enough to kill unwary tourists. Meat-eating lizards twenty times as heavy hunted in Australia in the recent geological past (silhouette shows size of the extinct giant compared to a modern Komodo Dragon).

odilians—several lacertilian families are equipped with modest natatory skills and have shellfish-crunching batteries. Hunting shelled prey in the Orinoco and other New World tropical rivers is the clam cracker par excellence, the two-foot-long caiman lizard so called because its deep tail and armor-studded hide recall the shape of the local alligators known as caimans. Bulging jaw muscles and nutcracker jaws make the caiman lizard nearly invincible in gustatory confrontations with Amazonian mollusks.

Monitor lizards have not limited their guild membership to the shellfish-eating clubs. On the Indonesian isle of Komodo is a monitor that kills and eats goats, water buffalo, and German tourists. The story of the Komodo dragon reads like the script for the original *King Kong* (a carefully crafted movie with excellent dinosaurs, molded by someone who read Al Romer's research paper). Rumors of a great lizard living on a tiny island, a real-life dragon called *ora* by the natives, reached explorers in the late nineteenth century. Expeditions brought the first skins and bones to museums in 1912, and, for once, legend paled before reality. Up to eight feet long and as heavy as a lioness, the adult Komodo dragon brandishes steak-knifelike teeth—sharp, recurved blades with serrated cutting edges. Showing the same sagacity found in veteran Nile crocodiles, fully adult dragons know their hunting territory from years of experience. They know where to lie along hilly game trails, awaiting the light footsteps of a deer. Attacks are instant successes or failures because the *ora* has no stamina, and if it misses on the first short rush, it has little sustained speed for a long pursuit. When attack succeeds, the cruel rows of slashing teeth cut fearful wounds on the rump and thigh of ambushed animals and the stricken prey may die of massive infection days later even if it manages to break free from the dragon's mouth. Tethered livestock suffer truly terrible cuts across the legs when an *ora* slinks into the compound under cover of the warm Indonesian nights. Several humans, both natives and European visitors, have died in savage daylight attacks. The victims simply had no warning sign that the *ora* was waiting patiently a few feet from trail's edge.

Fearsome though the Komodo dragon is, we must go much farther south, to mainland Australia, to find the full flowering of monitor evolution. The great Australian island continent is a down-under, topsy-turvy world in more ways than one. Instead of an interlocking guild system of small, medium, and large predators, filled mostly by mammals, such as we see in the Serengeti, the Australian predator guilds feature monitor lizards in many of the roles we are accustomed to believe were reserved for the Mammalia.

The badger role is played well by Gould's monitor, a digging predator specializing in buried prey. On other continents the brotherhood of furry hunters—weasels, ferrets, and mongooses—chase the small prey, but "down under" the long-bodied small predators are pygmy monitors. Tourists in minibuses gawk at leopards sleeping at midday in Kenyan game parks, but in the Australian outback the traveling lizard watcher can catch a glimpse of the seven-foot Perentie monitor, draped over a eucalyptus branch to escape the noonday heat. Native Aussie mammals take a decidedly second place to monitors in the freshwater guilds, too.

The greatest lacertilian hunter of this region is, however, missing today. A few thousand years ago a monitor Kong stalked the Australian landscape: *Megalania,* a massive half-ton lizard predator as big as a Kodiak bear. Fossil *Megalania* vertebrae and jaws with monstrous curved teeth were first discovered a century ago by pioneering Aussie naturalists and now are known from sites across Queensland, New South Wales, and Victoria. With the eyewitness accounts of Komodo dragons in mind, one must suppress an involuntary shudder at the image of a resurrected thousand-pound *ora* rushing out to tear apart the largest Australian mammal.

Dragons and half-ton monsters of Queensland's past should not sway us into believing that the vast lizard species-empire was built by brute force alone. Lizard adaptations include devices of greater subtlety—body ornaments designed for fraud, intimidation, display, and seduction. The Australian frilled lizard, one and a half feet long at most, is of a typical lacertilian temperament, slow to bite in earnest even when engaged in vigorous disputes over territory or potential mates. There's evolutionary wisdom in such restraint: Quarrelsome genes that give their owners a chip on the shoulder will get weeded out of the population, if constant brawling leaves the lizard scarred, crippled, and too exhausted to breed. Darwinian processes have operated on the frilled lizard to concentrate genes whose results are more theatrical than rowdy. Lining

Two-ton dragon lizard of ancient Australia. Fifteen feet long and as heavy as a bull rhinoceros, the extinct *Megalania* hunted giant kangaroos during the Pleistocene Epoch of the Age of Mammals, a few hundred thousand years ago. A scaled-up version of today's Komodo Dragon, *Megalania* died out quite recently, by geological standards, and for reasons that are totally unknown.

the lizard's mouth is tissue of the most brilliant vermilion. Hanging limp around the neck is a wide collar of folded skin. When the lizard must assert its presence, a direct biting attack is eschewed in favor of a grand thespian display: the mouth pops wide open, unveiling a sudden flash of red on palate and tongue, and the collar snaps erect, spreading a scaly corona about the neck like the frill around the Dutch Masters, increasing the apparent head size sixfold. Hissing and lunging forward, the frilled lizard goes through its act, a gaudy vaudevillian bit of behavior which transforms the little inoffensive lizard into an animated trick-or-treat mask.

Body ornament for intimidation produces some of the most decorative vertebral columns in lizardom. In most vertebrates the vertebral spines are strictly utilitarian and nonornamental. The bony prongs rise up from each vertebral segment to provide leverage for the back and neck muscles (the series of bumps down your back, between the shoulders, are the tops of vertebral spines). But the Australian water lizard grows spines so long they project far beyond the muscle contours and extend upward like a picket fence embedded in a thin sheet of tough skin. This lizard's intimidation technique, like that of most species, is broadside bluff. Turning sideways to its foe, the water lizard puffs itself up, standing as tall as possible, showing off its vertebral sail to best advantage, trying to prove that it is bigger and nastier than its rival. If your rival *looks* taller, then he might *be* bigger and stronger. This simple message is encoded in most lacertilian brains and plays out automatically during disputes, controlling the lizard's fight-or-flight response.

South American riverside forests are home for one of the best broadside bluffers, the Jesus lizard. Here the males sport among the most flamboyant vertebral crests known anywhere today. Sheets of bone protrude from the head and the picket fence rises from the torso to make the skinny lizard body look three times as big as it really is. The name "Jesus lizard" doesn't come from the puff-and-bluff display, but from the speedy getaway performed by the lizard when its tiny brain snaps over to the flight mode. Very long in the hind legs, the Jesus lizard can sprint so fast for a few dozen yards that its momentum carries it across the surface of lake or river, the long-toed strides propelling it far beyond the shore. After its walk-on-water dash the lizard can sink out of sight, a bewildering performance for most enemies.

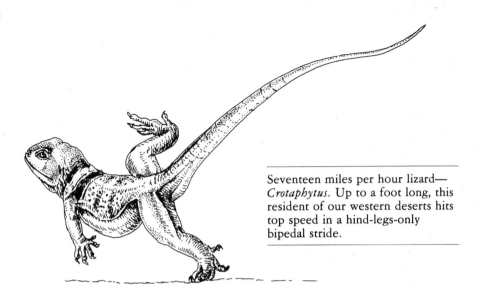

Seventeen miles per hour lizard—
Crotaphytus. Up to a foot long, this
resident of our western deserts hits
top speed in a hind-legs-only
bipedal stride.

All today's cold-blooded speed records are held by lizards. The best lacertilian sprinters are long-legged bipeds, species that at high speed tuck their arms under the chest and stride on hindquarter power alone. In our own American West the mountain boomer, a short-bodied, wide-headed predator that gulps down big desert bugs and other lizards, has been clocked at eighteen miles per hour. Lizard feats of arms and legs span the entire range possible for a land vertebrate, a complete evolutionary decathlon: burrowing by wormlike amphisbaenid lizards; sand-swimming by Kalahari skinks; snakelike grass-slithering by legless glass lizards; crocodilelike swimming by monitors; leaf-leaping by anolis lizards (Florida chameleons); claw-propelled digging; bipedal sprints; and the incredible slow branch stalk by the Old World tree chameleons. And there are even some lizards that can glide, using rib-supported wings.

Walking narrow branches is a tough high-wire act for most lizards, difficult to master because the basic lacertilian posture is a sprawl, with elbows and knees held far out beside the body and the paws held far apart. Gripping a narrow branch is awkward with such a wide-track gait. The prizes for the successful branch walker are enticing: hordes of insects and other juicy prey teem among the leaves, twigs, and stems. The Old World chameleons have solved this problem with a suite of limb adaptations rarely matched

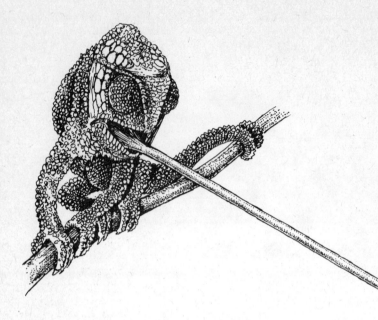

elsewhere. Most lizards have broad chests, which separate left and right shoulder sockets widely. Chameleons have deep chests, very narrow from side to side, like that of a cat. So the chameleon arms can swing fore and aft directly under the body. And the chameleon's forepaws can grip the narrowest of perches. Most lizard hands are rather crude five-fingered devices incapable of a precise grip. Chameleon hands are cleft—two fingers are separated from the other three at the wrist—and the chameleon can use the two as a sort of scaly thumb for gripping a branch. Hind limbs are similarly cast into a narrow-striding, gripping mode. With four precision grippers and a narrow stride, the true chameleon on the hunt makes all the slender vines and branches of the tropical forest unsafe for butterflies and beetles.

Our own mammalian order, the primates, prides itself on hand–eye coordination; monkeys, apes, and man are all good manipulators. But no mammal can rival the chameleon for eye–tongue coordination. The tongues of chameleons are explosive devices, lying loaded on the floor of the mouth, ready to fire forward as elongated, muscle-propelled missiles armed with a sticky, bug-catching warhead. Missile warheads are useless without their guid-

The Malay flying lizard, *Draco*. About six inches long, nose to tail tip.

ance systems, and the chameleon has a stereoscopic rangefinder and fire-control apparatus unique among vertebrates. Each chameleon eye is mounted in a scale-studded turret which can move independently, scanning the branches for insect targets. Once a beetle is located, eyes switch to attack mode—both turrets lock their stare forward on the target. Eyes feed the brain target data, distance, bearing, target size—the fire-control computations are swiftly made, automatically, without conscious thought, zap!—the tongue muscles contract, hurling the bony tongue base forward and

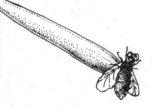

Best tongue show in lizarddom. Chameleons are lingual sharpshooters, firing their extensile tongues twice their body length to catch insects. This genus is *Microsaura*, one of the smallest varieties, only a few inches long head to hips. Other species reach a foot or more.

squeezing the tongue warhead at great speed out of its contracted state. Another Congolese beetle is swept into the high-tech chameleon jaws.

Lizards labor under the disadvantage of being the least publicized reptile clan, but their close kin the serpents bear the worst prejudice handed out by human society. This is unjust. Snake anatomy contains the most clever and intricately efficient feeding apparatuses to be observed anywhere among land vertebrates. Our human problem begins with our adaptive table manners; we're not accustomed to admiring creatures that can swallow something larger than their heads. All human parents, from Boston mayors to boomerang-wielding natives, warn their children not to stuff too large a hunk of food in their mouths. Human gullets are small and have only modest capacity for expansion. It's ecologically adaptive for human parents to discourage gulping big pieces of food, because choking is an uncomfortably common agent of human mortality.

Snakes, however, cannot chew. The evolutionary path they chose early in their career required unusual adaptations for swallowing huge hunks of food: (1) snakes are all predators, subsisting mostly on live prey; (2) they ambush by stealth, not by moving about scanning for victims, hence snakes don't meet a lot of potential prey each day; (3) therefore snakes have to make the most of each opportunity and should attack the largest potential prey. The Darwinian processes that favor the selection of big prey have also equipped serpents with their special organs for throttling and stabbing. Pythons have a crushing attack. They coil around large victims, constricting whenever their prey exhales, suffocating it slowly and with an economical expenditure of force. (Contrary to popular myth, big constrictors don't crush bones and pulp their victims into pudding; just enough force to asphyxiate seems to be the rule.) The poison attack evolved by several other snake families allows them to inject their venom with surgical precision through hollow fangs. Once the big victim is subdued by the constrictor's embrace or by a dose of poison, the snake must swallow it whole, because no snake has cutting teeth suitable for slicing the victim's body into bite-sized pieces.

Here is the nub of the problem: Snakes are long, narrow beasts with heads of very small width compared to many lizards and most frogs. Such a small, narrow head is a necessary component of the snake's fundamental mode of movement, sliding through narrow paths and down burrows. A giant tropical toad may have a mouth nearly as wide as its body is long, and so it can gulp down prey nearly as large as itself. But the poor puff adder, having successfully brought down a monkey offering enough meat to keep the snake going for a month, now faces an item of food at least *twice* the width of its own mouth. The solution to this gustatory dilemma has generated the most elegant cranial architecture in land vertebrates.

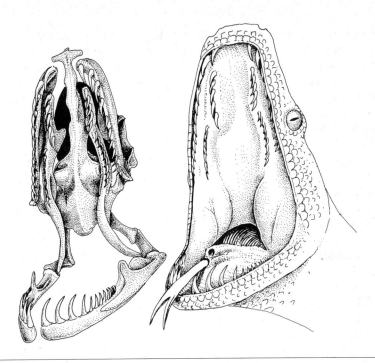

How to swallow something larger than your head—snake-style. A view directly into the wide-open mouth of a boa constrictor. All the upper and lower jaw bones are loosely connected by elastic ligaments, and each side of the skull has not one but two rows of curved teeth.

A great snake in the act of swallowing something larger than its head presents a marvel of reptilian engineering. The snake manipulates its prey's body with its mouth, until it faces the prey's head and the prey's limbs point away. Then the snake opens its jaws and begins to engulf the monkey. Not hurriedly, not with crude gulping and gnashing of teeth, but deliberately and precisely, the snake draws the monkey's head and shoulders into its mouth. We humans are limited by our rigid and brittle jaw, whose right and left sides are firmly joined at the chin so that the width of our mouth is fixed. The right and left halves of the serpent's jaw are joined only by an elastic ligament, so the "chin" can stretch as the monkey's head is swallowed. Within both right and left lower jaw, the snake possesses a hinge that allows even more expansion. Our human jaws move sideways only slightly where they meet the skull at the jaw joint, just in front of the ears (try moving your jaw from side to side with your finger resting on the jaw joint—you will feel only about ⅛″ of movement). But the right and left halves of the snake's jaw are hung on the skull by a long, folding strut, divided into two hinged sections like a carpenter's ruler. As the snake engulfs the monkey's shoulders, these joints swing outward on their flexible struts, enormously increasing the gullet's diameter to accommodate the outsized prey.

So far we have witnessed only the passive aspect of the puff adder's swallowing act—the hinges and elastic joints being pushed out by the prey's body as it is drawn into the snake's mouth. But the greater marvel is the way the snake powers its jaws to drag the prey down its throat. We think of swallowing as a minor muscular feat. We chew a few dozen times and gulp. Down goes a little masticated food accompanied by minor contractions of our tongue and esophageal apparatus. Our chewing muscles do most of the work; swallowing is not a major event. But since snakes don't chew, the entire body of the monkey is actively drawn into the snake's throat by the backward pull of fanged jaw bars, two above and two below. Unlike the soft roof of our mouth, the snake's palate possesses bars of bones, studded with backwardly curved teeth, on each side. The snake's jaw muscles can manipulate each palate bar backward by itself, the recurved teeth dragging the prey backward into the throat. After the bar has pulled as far backward as it can go, the jaw muscles lift it up and forward, while disengaging the curved teeth from the prey, and move the bar forward to start another stroke. The lower jaws can also be retracted independently, one side at a time, to aid in dragging the monkey down its throat.

To get a mental picture of the process as it might work in our heads, imagine that your jaw could expand at chin and jaw joint; imagine that you had two short hands, each holding a fork, attached to the roof of your mouth. You have a big monkey on your plate. You wrap your expandable jaws around it and your palate-forks stuff it down your throat in alternated strokes until the whole monkey carcass slides down. Finally, only the monkey's tail can be seen disappearing into your mouth.

No other land vertebrate today swallows more elegantly than the snakes. Serpent success—nearly three thousand living species—surely owes much to this sophisticated machinery for digestion, which allows snakes to exploit very large prey relative to their own body size. Human evolution produced a rather dull, simple jaw apparatus. Our brain size permitted us to compensate by inventing stone knives, steel carving sets, and Cuisinarts, so we can take a whole steer and swallow it, piece by piece. We should admire how evolution has solved this prey-bigger-than-your-head problem in snakes with entirely internal adaptations.

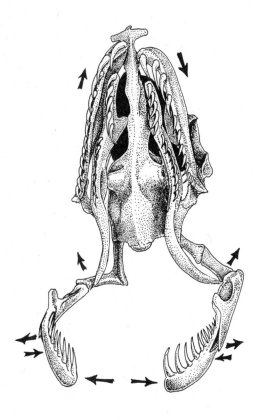

How the boa head works. In the roof of the mouth the two double-tooth rows move alternately—the left side pulls the prey backward down the throat as the right side reaches forward, and vice versa. To expand the gullet, all the cranial joints bend outward: The rear jaw strut swings out, the joint in each lower jaw flexes, and the right and left lower jaws stretch apart at the chin.

Panzercrocs—the exception that proves the rule.
During the Age of Mammals, very few cold-
blooded reptiles evolved large size and aggressive
habits and challenged the warm-blooded Mammalia.
An exception was the Panzercroc—*Pristichampsus*—
an eight-feet-long crocodilian that evolved long,
fast-running legs and hooflike claws for land
locomotion and steak-knife–like teeth for killing
and cutting up mammal prey (shown here is the
Dawn Horse, *Eohippus*). *Pristichampsus* hunted
during the Eocene Epoch, about 49 million years
ago, but it was very rare, much rarer than big
mammalian predators, proof that cold-bloodedness
was a great disadvantage.

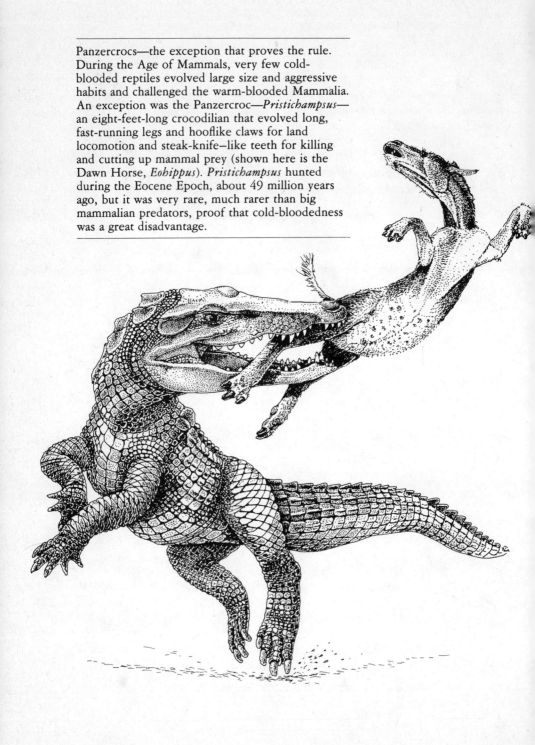

4

DINOSAURS SCORE WHERE KOMODO DRAGONS FAIL

Dinosaurs must be viewed as a giant evolutionary system, a vast conglomerate of species who shared a common adaptive plan. No adaptive plan is perfect—neither warm-bloodedness nor cold-bloodedness, for example, works best all the time. If we probe the nature of the dinosaurs' success, we can feel out the basic strengths and weaknesses that existed within the dinosaurian organization. And that will allow us to understand more about precisely what kind of animals they were.

Orthodox theory has it that dinosaurs were merely "good reptiles," essentially scaled-up versions of modern lizards and crocs whose metabolism was pitifully low compared to mammals'. So we can begin our inquiry into the nature of the dinosaurs' success by asking, What are the limitations of the cold-blooded reptiles—where do they fail today? As we saw in the previous chapter, reptiles and amphibians do overwhelmingly outscore mammals in total species count. But it must also be said that there are ecological categories where the cold-blooded league is almost entirely shut out. If the basic organization of the dinosaurs really was reptilian, then the pattern of deficiencies we observe in today's Reptilia should match the picture we get from the dinosaurian world. But what if the dinosaurs' successes turn out to be totally different from those of modern reptiles? What would that mean for the orthodox theory? If we discover that dinosaurs succeeded where modern reptiles fail, and vice versa, then such a theory would be totally incorrect.

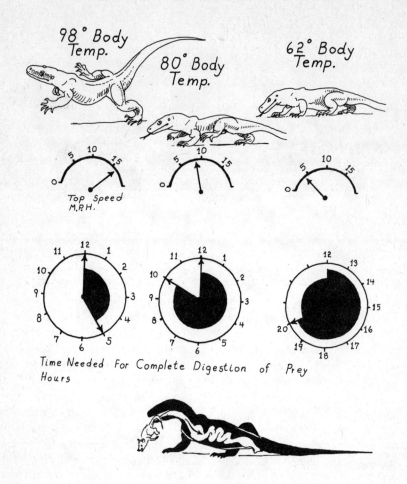

98° Body Temp.

80° Body Temp.

62° Body Temp.

Top Speed M.P.H.

Time Needed For Complete Digestion of Prey
Hours

Why body temperature is so important. All physiological performance peaks at one narrow temperature range, and the whole body machinery slows down when body temperature falls. Many lizards are at peak form at a body temperature close to a human's—about 98 degrees F. But when body temperature drops 10 degrees C (18 degrees F), performance drops to half—running speed is half as fast and digestion takes twice as long. When temperature drops another 10 degrees C, performance falls to only one quarter of the peak levels.

Super-giant tortoise—*Colossochelys*. Today a big Galapagos tortoise can reach fifty inches long (measured front to back on the bottom shell) and five hundred pounds. But a few million years ago *Colossochelys* grew to eighty inches and four tons or more. Shown here is the profile of a five-hundred-pounder with rider and, in silhouette, the giant *Colossochelys*.

In this chapter we can begin by considering the reptilian giants that came after the end of the Cretaceous, after the end of the dinosaurs. These cold-blooded monsters evolved during the Age of Mammals. Their story teaches many lessons about reptilian failure. *Colossochelys,* the king of the giant tortoises, presents one such lesson. Bones from the two-million-year-old sediment of the Siwalik Hills in India contain fragments of elephant, hyena, hippo, bear—and of *Colossochelys.* The fossils of the tortoise king has been found everywhere in the Old World tropics, from Kenya to Cape Province to Java (fragments suggest its presence in Florida, too, during this age). Everywhere it was found, there was an accompanying rich fauna of big, modern-type mammals. Complete *Colossochelys* shells are one of the most breathtaking displays in all of terrestrial turtledom. They look like fossilized Volkswagen Beetles, enormous bone domes six feet long and three feet high. When alive and fully grown, *Colossochelys* dwarfed even the largest giant tortoise alive today.

Tortoises are paradoxical reptiles. Their history is a success story, but they also betray the basic flaws in the economic organization of the Reptilia. Tortoises constitute a family of the turtle group, and represent the acme of turtle adaptation to dry-land habitats. Their feet are super-compact, with short toes and elephantlike cushion pads, and their beaks enable them to crop grass like a cow (most other modern reptiles are carnivores). Tortoises aren't an ancient tribe at all; the first tortoise didn't evolve until the Eocene Epoch of the Age of Mammals, fifteen million years after the last dinosaur died. So tortoises are one of the very last big-bodied reptiles to make their appearance. And their dome-shelled clan scored major ecological successes for forty-five million years despite the potential danger from mammalian meat-eaters and plant-eaters.

Giant tortoises could defeat any mammalian predator except one—man. Because of man, two-ton tortoises are totally extinct today and even the three-hundred-pounders are very rare, restricted to a few desert islands—the Galapagos off Ecuador and the Aldabras in the Indian Ocean. The demise of the giant tortoises is thus a very recent event for which our own species is probably to blame. One human hunter couldn't kill a giant tortoise easily, but six together could use a branch to tip the tortoise on its back, then build a fire under its shell, and stew the poor beast in its own carapace. Human hunters were a very late development in the Age of Mammals, and they started multiplying significantly only in the last two million years. But once they got going, our primordial forefathers cut a wide swath through both the Old World and the New, exterminating dozens of big species of mammal—mammoths, mastodons, saber-toothed cats, giant ground sloths, to name but a few. And they killed giant tortoises. That must not allow us to forget the remarkable success of giant tortoises up till the advent of human hunters. The best nonhuman predators couldn't kill off the big tortoises; saber-toothed cats, giant bears, oversized hyenas, clever wolves, all in their heyday failed to suppress *Colossochelys*.

These giant tortoises demonstrate that cold-blooded reptiles could handle mammalian enemies (nonhuman ones). But their manner of success also reveals the limitations of the reptile's adaptive equipment. Tortoises were "cold-blooded" in the narrow, physiological sense of the word: they had very low metabolism compared to that of a mammal of the same size. A tortoise could heat up its body tissue only if it had access to abundant solar energy in the form of direct sunlight or solar-warmed sand or rocks. Tortoises were also typically reptilian in the small size of their heart and lungs. They couldn't keep up a level of activity anywhere nearly as high as a dog, bear, or hyena could. How, then, could the tortoise overcome the debilitating effects of its low metabolic performance? Armor. Quite simply, tortoises succeeded because they didn't confront mammals in direct tests of strength and coordination. Tortoises didn't have to flee the lions the way wildebeest do today, with a burst of high speed. And tortoises didn't have to defend themselves the way wild boar do, with aggressive counterattacks. When threatened, the tortoise simply pulled in all its appendages—head, tail, and legs—and waited out the danger, with vulnerable body tissue withdrawn into its incredibly strong bony shell. If it had to, a giant tortoise could wait for hours, for days, even for months, because its low metabolism allowed long fasts. Tortoises beat mammal attacks by a totally passive defense. There's a lesson here, one that orthodox paleontologists ignore: When warm-blooded mammals abound, reptiles can't evolve large size on land unless very special adaptations permit the reptiles to avoid direct confrontation.

In the very same tropical woodlands and bush where giant tortoises flourished for so long, another exceptional reptilian evolved—the giant land snake. Big snakes labor under the same limitations as tortoises when faced by mammal predators. Giant pythons have a low metabolism so they can't keep their bodies warm if solar heat isn't abundant. Their heart and lungs are low-powered affairs compared to the typical mammalian design. And snakes can't compete in prolonged contests of violent activity. How, then, do giant snakes survive among the lions and hyenas? By stealth and patience. Giant snakes don't try to compete with hyenas in running down antelope over long chases. They don't prowl over hundreds of acres the way lions do. A twenty-foot python glides silently out of its hole near a waterhole's edge and lies in wait, concealed by its camouflaged hide and long, low silhouette. Here again, the low reptilian metabolism permits the giant snake to wait as no mammal could, for weeks if necessary. Finally, an unwary antelope comes to drink and steps too close to the python. In one motion two hundred pounds of snake coil around the antelope's chest. The snake kills by subtle alterations of its grip, not by violent contractions. The snake's body musculature can't work a tenth as hard as a lion's over an hour's time. But by contracting every time the antelope exhales, the snake's coils can finally tighten into a suffocating straitjacket.

Big snakes, like big tortoises, deserve credit for their success in the Age of Mammals. Giant land snakes have hunted mammals in all tropical continents for the last thirty million years. But we can see that their success comes only by avoiding mammal-style hunting tactics. The great serpents succeed by being something a warm-blooded mammal could never be—a hunter of infinite patience, with a legless body designed for maximally cryptic locomotion and ambush.

Is there any reptile which has successfully challenged large land mammals for the role of normal, four-footed predator? The Komodo dragon lizard is a possible candidate. Orthodox paleontologists often point to the Komodo dragon as the perfect modern analogue of the dinosaurs—a big terrestrial reptile that succeeds in dominating a warm tropical ecosystem. But the Komodo dragon is a red herring.

The truth about it helps demonstrate that the dinosaurs' success couldn't possibly be the result of a lizard-style metabolism. Komodo dragons, it is true, can kill the largest land mammals on Komodo—even adult horses and water buffalo. But the dragon rules a kingdom of tiny extent. No dragons survive on the nearby big islands of Java or Sumatra. Dragons swim well and could easily get to these bigger areas, yet their entire breeding population remains restricted to a handful of tiny islands. And unlike the tortoises, these dragons in the past have never extended to the big islands and mainland areas. There's an obvious explanation for these geographical limitations. The dragon succeeds only where it's free from interference from large mammal predators. Leopards, tigers, and sun bears prowl the big Indonesian islands, and there were large hyenas too until a few million years ago. The mainland of Southeast Asia has hosted big cats, wolves, and hyenas in dangerous profusion. On Komodo Island not one large mammal predator has ever existed in the wild. (Natives keep dogs on Komodo, but these canines are a wretchedly scrawny lot, hardly a threat to the *ora*.)

The conclusion is inescapable: Giant predator lizards can't evolve in the presence of big mammal predators. So the lesson is that mammals suppress much of the evolutionary potential of modern lizards. Is the Komodo dragon a good working model of how dinosaurs succeeded? Absolutely not. Dinosaurs suppressed the evolutionary potential of mammals, not the other way around. And dinosaurs carried out this suppression everywhere, on all the continents, not merely on a few tiny tropical isles. Dinosaurs succeeded where Komodo dragons fail.

Crocodiles today teach much the same lesson concerning the limitations of reptiles. They certainly are dangerous to big mammals, but croc hunting tactics are yet another admission of reptile inferiority in direct confrontation. Nearly all the large mammals killed by Nile crocodiles are caught near the water's edge. Modern crocodiles don't go hunting much over dry land, and don't challenge mammals in the role of terrestrial meat-eater out on the savannah or in the woodlands. Croc tactics are variants of the basic reptilian theme: avoid confrontation with big mammals on land, ambush from special sites that give a reptile an edge. Their low metabolism allows crocs to stay underwater much longer than a mammal or bird could, and thus tropical rivers and streams have remained the locales for an evolutionary proliferation of big crocodilian predators all through the Age of Mammals. But on land, crocs don't score. (There was a mammal-killing croc on land in the Eocene Epoch [forty million years ago], but it was rare except in swamps.)

All these facts of modern reptilian failure are damaging to the orthodox theory of dinosaurs, which consists of one central credo: Dinosaur metabolism was nothing unusual, merely the standard lizard-style system blown up to accommodate multi-ton monsters; dinosaur hearts and lungs were as inferior to the big mammals' as giant tortoises' were. If this credo is correct, then the dinosaurs' successes and failures should follow the identical ecological pattern to that of the modern Reptilia. But the entire history of the dinosaurs is totally and indisputably the opposite of the tortoise–lizard–croc–turtle history today. Let's summarize the ecological box scores:

Modern reptiles score very high, higher than mammals, as small-sized species. But dinosaurs produced no really small species, not one with an adult weight of less than two ounces (the average for lizards), and very few of less than ten pounds. So dinosaurs failed miserably where modern reptiles succeed magnificently.

Modern reptiles dominate the role of large freshwater predator. But dinosaurs didn't produce any swimming predators at all. All the dinosaurian meat-eaters—*Tyrannosaurus, Allosaurus,* and their ecological colleagues—were basically dry-land types. Again, dinosaurs failed where modern reptiles succeeded.

Modern reptiles and their cold-blooded cousins the Amphibia score very high as *small* freshwater predators—the lakes and streams abound with little swimming frogs, snakes, and turtles. But not one dinosaur was specialized for this type of role. Yet again, dinosaurs failed in roles where modern reptiles and amphibians succeed.

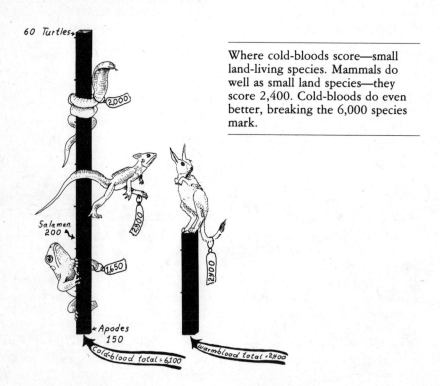

60 Turtles

2,000

Salaman. 200

2,900

1,650

Apodes 150

Cold-blood total = 6,100

2,400

Warm-blood total = 2,400

Where cold-bloods score—small land-living species. Mammals do well as small land species—they score 2,400. Cold-bloods do even better, breaking the 6,000 species mark.

13 Crocs
50 Snakes
20 Lizards

160

120

960

50

Where cold-bloods score—
small species in freshwater.
Only 50 small mammal
species make their living in
streams, lakes, and ponds.
But nearly 1,300 species of
cold-blooded reptile and
amphibian fill out these
ecological roles.

Where cold-bloods score—
big-bodied species in
freshwater. There's only 1
really big mammal today in
the semiaquatic niche—the
hippo. But there are 15
crocs, turtles, and snakes in
this ecological category.

coldblood total : 1290

warmblood total = 50

13

1

1

coldblood total = 15

warmblood total = 1

Modern reptiles fail nearly completely as big, active land predators wherever land predators roam, and mammals clearly suppress the evolution of big Komodo dragon–type hunters. But dinosaurs excelled at being big, land predators, and the dinosaurs suppressed the evolution of large mammals. Therefore, dinosaurs succeeded where modern reptiles fail.

Modern reptiles can evolve large body size only if they possess special adaptations—tortoises have their armor and giant snakes their stealthy shape and habits. But only a few dinosaurs were heavily armored, and every dinosaur had relatively long legs. Dinosaurs didn't slither about, trying to hide. They succeeded gloriously as big, active land critters, roles where the Reptilia fail.

In the presence of these facts, is there any way of saving the orthodox theory of dinosaurs? Can the idea of *Tyrannosaurus* and *Brontosaurus* as giant cold-bloods be salvaged? A number of paleontologists believe so. They rest their belief on a theory called "mass homeothermy." This theory maintains that dinosaurs succeeded as cold-blood reptiles, and didn't require a high metabolism because they kept their body temperatures high and constant simply by evolving gigantic body size. "Homeothermy" literally means constant temperature, and "mass" refers here to body mass. In a word, mass homeothermy means keeping warm by being huge. Yale Professor Richard Swann Lull was the first to spell out this

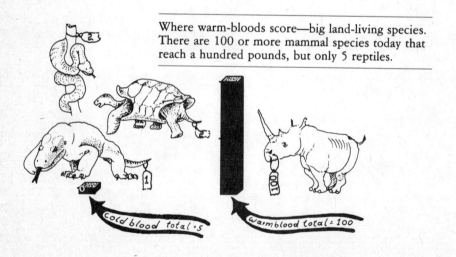

Where warm-bloods score—big land-living species. There are 100 or more mammal species today that reach a hundred pounds, but only 5 reptiles.

Cold blood total = 5

Warmblood total = 100

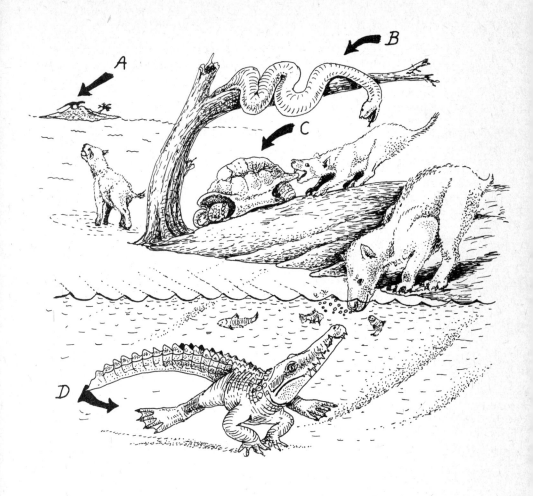

How big reptiles cope with big mammals. During the Age of Mammals, big cold-blooded reptiles evolved four different ways of surviving: a) Live on a remote island too small for big mammals (the Komodo Dragon took this route); b) Evolve a cryptic, camouflaged body form (giant pythons and boas are examples); c) Evolve stout body armor (giant tortoises); d) Evolve aquatic habits in order to stay under water much longer than a mammal can (a tactic used by big crocodiles and turtles). Which of these four methods did big dinosaurs use? Answer—e) None of the above.

idea, back in the 1920s, though the general notion had been suggested long before. The idea is popular because it focuses on ecology's most important working principles: The principles of how the performance of every bodily organ, from brains to intestines, is altered by the ebb and flow of body heat, and of how body size controls the way in which bodies gain and lose heat.

Mass homeothermy recognizes, quite correctly, that "good reptiles" and "good mammals" have totally different solutions to the problems of heat. The Reptilia have a fundamentally laid-back, nonconfrontationist approach to ecological action and reaction. Mammals, on the other hand, are aggressive and compulsive about food, and seem positively frenetic compared to their reptilian

---

How cold-bloodedness works. When the sun's rays are warm but not too hot, the ten-pound lizard's blood is every bit as warm as the ten-pound pig's. But when the sun's rays are blocked by clouds and rain, the lizard's metabolism is much too low to keep its body temperature up and its mental and physical condition slips into a somnolent torpor. If the sun is too hot, the lizard can't sweat or pant the way a mammal or bird can and the poor lizard's brain heats up until it addles.

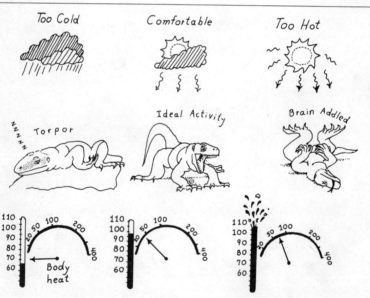

neighbors. Consequently, reptiles have very low yearly metabolic needs compared to most mammals' and, on average, a reptile doesn't need to find food every day.

A three-ounce mammal (chipmunk size) has to scurry about every day to gather nuts and berries to stoke its metabolic furnace. The mammal therefore is forced by its metabolism to be a confrontationist; it must go out and confront the weather and predators and competitors daily. But a three-ounce lizard can stay tucked snugly in its burrow for weeks, waiting until all is safe before it scuttles out to forage for food. High metabolism does give the chipmunk some advantages. The constant supply of body heat lets the mammal keep its temperature high and constant most of the time despite fluctuations in the weather. Everything else being equal, constant body temperature is beneficial because enzymes—the chemicals that keep bodily processes working—reach peak output within a narrow range of temperatures. And so perfect homeothermy allows evolution to fine-tune any creature's physiological mechanisms.

Hot-blooded metabolism buys freedom in time and space. If a species has a high heat production, it can forage around for food at peak efficiency in the shade. But a cold-blood must shuttle back and forth, basking in the sun to warm up before chasing prey in the shade.

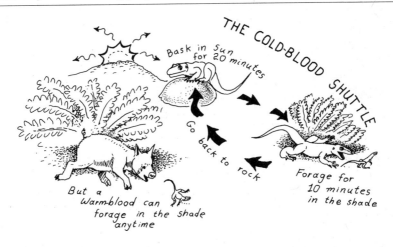

If body temperatures fluctuate wildly, on the other hand, then internal body chemistry can never settle into an optimal mode. If tissues get too cold, metabolism will slow to stalling speed. If tissues overheat, the enzymes can denature and the creatures' innards addle (the central nervous system, for example, is the most sensitive in humans; brain death takes only a few dozen minutes at 108°F).

How warm-bloodedness works—Part 1. Typical mammals and birds have super-high body-heat production nearly all the time. When the weather is warm, blood flow to the skin increases, so more body heat escapes into the air. When the weather is cool, blood flow to the skin decreases, so more of the body heat is kept in the body.

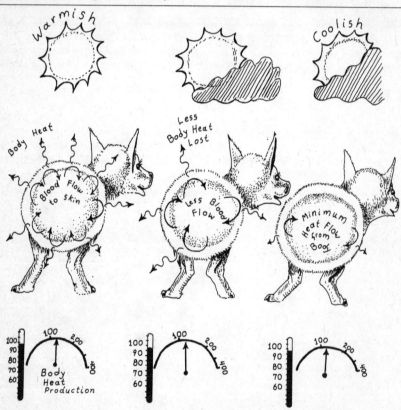

A useful rule to help us understand all this is that $Q_{10} = 2$, which means that for every ten-degree change in body temperature (measured in Centigrade), the rate of a physiological process changes twofold. According to this formula, a lizard which enjoys peak enzyme activity at 38° Centigrade (normal human body temperature) would suffer a decline to one half of optimal rates at 28°C and to one quarter at 18°C. A chipmunk, with its high metabolism, can keep its internal chemistry operating optimally even when its habitat cools. Therefore the chipmunk can run at top physiological efficiency even when it spends hours foraging in deep shade and in other locales lacking warmth. The three-ounce lizard is much more severely constrained geographically. Its metabolism isn't strong

How warm-bloodedness works—Part 2. Birds and mammals have extra physiological adaptations for extreme weather. If it gets too hot, sweating or panting will increase the heat loss from the body. If it gets too cold, shivering will increase the body-heat production two or three times.

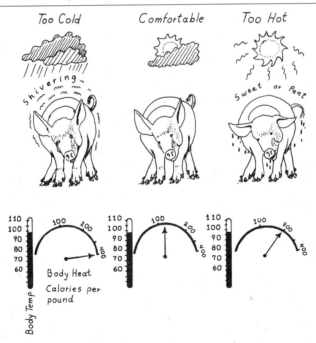

enough to keep its body temperature constant in cool, dark places. High body heat also gives the chipmunk more flexibility in time than its lizard neighbor. The mammal can keep its temperature high even during the cool parts of the day, during the early morning and evening of summer, or all day during winter. Yet the reptile has some compensating advantages. Since it doesn't have to fuel its metabolic fires as continuously, it can afford to wait until conditions are just right before it risks confrontation with dangerous neighbors.

These are the principles that define the boundaries of the reptile's modern ecological successes: physiological guerrilla warfare, conflict by hit-and-run, wait-and-hit. These reptile rules work perfectly for relatively small species. Small snakes and lizards can hide in hollow logs, burrows, or up in the trees when enemies threaten. Eight thousand living species of land reptile and amphibian follow variants of the wait-and-hit strategy. All are small enough to stay protected in their habitat lairs, waiting for the opportune time to emerge. There's nothing cowardly or disreputable about this reptile strategy; their physiological equipment simply represents an alternative mode of adapting compared to the constant hyperactivity of most mammals.

Wait-and-hit strategy works only if the reptile has a safe place to wait. And there's the great problem for big land reptiles: finding a hole to hide a two-hundred-pound lizard is difficult. Komodo dragons seek caves or other lairs, but the bigger the lizard grows the fewer the lairs that fit. Tortoises solve this problem by carrying their own cave with them wherever they go. No other big reptile has solved the problem so well.

But where could a two-ton *Allosaurus* hide? The theory of mass homeothermy maintains dinosaurs didn't need such holes to hide in because they were so big their bodies never cooled to dangerously low temperatures. Two laws concerning body heat and body size serve as the foundations for this argument. First, bigger bodies produce less body heat per pound per hour. Second, bigger bodies lose less body heat per pound through the skin. Together, these two laws mean that it's easier to keep warm in a big body than in a little one.

Physiologists describe these laws as examples of the "mouse–to–elephant phenomenon." All through the animal kingdom, the

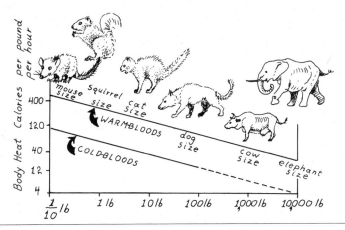

Big or little, every cold-blood puts out much less body heat than a warm-blood of the same size. If you have a lizard warmed up to 98.7 degrees F, its body heat is about one fourth as high as a typical mammal of the same body bulk. And both warm-bloods and cold-bloods produce less heat the bigger they get. If we increase body size ten thousand times, the heat production drops to one tenth.

production of body heat drops in a very regular way as body size increases. A simple mathematical shorthand defines this phenomenon: $M = k/W^{-\frac{1}{4}}$; M is metabolic heat production, W is body weight, and k is a constant. A bunny weighs about one pound; a five-ton elephant is 10,000 times heavier. So the elephant's production of body heat is $(10,000)^{\frac{1}{4}}$ times less per pound, or ten times less per pound than that of the bunny.

An old saw perfectly illustrates this mouse–to–elephant phenomenon: "What will keep you warmer on a cold night at the zoo, snuggling up with a five-ton bull elephant, or with 10,000 bunnies who altogether weigh five tons?" Answer: The bunnies. They put out ten times as much heat.

Producing less heat per pound, however, doesn't mean a big animal is colder than a small one—just the reverse. When a vertebrate body is at rest, it loses heat to the environment mostly through its skin. If it has a lot of flesh per square inch of skin, it saves heat. If size goes up, the skin area per pound of flesh goes down. Hence the big animal keeps warmer more easily because the area of its skin surface is less, relative to its heat output. (The mathematical shorthand for this corporeal geometry is $A = k/W^{-1/3}$; A is skin area per pound, W is body weight, and k is a constant. This relationship holds true only as long as body shape stays similar. So we can't use the same formula for snakes and turtles.)

Now let's compare the bunny to the bull elephant again. The elephant is 10,000 times heavier than a bunny. Bunny and elephant have roughly similar shapes—a compact body and one set of skinny protuberances (trunk for elephant, ears for bunny). The elephant has much less skin per pound—about 22 times less than the bunny. So the elephant has proportionately much less skin area through which to lose its body heat. The two mouse–to–elephant thermal laws therefore combine to give the big animal better heat-conserving properties. The elephant's ratio of 22 times less skin per pound compensates for its ten times less heat production per pound, granting it a net advantage of $22 \div 10$, or about 2.2. The elephant produces body heat 2.2 times faster per pound per square inch of skin than the bunny.

The biothermal bottom line here is this: It is easier to chill a mouse than an elephant. Zoo keepers know this from experience. Big mammals, even species from tropical homelands, adapt to winter outside in northern zoos better than do small species. Open the cage window, let the cold draft in, and the elephant doesn't feel much. But the poor mouse starts shivering immediately and literally will shiver itself to death in an hour or so. (Shivering is the mammalian body's way of increasing heat production to meet the thermal crisis. Shivering burns up the calories at enormous rates, up to five times the standard metabolism, and a shivering mouse literally can burn itself out quite quickly because the body is so small.)

Today's Reptilia share their own version of this mouse–to–elephant formula. Big reptiles, like giant tortoises, produce less body heat per pound than little ones, and big reptiles also have much less skin area per pound. So it happens naturally that a giant lizard or tortoise doesn't chill as quickly as a little one. If you take heat lamps and warm your three-hundred-pound tortoise and your three-pound box turtle to 90°F body temperature, and put both of them outdoors on a cool cloudy day, the big tortoise will lose its body

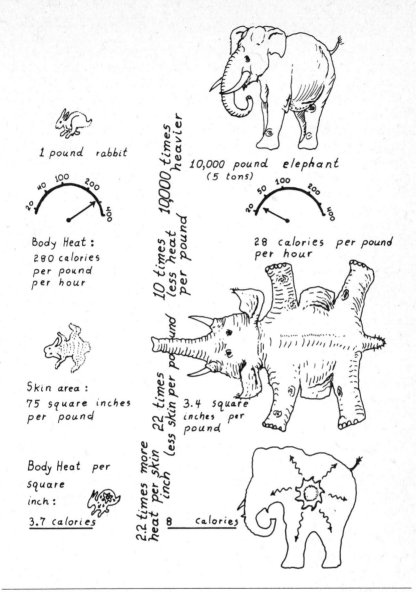

1 pound rabbit

10,000 pound elephant
(5 tons)

Body Heat :
280 calories
per pound
per hour

28 calories per pound
per hour

10,000 times heavier

10 times less heat per pound

22 times less skin per pound

Skin area :
75 square inches
per pound

3.4 square
inches per
pound

Body Heat per
square
inch :

3.7 calories

2.2 times more heat per skin inch

8   calories

Why big animals stay warmer easier. An elephant is ten thousand times heavier than a rabbit and produces body heat one tenth as rapidly. But still the elephant keeps warmer because it has much less skin area per pound and much more body heat per square inch of skin.

heat much less quickly. However, no matter how big the reptile is, its metabolism will always be lower than that of a "warm-blooded" mammal or bird of the same weight—about four times lower. So a three-hundred-pound pig keeps warm more easily than a three-hundred-pound tortoise (in terms of the standard metabolic formula, $M = k/W^{-1/4}$; that means the reptile k is one fourth the mammal k, and the reptile metabolic rating is $1/4$ that of the mammals).

The theory of mass homeothermy starts by assuming a point of view. Assuming that dinosaurs were cold-blooded and had low metabolism, can we explain their success? The theory *begins* by assuming "cold-bloodedness." Stuck with the model of hypothetical dinosaurs who produce low body heat, the only way to make their body temperature stay reasonably high and constant is to make their bodies as huge as possible. So the theory says that dinosaurs were successful because they evolved gigantic body size and consequently their body temperature was maintained without the need for excessive metabolism. This theory would work best in a warm, tropical climate. And fossil plant evidence shows the Mesozoic world was, on average, much warmer farther up toward the poles than it is today. So the environmental context would seem perfect for giant homeotherms with low metabolism.

Orthodox paleontologists have rallied round the standard of homeothermy, triumphantly proclaiming it obviates any need even to consider the hypothesis of high metabolism in dinosaurs. But their enthusiasm is ill-founded. The theory doesn't work and it's fairly easy to demonstrate its flaws. A few years ago a young biophysicist, Jim Spotila, worked up a computer program to show how a two-ton reptile with low metabolism might regulate its body temperature. Jim placed the hypothetical beast into a nearly ideal climate—present-day southern Florida—and never allowed rainfall to influence its body heat. Jim's theoretical two-tonner did pretty well. It could maintain its body temperature between 30°C and 38°C for most of the year. And its body temperature never rose to where it might addle its brain, nor fell to where it might get frostbite.

The advocates of mass homeothermy seized upon Jim's computerized two-ton lizard as proof positive that dinosaurs didn't need high metabolism. But they missed an important point: The two-ton lizard might certainly regulate its temperature better than a little two-ounce lizard, but a two-ton dinosaur with high metabo-

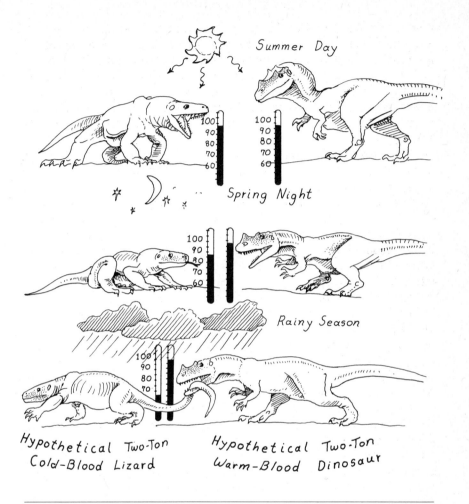

Summer Day

Spring Night

Rainy Season

Hypothetical Two-Ton Cold-Blood Lizard

Hypothetical Two-Ton Warm-Blood Dinosaur

Why the mass homeothermy theory doesn't work. Compare the performance of a hypothetical two-ton cold-blooded lizard with a two-ton warm-blooded dinosaur. In a warm climate with clear skies the two-ton cold-blood can keep its body temperature high and constant during the daylight hours. But on cool nights the lizard's temperature will slip ten degrees below the warm-blooded dinosaur's, so the dinosaur would have a distinct advantage. And during the rainy season, when the sun is blotted out for weeks on end, the two-ton cold-blood simply won't have the body heat to prevent a disastrous fall in body temperature.

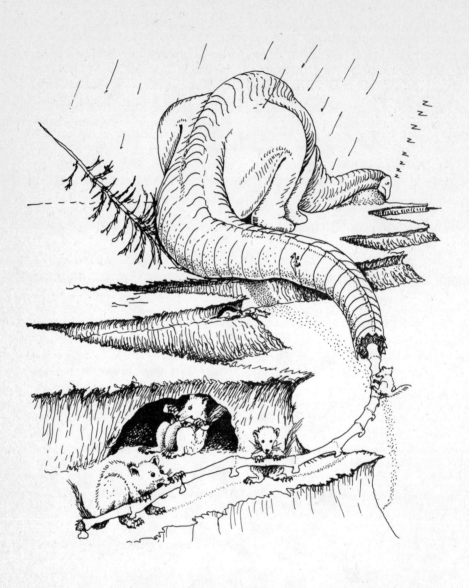

Mesozoic nightmare—being a cold-blooded dinosaur during the rainy season. If big dinosaurs really were mass homeotherms, then the rainy season would have sapped their body heat and left them torpid and vulnerable to the warm-blooded mammals.

lism would do much better than that, much better. If Jim's computerized beast had possessed the high metabolism typical of mammals, it could have kept its body temperature between 38°C and 38.5°C all year, 6 or 8 degrees less variation than the low-metabolism model. And if Jim's computer had allowed rain to fall on the beast, the high-metabolism version would greatly surpass the low-metabolism model, because the high-metabolism model could shiver to raise heat production so high that rain wouldn't lower body temperature at all (no reptile can shiver like a mammal or bird).

What sort of advantages would a dinosaur with high metabolism garner from temperature regulation that keeps variation up to eight degrees less than one with low metabolism? As we have seen, a drop of eight degrees in body temperature implies a drop of 20 to 50 percent in physiological prowess (the exact drop, remember, depends upon Q_{10}). If two-ton, high-metabolism dinosaurs met two-ton low metabolism dinosaurs in southern Florida, sooner or later the animals with low metabolism would find themselves outclassed in all the ecological contests necessary for survival—running, fighting, digesting, mating, growing. In direct confrontation, high metabolism always conquers low metabolism, even when bodies are huge and climate warm. Even if the advantage of high metabolism was only 10 percent, the laws of evolution would force the low-metabolism model to extinction. Geneticists in the 1930s proved that even a tiny net advantage, say 5 percent, would imply that one adaptive system would replace another over hundreds of generations.

Applying these ideas to the questions of the Mesozoic, how could dinosaurs have suppressed mammals for over a hundred million years? Merely by being big? No, it wouldn't work. Dinosaurs could not have maximized their physiological output simply by being big. Those Jurassic and Cretaceous mammals must have had some sort of high metabolism—since all the most primitive living mammals do. If the dinosaurs were equipped only with low-metabolism biothermal weaponry, they couldn't have prevented the Mammalia from evolving to fill all the large-bodied niches. Low-metabolism dinosaurs would have survived only by staying small, hiding in their holes, nipping out to forage when the big mammals weren't looking—just as modern reptiles do.

Mass homeothermy falls into another error: it entirely ignores the small- and medium-size dinosaurs. There weren't any tiny (less than two-ounce) dinosaurs—or at least none have been found. But there were ten-, fifty-, and hundred-pounders, and this size range contained the dinosaur species that would have interacted with the Mesozoic mammals. A twenty-ton *Brontosaurus* probably didn't interact with the two-ounce mammals of Como Bluff. The brontosaur ate tree leaves and would have swallowed a mammal only by accident, as we might swallow a caterpillar hiding in a chef's salad. *Ornitholestes* was another story. It is a twenty-pound Como dinosaur with big eyes, sharp teeth, and quick legs for darting through the underbrush and hunting Jurassic small prey. *Ornitholestes* must have hunted mammals—the Como furballs were just the right size to fit the predator's jaws. Before *Ornitholestes,* the earliest dinosaurs of all, *Lagosuchus* of the Triassic Period, were also small, lively hunters. Hence all through the Mesozoic, the dinosaurs supplied mid-sized predators that must have continuously confronted the Mesozoic Mammalia.

These small, mammal-hunting dinosaurs were far too little to reap the theoretical benefits of big body size in keeping temperature constant. The only way *Ornitholestes* could have kept its body temperature high and constant was by having a high constant metabolism. The fact that *Ornitholestes* and its brethren succeeded in keeping the Mammalia small for over a hundred million years is a powerful argument that these dinosaurs possessed basic physiological equipment equal to or better than a mammal's.

Ornitholestes was an impressive little dinosaur, and even the diehard defenders of orthodoxy yield a little to admit that perhaps *Ornitholestes* and its kin might have had high metabolism. Such a concession, however, would lead to yet another inconsistency in the theory of mass homeothermy. Big dinosaurs, all of them, evolved from small-dinosaur ancestors. The idea that little ancestors had high metabolism and their bigger descendants didn't, would be tantamount to arguing that evolution reversed itself. (In mathematical terms, that means the constant k would get smaller as size got bigger.) Modern elephants and rhinos evolved from small ancestors with high metabolism, without reversing their metabolic rating (and without changing their k). If big mammals didn't lose high metabolism, why should big dinosaurs have? In fact, there's

The Late Jurassic forty-pound predator *Ornitholestes* terrorizes a mammal.

not a single documented case of a large descendant completely abandoning a high-energy heritage handed down from a small ancestor. It could happen—in theory—but given the evidence, it is more logical to assume that a small high-metabolism dinosaur would produce big, high-metabolism descendants.

The assumption that big dinosaurs didn't require high metabolism also ignores the fact that each dinosaur community co-evolved with others—the big plant-eaters interacted with medium-size meat-eaters, which interacted with small meat-eaters, and so on throughout the ecological web of relationships. *Ornitholestes* was a close relative of big *Allosaurus,* a predator that reached a ton or more in size. It's difficult to believe that *Allosaurus* had a physiological structure very different from its little cousin—the bony architecture is startlingly similar. So if *Allosaurus* was 100 percent warm-blooded, with a high metabolism, then the plant-eaters that had to cope with it would have required matching physiological adaptations.

Another weakness in the theory of mass homeothermy is its assumption that dinosaurs succeeded only where the climate was warm and tropical. It's true that most of the best-known Cretaceous graveyards—the Judith River Delta in Montana and Alberta, for example—yield strong evidence of warm habitats with mild winters. Big fossil crocodiles and soft-shelled turtles can be found there, and these clearly reptilian types required year-round warmth. Fossil leaves from these sediments represent plants of tropical aspect (the leaves of dicots in the tropics tend toward "whole margin" shapes, with the leaf edge smooth and not sculptured into complicated edges; leaves from habitats with cold winters, on the other hand, tend toward complex shapes like those of our New England oaks). The chemistry of fossil seashells from the nearby marine beds also show that winters were warm (the ratio of the oxygen isotopes O^{16} and O^{18} in the lime shells indicates the temperatures of the seawater when the animals were alive). It's also true that the best-known Jurassic dinosaur beds yield fossil plants that indicate warm habitats (Jurassic flora at Como feature many tropical-type ferns). And finally, all the evidence from plants, fossils, and geochemistry demonstrates that tropical conditions prevailed farther up toward the poles all during the Mesozoic than they do today, so that tropical conditions were present even as high as latitude 45°.

But—and this is a big but—dinosaurs were also the dominant big-bodied land life form in less well publicized sites where the climate was much cooler in Mesozoic days. A good example is in South Australia, where Early Cretaceous dinosaur bones are found in lake beds deposited at 70° south latitude. Fossil plants and geochemistry show that winters here were cold while those dinosaurs were alive—cold enough so that frost formed and lakes froze over. How could a cold-blooded giant dinosaur have survived those chilly Australian winters? Giant tortoises and crocodiles can't cope with such winters today.

The final major shortcoming of the orthodox theory of mass homeothermy is that it ignores what really happens to genuinely giant reptiles with low metabolism. The theory holds that all the benefits of constant body temperature can be enjoyed in a tropical climate, without high metabolism, if body size is large enough. But if being a low-metabolism giant reptile were so efficient, why aren't today's tropics overrun with two-ton lizards and frogs? How many species of multi-ton reptile lurk today in warm terrestrial habitats? None. On the other hand, how many species of tropical high-metabolism mammal presently grow to one ton or larger? Quite a few—three rhinos, a hippo, two elephants, giraffes, some races of water buffalo.

So mass homeothermy doesn't work in today's ecosystems. The message from the tropics is unambiguous: To be a successful big land animal, you must cope with mammals, and to cope with mammals you must be a mammal yourself, or at least have metabolism as high as a mammal's. And big mammals have suppressed big reptiles in our tropics for the last sixty-five million years. So how can the dinosaurs' success over mammals' be explained? By assuming that dinosaurs had low-energy metabolic styles? Not very likely.

To understand the dinosaurs, we need a new theory, a heresy. Or rather we need the renaissance of an old nineteenth-century view, which believed that the dinosaurian system possessed key elements not found in the adaptive tool kit of modern Reptilia.

PART 2
THE HABITAT OF THE DINOSAURS

5

THE CASE OF THE BRONTOSAURUS: FINDING THE BODY

At Como there is a limestone ledge called Cam Bench, named for a *Camarasaurus* skeleton that lies there, eroding out bit by bit. The dinosaur's pale gray, weatherbeaten fragments are slowly disintegrating beneath the endless blows of sun and rain. *Camarasaurus* was a smallish brontosaur, with rather long neck and tail, probably no more than eight tons alive. It has been left to decay because the skeleton isn't complete enough to justify the two weeks of quarrying necessary to chip it free. It is nonetheless quite important because preserved around it is a trail of fossil clues that stare out at us from the day, 140 million years ago, when this animal died. This body can tell us something.

How did this *Camarasaurus* die? If this question can be answered, it will provide unique insights into how it lived, what enemies it feared. This mangled carcass can in fact help test the widely believed theory that brontosaurs were swamp dwellers, sloshing around lakes and rivers up to their armpits to keep their impracticable bulk buoyed up by the tepid waters. If that theory is true, then this *Camarasaurus* very likely died in its favorite watery habitat. Can that idea be tested? Very easily. If the body sank in an aqueous grave, it should be resting on the type of sediment laid down on lake or stream bottoms.

Along this limestone outcrop where the bones are eroding out, there is only this one carcass—scattered vertebrae from the neck

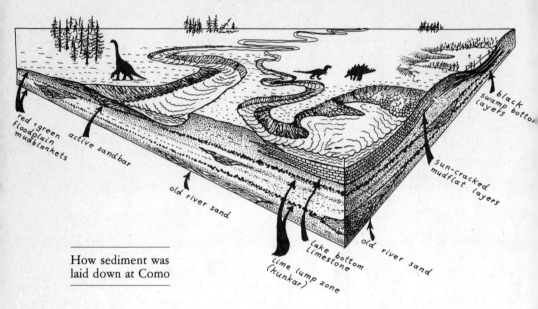

red + green
floodplain
mudblankets

active sandbar

old river sand

How sediment was
laid down at Como

lime lump zone
(kunkar)

lake bottom
limestone

old river sand

sun-cracked
mudflat layers

black
swamp bottom
layers

and back, a shoulder blade, a thigh bone—all the right size to fit together and make one skeleton. The bones aren't jumbled on top of each other. They lie close together in a layer one bone deep. Clearly this body fell apart, ligaments rotted, knee detached from thigh, neck vertebrae separated from one another. Such rotting of the ligaments *could* have happened underwater while the body lay in the muck. Crayfish, turtles, and other bottom scavengers would have crawled over it, tugging and biting at every shred of flesh. Or it *could* have happened on land, where dryshod predators could have pulled the meat and gristle apart. Which scenario is correct?

The surrounding rock of the bone layer is a dark gray mudstone containing little streaks of sand grains here and there. No distinct layers, though; no fine horizontal bedding. And that is a clue. Mud settling through standing water—a pond or swamp—usually deposits clear-cut layers, one piled on top of the other, because the flow of mud particles is almost never constant. Usually, the flow of mud into the lake varies with rainfall and flooding. A spring flood sweeps coarser sand into the pond, the sand sinks, a layer is formed. Gentle showers wash fine mud into the pond, the mud sinks on top of the sand, another layer. Such layers of mud undisturbed often form delicate sheets, because the fine clay particles which compose them are in the shape of microscopic plates that lie flat on top of each other. The lack of layering thus *suggests* this body didn't lie on a deep lake bottom. Very deep lakes usually possess bottom mud with very clearly defined layering.

Protruding from under the bones is the limestone of the Cam Bench. Why does this layer stick out from the eroded bank? Obviously because it's harder than the bone-bearing layer above it. But why is the lower layer harder? Because it's a limestone—and limestone resists erosion in this dry Wyoming climate. But this is not an ordinary limestone. Most limestones form underwater, when lime (calcium carbonate) precipitates out of solution. This limestone is made up of little balls, from pinhead size to golfball size, packed together, jammed onto each other.

A closer look at these broken lime balls reveals that some have a tiny grain of sand at the core, others a few mud streaks. Such lime balls can form in gently agitated warm water—the action of the waves rocks small particles as they clothe themselves with layer after layer of lime. But these aqueous lime balls, called ooids ("oh-oyds"), are usually all of nearly one size, not at all like these in the Cam Bench, where pin-sized balls lie adjacent to others a hundred times bigger. Furthermore, ooids usually show internal layering, like an onion. These lime balls below the skeleton don't. Instead, the Cam Bench lime balls look like the ones called kunkar that grow in well-drained soil today in tropical India. So the camarasaur *could* have died on dry land.

Now, do such irregular lime balls as these grow anywhere today—and if they do, in what kind of habitat? An answer to that question might tell us where this one dinosaur body lays, and therefore where it lived and died. Soil scientists in Australia and India have found exactly the right thing. In tropical soils where the particles are well drained for most of the year, a trench dug into the soil layers will reveal a zone of lime balls a foot or so beneath the surface. Lime in solution washes down from the soil's surface during the rains. As the water dribbles and oozes downward, some lime drops out of solution and tiny lime pellets form, growing bigger each year. In many tropical landscapes these lime

Death of a camarasaur. Adult camarasaurs were too big to fear most predators—the camarasaur's fifteen-ton bulk was immune to attack by the average one-ton *Allosaurus*. But when sickness weakened their resistance, even a full-grown camarasaur could fall victim to the steak-knife teeth of the dinosaurian hunters.

balls lack internal layering—just like the ones under the fossil dinosaur. Aussie soil scientists call this type of lime ball "kunkar." If the lime balls under the skeleton are truly kunkar, that would be grounds for considerable excitement. Kunkar nodules would prove that this soil was originally not swampy and wet, because swampy soil water is usually acid and dissolves lime balls as quickly

as they form. In such acid soil a conspicuous kunkar layer never forms. If brontosaurs died on kunkar-growing soil, it means they probably lived on dry, firm ground, and not in swamps at all.

Layers of lime balls growing today in Indian soils have one characteristic signature: the balls get bigger toward the top of the layers, because the top balls receive more lime from the water percolating downward. A trench hacked into this limestone out-crop exposes unweathered rock, and the balls do indeed get big-ger toward the top! It certainly looks as if this particular brontosaur died on a land surface, just above a layer of irregular-sized lime balls growing in the soil.

Could we prove, however, that this was a land death? What's on trial here is not a murder suspect, but a suspect theory. Tra-ditional theory maintains that the big dinosaurs were water crea-tures—and old theories are tough antagonists in court. The scientific establishment tends to believe that old, accepted views are correct unless shown to be wrong beyond any reasonable doubt.

Here at Como we need the equivalent of the spent bullet, lodged in the carcass, that can be securely traced to the murder weapon. We need to find one more independent piece of evi-dence that this multi-ton giant met its end on land. Further inves-tigation around the eroding carcass reveals that some of the bones are scarred by deep, knifelike wounds. These could be teeth marks of the predator that killed the brontosaur or of the scavengers that stripped the body after death. In the dust a gleaming piece of tooth enamel catches the sun: a tooth, four inches long, pointed, sharp-edged, with sawlike serrations along front and back. Not the *Ca-marasaurus*'s tooth—the victim was a vegetarian. The tooth is from a big *Ceratosaurus,* a bipedal predator.

The *Ceratosaurus* tooth is the spent bullet. This predator's tooth clearly broke off as the flesh-eater bit into the brontosaur. At the base of the tooth, where the root should be, is a deep pit where the root had been dissolved by the *Ceratosaurus* gums. We hu-mans think of tooth loss as a tragedy, because once gone, our adult molars leave nothing but a hole in our jaw. But *Ceratosaurus* and all the other dinosaurs had an endless supply of teeth forming at each socket. As a new tooth grew in the socket, its pointed tip pushed out the old tooth. The new tooth grew upward as its root grew longer—the old tooth's root was dissolved to make room for the new. An X-ray of a ceratosaur jaw reveals four or five teeth, all in a row from top to bottom, each growing upward, pushing the one ahead.

Fractured edges around the root remnants of the ceratosaur tooth show that it broke off during life; the big predator was bit-ing into something hard enough to break the tooth off its jaw.

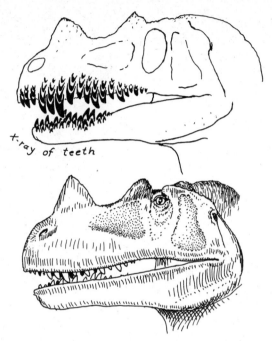

x-ray of teeth

How new teeth grow out
and push old teeth from
their sockets (shown is the
predator *Ceratosaurus*)

There is another *Ceratosaurus* tooth, and another. All three are the right size to come from the same animal, and all three broke off during life. This is the killer—or at least a beast that bit into the dinosaur carcass after death. And *Ceratosaurus* was a land predator, not a swimming meat-eater like the crocodile. *Ceratosaurus* hunted for prey on land, either killing the young and weak brontosaurs or searching for brontosaurs that had died of other causes. So broken ceratosaur teeth supply our sought-for clue. The case for death on land is solid.

If a brontosaur carcass lay on a pond bottom or floated on a lake surface, it wouldn't attract land predators, it would attract crocodiles, which swarm over any available meat. Crocs shed their teeth just as dinosaurs did, and when crocs bite into a big carcass, a few teeth usually detach and stick embedded in the body. If our brontosaur carcass had ever been in water for a long time, we would expect to find some croc teeth with it. But in digging around the site for five days, not one croc tooth turns up. A few feet above the lime-pellet bench there is a stream deposit full of broken and shed crocodile teeth mixed with clamshells and turtle bones. This deposit clearly preserves the record of water-living predators and their victims, a record which is absent in the case of our *Camarasaurus*.

The case is looking good—the vegetarian *Camarasaurus* died on land and was chewed up by a *Ceratosaurus*. But now a complication. The plot thickens when the shed teeth of a second big predator, *Allosaurus,* a very different species, turn up among the *Camarasaurus* bones. And then yet another predator's teeth—this time a very small killer, *Coelurus,* only the size of a very big turkey.

A moment's reflection produces a clear solution to this case. In land ecosystems today, a big carcass is an enormous amount of protein waiting to be used by any and all predators. Whether a big water buffalo dies of disease or from a lion's attack, the sight and smell of the dead hulk will attract lions, hyenas, jackals, and vultures from a wide radius. After the biggest, most aggressive flesh-eaters have eaten their fill, the smaller jackals and foxes nip in to tear off their share. And thus one buffalo carcass gets chewed and pulled apart by successive crews of large and small predators.

Only one more piece of evidence of how the Cam Bench brontosaur died and was preserved is necessary to complete the case. The chewed-apart *Camarasaurus* carcass had somehow to be covered by a layer of sediment. But the layer of rock entombing the skeleton is unambiguous at this point. It was deposited as a blanket of mud when a turbid sheet of water inundated the floodplain. Floodplains are special places. They can remain dry and grow a rich carpet of bushes, young trees, ferns, and ground pines for half a year, ten years, or as much as a thousand years. It's during these intervals that the kunkar lime balls form in the soil below their surface. But floodplains get their layers of sediment when neighboring rivers and streams overflow. The floodwaters that have been rushing through river and stream channels slow down dramatically when they spill over and spread across the flat lowland. Homeowners in Cincinnati and other floodplain cities today know firsthand how such sheets of floodwater can bury even large objects in mud—sometimes objects as large as a station wagon or a one-story house can disappear in a few days. Our *Camarasaurus* carcass, chewed and pulled apart, was gently and thoroughly covered by just such a mud deluge. Then the floodwaters receded. The new layer of mud dried out. Plants began to germinate, ferns sprouted. A new soil surface developed atop the new blanket of mud, supporting the very same type of plants that had fed the camarasaur during its lifetime. Once again, kunkar nodules began to grow below the surface and among the camarasaur bones.

But other arguments are possible here. The Russians have given a name to this type of paleontology, "taphonomy," coined from the Greek word for "burial" and the word for "laws." Taphonomy is when a paleontologist reconstructs the corpse's burial so it can reveal how and where it lived. It is important not to be fooled by first impressions. A fossil body's location may not be where the death occurred. Fossilization can be misleading. A dead brontosaur like the *Camarasaurus* lying on a floodplain probably means the death occurred on land. But maybe a flood washed the body from a river onto the land. It could happen.

Worse yet, there's the possibility the body was dragged from a watery death site by the actual killer. Meat is meat and in short supply in most habitats. So a land predator, such as a lion, might pounce on a crocodile in shallow water if the croc were unwary. Without the least concern for whether it were messing up a potential fossil, it would drag the dead croc hundreds of yards from the shore to some secure spot on dry ground where the big cat could enjoy its scale-wrapped dinner in comfort. Large land-hunting dinosaurs could have done the same, dragging aquatic prey into their terrestrial habitat.

Taphonomy is therefore a science requiring subtlety, broad knowledge, and a good eye for detail. The diligent carcass sleuth must be alert for signs that the dead body has been moved from water to land, or vice versa. Crocodiles are usually as hungry as lions and nearly as crafty. A big croc will wait in the reed-choked shallows for an unwary zebra to come and drink. In a minute the zebra disappears beneath the river's surface. And an unwary paleontologist, excavating the scene a million years later, might be similarly ambushed—he might conclude that the zebra was aquatic because he found its mangled bones buried in river sand.

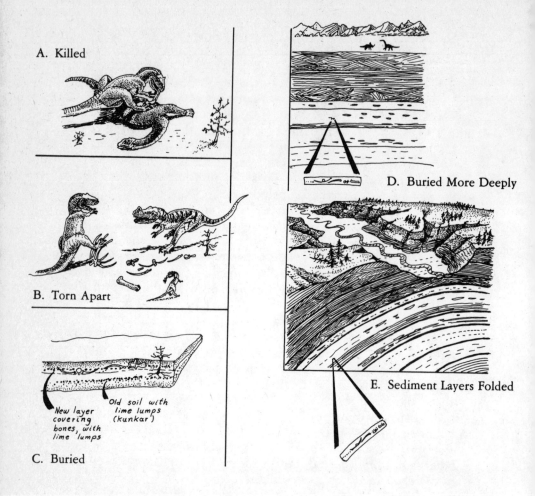

A. Killed

B. Torn Apart

C. Buried

New layer covering bones, with lime lumps

Old soil with lime lumps (kunkar)

D. Buried More Deeply

E. Sediment Layers Folded

Rivers are moreover terrible deceivers. When modern rivers overflow, they wash all sorts of living and dead matter off the floodplain into the river channel. Consequently, sediment buries dead squirrels, unfortunate cows caught in the flood, lawn furniture, shopping carts, and other terrestrial debris, along with the fish bones and clamshells that belong there. If hundreds of dinosaur skeletons from a single species are found preserved in river-channel sediment, could it safely be concluded that the species was water-loving? Not at all. Big floods along the Missouri wash

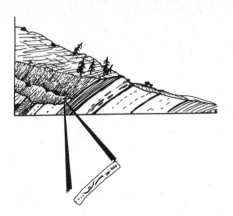

F. Erosion Exposes Bones

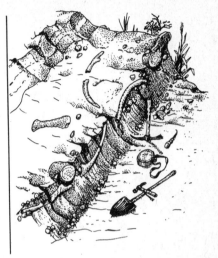

G. Discovery!

hundreds of Hereford steers into the rivers, where they eventually become buried in sandbars. Does that prove Herefords are an aquatic species? Dinosaurs of all kinds have been found in stream- and river-laid sandstones—predators like *Allosaurus,* armored stegosaurs, and giant brontosaurs. A few years ago, a young Canadian paleontologist was misled into concluding that duckbill and horned dinosaurs were aquatic because in Alberta their skeletons are concentrated within the river sandstones. But Mesozoic rivers were as deceptive as modern ones. Dinosaurs in rivers prove nothing conclusively.

If rivers can't be trusted, where can paleontologists turn for a truthful account of the brontosaur's habitat preference? Lake bottoms and floodplains are the most trustworthy locales. Some fish and crocs do get washed and dragged up onto plains, and some zebras and lions do get washed or dragged into lakes. But, on average, lake-bottom mud preserves mostly water creatures, and floodplain mud preserves mostly landlubbers. What is needed, then, beyond the single Cam Bench *Camarasaurus,* to solve the mystery of the brontosaur's habitat is a broad statistical survey. Thanks to a *National Geographic* research grant, a group of brontosaurophiles called the "Morrison Dinosaur Habit Research Group" was able to make just this type of survey between 1974 and 1977.

A study trench was dug up through the whole three-hundred-foot thickness of the Morrison Formation at Como. It laid bare cycle after cycle of life, death, burial, and new life. Twenty different kunkar layers were exposed, each marking a time when a floodplain was dry and green and its vegetation nourished dinosaur life. And just above each kunkar layer were zones of fossils —chewed dinosaurs—whose bodies had been written into rock history by entombing layers of mud. When the statistics were tabulated, eighty percent of the brontosaur graves were in floodplains, twenty percent in river channels and zero percent in lakes or swamps.

But the quarries at Sheep Creek, Wyoming, twenty miles north of Como, posed a special mystery. Here was a big quarry full of brontosaurs—and their carcasses had been entombed in lake-bottom limestone. Had lake-dwelling brontosaurs been found at last? Did the Sheep Creek skeletons represent bottom-walking behemoths that fed on soft lake plants just as brontosaur orthodoxy had preached for eighty years? If the Sheep Creek limestone was deposited in a deep lake, then perhaps the brontosaurs found here were the classic dwellers in aquatic habitats.

A careful investigation of the quarry made it clear that something was wrong with the deep-lake theory. As a carcass sleuth looks for clues, he or she must be alert for negative evidence; sometimes what's missing reveals more than what is present. And negative clues were everywhere at Sheep Creek. No fish bones or crocodile bones and almost no turtle remains (only one fragment of shell) were ever found in the Sheep Creek limestone. What sort of lake had no fish or crocodiles or aquatic turtles? Nearly all tropical lakes today are quite full of these swimming creatures. Snail shells were found, but only of the type usually present in ponds and along lake margins.

More perplexing clues turned up. Kay Behrensmeyer, among the best young American taphonomists, found giant dinosaur footprints in the same limestone which contained the dinosaur bones. Bottom-walkers can make shallow prints in the mud under deep water. But these prints were far too deep to have been made by a brontosaur buoyed by twelve feet of lake.

Finally, a key piece of the puzzle fell into place. "Fossil sunlight," in the form of mud cracks in the lake bottom, turned up. When mudflats are exposed to air, the mud surface dries and contracts, cracking itself into a mosaic of hexagons separated by fissures. In a sunny, dry climate, mud cracks can grow to depths of a foot or more quite quickly. And such cracks can fossilize when a subsequent flood washes a layer of sand over the cracked mudflat, filling the fissures with sediment. Usually the soil filling the fissures has a different texture from that of the mudflat. So when a fossil mudflat and its crack-filling are exposed by erosion, the ancient dried surface faithfully preserves the record of sunrays millions of years old.

Mud cracks in the lake-bottom limestone proved that the lake had dried up repeatedly. Such sun cracks were found on several layers piled on top of one another on the ancient lake bottom. If the Sheep Creek dinosaurs had died in a shallow lake, then the strange negative clues could be easily explained. Shallow lakes are subject to cycles of drying up and wetness; some lake beds in today's tropical Africa are dry most of the time and fill up only during exceptional floods. Small snails can live in such shallow water, but bigger swimming creatures—fish, turtles, and crocodiles—obviously cannot. The best explanation for the peculiarities of the Sheep Creek fossils produced an unexpected twist to the theory of swamp-living brontosaurs. These brontosaurs probably did die on a lake bottom, or at least near it. But there probably was not enough water in the lake to float a small crocodile, let alone a multi-ton dinosaur.

The ultimate irony at Sheep Creek is that the dinosaurs may have died there during a drought. When dead bodies lie on dry land for awhile, the air and sun desiccate the muscles and back ligaments, contorting the entire body so that the neck and tail are twisted up above the level of the back. Did drought kill the brontosaurs? There was strong evidence for dry seasons during the brontosaur's heyday in the late Jurassic. Beds of lime pellets show up in dozens of layers, and each kunkar zone recorded a time of repeating dry seasons. But the contorted state of the brontosaur bodies provided the strongest proof of terrible drought. At Sheep Creek, the eighty-foot body of a brontosaur was twisted precisely in the manner of a drought victim. And at Dinosaur National

Monument in Utah, a four-hundred-foot-wide stream bed preserves dozens of gigantic dinosaur bodies, all twisted with their huge necks and tails over their backs. Similar scenes of Jurassic death by drying are repeated in quarry after quarry.

Ordinarily, carcasses are pulled apart by scavengers right after death. If the weather is mild, lions, hyenas, and jackals can reduce a dead elephant to a scattered mass of disjointed bones in a few days. But drought kills animals faster than the scavengers can dismember them, and severe drought kills the scavengers too. So the landscape becomes full of bodies drying up under the sun. Scores of brontosaur bodies found in the Morrison beds show the telltale signs of mummification under the merciless sun of Jurassic droughts.

Our scenario of what occurred at Sheep Creek contains the script for a Jurassic tragedy. Around the shores of a drying lake bed gather the beleaguered giants—*Diplodocus, Brontosaurus, Stegosaurus*. With their huge elephantine feet they dig into the mud, trying to reach the water table fast receding under the desiccating Mesozoic sun. For a few weeks the giants survive, drinking from the muddy water which seeps into the deep holes excavated by their feet. Smaller dinosaurs sneak in to drink when the giants are unaware, though the biggest brontosaurs angrily defend their dwindling water stores. Finally even the strongest cannot survive. Twenty-ton bodies collapse on the lake shore and twist into sun-dried mummies. After six months or a year or two, the monsoon rains return. The level of the lake rises, spreading a soft blanket of limy mud over the sun-dried bodies lying on the parched clay.

The final box score tabulated for the survey of brontosaurs was emphatically in favor of land habits. Few brontosaurs were buried in deep lakes. Many died and became entombed on floodplains where the water creatures—crocs and turtles—were rare or absent. Altogether, the brontosaurs showed the same burial pattern Kay Behrensmeyer has observed for land-living elephants in East African sediments.

We then reconstructed the overall environment from the patchwork of different individual habitats found throughout the landscape during Morrison times. This broader level of sleuthing yielded a picture of the Jurassic world in the Western United States. It consisted of a system of broad, flat floodplains, small rivers, shallow ponds, and occasionally deep lakes, all subjected to cycles of killing droughts.

After the results of our carcass-sleuthing in the Morrison, a nagging question remained. Why did the brontosaurs die out in Wyoming and Colorado at the end of the Jurassic Period?

After achieving extraordinary success right up into the last

Late Cretaceous brontosaurs avoided swampy forests. The Alberta delta was wet year-round most years, and brontosaurs weren't there. But in North Horn, Utah, there was a distinct dry season (producing kunkar) and the brontosaur *Alamosaurus* enjoyed the climate.

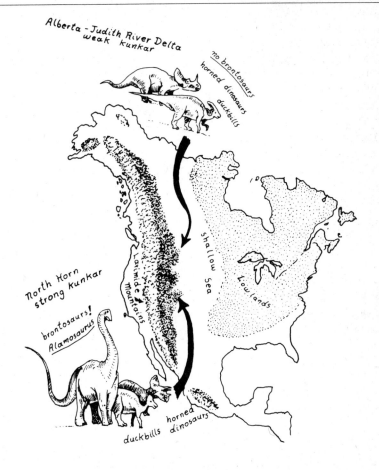

zones of the Jurassic, the brontosaur clan declined suddenly in Wyoming and Colorado before the waves of new dinosaur families flooding the land ecosystems at the beginning of the Cretaceous. These new dinosaurs of the Cretaceous were all members of the beaked Dinosauria. And the entire Cretaceous Period records the proliferation of beaked dinosaurs into several rich families of duckbill dinosaurs, armored dinosaurs, and horned dinosaurs. This transition from the Jurassic to the Cretaceous marks the most profound shake-up in the history of dinosaurian families.

Why did brontosaurs lose their ecological preeminence in these areas? According to most books on dinosaurs written for children, the swamp-loving brontosaurs would have died out as their marshy haunts drained away. Our comprehensive survey of brontosaur quarries has already shown that this view simply won't hold—Jurassic brontosaurs were living and breeding for millions of years in habitats with a distinct dry season.

There is, however, another good explanation for the absence of the brontosaurs from the areas they had formerly dominated, an explanation that takes dinosaur orthodoxy and stands it on its head.

To understand this disappearance of the brontosaurs properly, we must turn to the great deltas of the Cretaceous. Deltas are formed where rivers meet the sea. Where the modern Mississippi flows into the Gulf of Mexico, for example, the mud-laden river water dumps millions of tons of sediment every year along the boundary between land and ocean. This influx of sediment is building a huge mud platform out into the Gulf. In Late Cretaceous times, a series of short rivers flowed eastward from the rising young mass of the Rocky Mountains into a broad and shallow sea that covered most of what is now the High Plains country of eastern Alberta, Montana, and Wyoming. Together, these Cretaceous rivers built a continuous sequence of overlapping deltas from Canada south to New Mexico, and these deltas are full of dinosaur skeletons.

These Cretaceous deltas are among the best-studied dinosaur habitats of the entire Mesozoic, because compelling economic interests drive geological exploration—the deltaic sands are often full of oil and the deltaic swamp deposits are full of coal. Thanks to the oil and coal companies, which have poured millions into research, and to American and Canadian museums, which have consequently excavated hundreds of dinosaurs, the climate and landscape of the Cretaceous deltas can be reconstructed in minute detail. Conditions had clearly changed from Jurassic days. Gone were the dry meadows and kunkar-layered soils. In their place stood stagnant bayous, estuaries, and cypress swamps like those of present-day Louisiana. Beneath these broad bodies of delta water, organic residue from innumerable rotting leaves and branches was continuously pressed into a compact, carbon-rich coal.

These deltas began yielding dinosaurs in the 1880s. The rock layers they contained were named the Laramie Beds after the Wyoming frontier town. Now, if the orthodox story of brontosaur habits is to be believed, then the Laramie Deltas should have been prime brontosaur country. But the brontosaurs aren't there. Dinosaurs in profusion are found in the Laramie Deltas—beautifully preserved skeletons, with every joint in place, of duckbill dinosaurs, horned dinosaurs, armored dinosaurs. But after a century of thorough exploration, not one brontosaur has turned up.

Yet brontosaurs did still exist during this time. Late Cretaceous beds in Brazil, Argentina, India, Mongolia, and even in closeby New Mexico have yielded dinosaurs. Why then did these latter-day brontosaurs avoid the Laramie Deltas?

Perhaps a consummately heterodox suggestion answers that question: Maybe the brontosaurs did not need soggy terrain, maybe they positively hated it. Some of today's large animals—zebra, wildebeest, and lions, for example—dislike swamps intensely. Brontosaurs may have disliked the mushy soil so much that they avoided the swampy deltas entirely.

This idea could be tested in the quarries of the North Horn Mountains of Utah, the northernmost locales that contain Late Cretaceous brontosaurs. If the North Horn quarries contain evidence for dry, well-drained floodplains—the type of habitat missing further north in the area of the Laramie Deltas—then a strong case for hydrophobic dinosaurs can be made, and their decline in Wyoming and Colorado in Cretaceous times consequently explained.

Alamosaurus, "lizard of the Alamo," is the name given to the Cretaceous brontosaur found in Utah. "Alamo" in this case refers to the Ojo Alamo Mountains of New Mexico where the beast was

The Alamo brontosaur in its dry Utah home. Two *Alamosaurus* (at left) watch as a meat-eating *Albertosaurus* tries to attack the spiny-frilled *Styracosaurus* (at right). A trombone-crested duckbill (*Parasaurolophus*) flees the scene. During these Late Cretaceous times, the brontosaur clan avoided the humid habitats of the northern deltas in Wyoming and Montana and Alberta, but farther south, where summers were dry and hot, *Alamosaurus* reigned supreme as the biggest plant-eater.

first found in 1922. When that first *Alamosaurus* was found, it jolted American paleontology; until then, no one had found a single scrap of Late Cretaceous brontosaur in the United States. And it had been assumed that all the North American brontosaurs had died out long before Late Cretaceous times.

The best specimen of *Alamosaurus* was excavated from a beautiful cliff face within the North Horn Mountains of Utah. Dinosaurs from the North Horn constitute an intriguing mix. Scientists from the Smithsonian Institution have disinterred horned dinosaurs, duckbills, and flesh-eating tyrannosaurs—all three groups that dominate in the Cretaceous deltas of Wyoming-Alberta. But they have also discovered several gigantic brontosaurs, all belonging to *Alamosaurus*. And the North Horn quarry did not at all resemble a typical Cretaceous delta. Instead, it looked precisely like a Jurassic quarry from Como. A walk up the cliff face yielded a count of seven distinct layers of lime pellets. So the North Horn habitat in Late Cretaceous times was much less soggy, on average, than the contemporaneous locales on the deltas in Wyoming. Like its Jurassic forebears, therefore, *Alamosaurus* lived and died in a landscape of dry, well-drained floodplains.

The ultimate in heterodox thinking seems justified. And brontosaur orthodoxy had it completely incorrect. Brontosaurs didn't require deep swamps to buoy their bulk; they didn't even like to be near swamps. *Brontosaurus* and its kin of the Jurassic Age favored truly terrestrial haunts with dry soils. And when great swamps did spread across vast areas of the Cretaceous world, as in Wyoming and Colorado, the brontosaur clans simply eschewed this soggy terrain and moved their evolutionary centers elsewhere, to locales where the brontosaurs could feel the reassuring texture of dry floodplains beneath their feet.

6

GIZZARD STONES AND BRONTOSAUR MENUS

Seen from a distance, a live *Brontosaurus* would appear not to have any head at all. Both neck and tail would just seem to taper gradually to a point both fore and aft. Up close the head would appear, of course—about the size of an average horse's. Less than two feet of brontosaur head to go with seventy feet of neck, body, and tail. A two-foot horse's head, with a mouthful of big molar teeth, can feed an eight-hundred-pound horse body. A two-foot brontosaur head, with only a handful of pencil-size front teeth, had to feed twenty or thirty tons of body. Obviously, the standard orthodoxy has it, the brontosaur's extreme microcephaly imposed severe dietary restrictions. Only the most nutritious and softest of water vegetation would have met the stringent requirements. And even with a superabundant supply of such green mush, the brontosaur's metabolism would still had to have been incredibly low—somewhere between the level of a tortoise's and a cactus's—for the great beast to survive at all.

This argument has been repeated hundreds of times by schoolteachers and Ivy League professors alike. A recent issue of *National Geographic* featured a long piece by a respected curator at a university museum. Typically, this author scoffed at the idea of any brontosaur's having a high metabolism. He dismissed any such notion with a single fact: its head was too small. In a 1984 article in a technical journal, a young paleontologist presented a mathematically reasoned argument that proved beyond the least

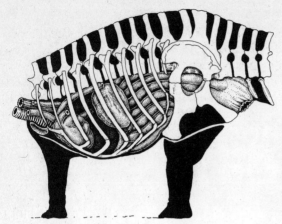

Yard-wide gizzard of a *Brontosaurus*. With its thick muscular walls and lining of hard rocks, the brontosaur gizzard could grind enough tough leafage to fuel a warm-blooded body.

doubt that the big brontosaur's meager cranial apparatus was hopelessly undersized to provide for any sort of high metabolism. Several years before, a graduate student from Yale lecturing before an enthralled audience at Harvard used the rate at which moose chew water lilies to prove irrefutably that a twenty-ton brontosaur simply could not support anything but the most subdued and sluggish life style. Documentary proof. Irrefutable logic. The giant brontosaurs could only have spent all their lives in a somnolent state of semi-torpor, just barely moving their long necks to reach into the lukewarm water, poking slowly about for the softest part of the Jurassic swamp salads.

But all these arguments, both popular and professional, leave out important pieces of the brontosaur puzzle: gizzards, stones, and moas.

A white mouse sacrificed to a hungering alligator posthumously provides a most important clue. The bones of the mouse show up quite clearly in the alligator's stomach on the laboratory's television X-ray monitor. But the mouse's bones are not alone. The alligator's after-stomach is lined with hard, dense objects—gizzard stones. The gizzard stones are convulsed by sudden muscular contractions of the gizzard's walls. The monitor clearly shows the mouse is being chewed, not by teeth in the mouth but by stones in the gizzard.

Naturalists who study big 'gators and crocs in the wild find huge masses of gizzard stones when they cut open the animals to study their feeding habits. The stones are found only in one chamber of the stomach—the gizzard—and this one chamber has walls with grooves and folds to permit expansion and contraction. Even without X-ray monitoring, it is obvious that this stomach chamber is a churning compartment designed to crush and pulp the prey's body after the gastric juices begin their preliminary chemical treatment. Crocs usually select very hard stones—quartz and granite pebbles, for example—to line their gizzards. If such materials are lacking in their native streams, they may use angular bits of hard wood, pieces of glass bottles, or whatever else is available. I have also seen one or two near-perfect fossil alligator skeletons containing a neat bundle of hard pebbles clustered between the ribs precisely where the gizzard was in life. These fossilized gastric mills demonstrate plainly that gizzard stones have been an essential functional component of crocodilian food processing for many millions of years. And the study of crocodilian gizzards leads to some intriguing conclusions about evolution both in birds and in the Dinosauria.

Zoos mislead their visitors by the way the species are housed. Birds are in the Bird House, of course, and crocodiles are always segregated to the Reptile House with the other naked-skinned, scale-covered brutes. So the average visitor leaves the zoo firmly persuaded that crocodilians are reptiles while birds are an entirely different group defined by "unreptilian" characteristics—feathers and flight. But a turkey's body and a croc's body laid out on a lab bench would present startling evidence of how wrong the zoos are once the two stomachs were cut into. The anatomy of their gizzards is strong evidence that crocodilians and birds are closely related and should be housed together in zoological classification, if not in zoo buildings.

Both birds and crocs have the identical plan to their specialized gizzard apparatus, and this type of internal food processor is absent in the other "reptiles"—lizards, snakes, and turtles. In both birds and crocs, the gizzard is a thick-walled, muscular, crushing compartment with two great tendons reinforcing the walls of muscle (these are the shiny sheets of tough tissue you cut off the turkey gizzard before cooking it). In both birds and crocs, the muscular gizzard is just aft of the thin-walled glandular stomach where food is softened by gastric juices.

This croc–bird digestive system makes a lot of mechanical

sense. We humans chew our food first, then pass it to the glandular stomach, where it is softened by stomach juices. Our system makes our teeth do the heavy work; they must crunch up the food as it comes directly through the lips. If the human diet is a civilized one, full of soft TV dinners and tender cuts of meat, our teeth don't wear much. But in primitive human societies the natural foods are often tough and gritty—the Anasazi Indians of ancient New Mexico wore their teeth down to the gums because tiny bits of sand got mixed into their cornmeal when it was ground on stone matates. Even horses wear out their huge molars if they have to feed on grass growing in gritty soil. But consider the advantages of the croc–bird system. They swallow without chewing and pass their food directly into the glandular stomach, where the food rests, softened by the gastric biochemistry. Then sphincter muscles act as gastric gatekeepers, letting the food pass on to the gizzard where it is chewed. The "teeth" of this system (the gizzard stones) don't begin their crunching work until the food has been rinsed, soaked, and softened.

Crocs have powerful digestive processes. However, no croc species eats vegetation purposely; sometimes weeds are swallowed accidentally when the croc swallows turtles or fish. So crocs don't provide a complete picture of how a gizzard might work in an herbivorous dinosaur like *Brontosaurus*. Fortunately many species of bird are plant-eaters, and vegetarian birds perform some truly spectacular gastric feats with their rock-lined gizzards. Ducks and geese shovel up hard nuts and grains and even live clams, chug them down to the gizzard, and crunch them up with the gizzard's lining. Clamshells, acorns, and corn kernels are all equally cracked into small pieces by this formidable gastric mill. Fruit pigeons do even better; their gizzard is especially tough and contains horn-covered "teeth" growing from the inside lining. Even the hardest of tropical nuts are swallowed whole, passed into the gizzard, and cracked with an audible thunk. Ostrichs shot in the wild have gizzards lined with the hardest rocks—usually those rich in quartz—available in the countryside. And a large bird can carry around as much as a double handful of these stong gastric tools.

Now the problem of tooth wear in nature is not a minor one. When wild species wear out their adult teeth and can't replace them, they die. Elephants possess huge adult teeth, the largest ever evolved. But every elephant eventually wears out its last molar and wastes away along some swampy shore, attempting to gum soft water plants for nourishment. Having a continuous supply of teeth in each socket, as was the case for dinosaurs, eases the tooth-wear problem but doesn't remove it entirely. The basic adaptive difficulty is that the hardest material in a tooth—the enamel—is still much softer than the grit that covers most foods in nature. Wind-blown dust generally contains tiny specks of silica. Silica is natural glass, a very common material in rocks and soil. Plants growing in natural soils become coated with windblown grit and with dirt containing silica particles.

Not only do soil and wind tend to make plant food gritty, but plants themselves sometimes evolve silica armor to discourage the plant-eaters. Horsetails are one such armored type of plant, an ancient group dating back to long before the dinosaur. Modern horsetail species are sometimes called "scouring rushes" because peasant housewives used to scrub pots with horsetail stems. They scour well because evolution has provided them with special cells that concentrate silica from the soil. The silica cells armor the entire stem with row after row of glass-hard microlumps. A plant-eater learns quickly that a diet of horsetails will erode its teeth down to the gumline.

Gizzards not only give plant-eaters an edge in their evolutionary struggle with plants. They also confer the freedom to do other things besides constant chewing. Pity the poor plant-eater with neither gizzard nor ruminating stomach—a zebra, for example. The zebra must chew each lump of grass directly, without soaking or softening. Zebra heads are large for their bodies and are provided with huge molars—twelve on each side of the mouth (twice the number humans have). Even with this dental armory, when grass is tough and sparse, zebras are nonetheless forced to spend nearly all their working hours plucking and chewing. All this chewing demands that the zebras remain out on the plains, exposed to rain, wind, and constant danger from lions and hyenas.

What would happen if a zebra were supplied with a hypo-thetical gizzard? Such a zebra could pluck up grass quickly, without masticating, fill its forestomach chamber, and retreat to the shade and safety of a bush-covered hill to let its gizzard do all the work of mastication. Gizzards also free the animal's mouth for other activities—such as sex. With its gizzard doing all the work of chewing, the zebra could use its mouth to snort and whinny and make all sorts of elaborate noise display to attract mates and frighten sexual rivals. Ever wonder how tiny songbirds can afford to spend so much of their time singing? Little birds are notorious for their high metabolism, but when do they find the time to chew? They don't. As the warbler sings, its gizzard and forestomach are doing the food processing without interfering with the music.

Cud-chewing mammals have evolved a soak-and-soften mechanism almost as good as the gizzard. A cow or deer plucks a mouthful of gritty grass, swallows it without chewing, and passes the lump of grass to a series of special stomach chambers. These chambers are fermentation vats where gastric juices and yeastlike microorganisms clean the wad of food and break down the tough plant fiber. Only after the lump of grass has soaked and softened is it passed back up to the mouth to be chewed by the molars. The technical name for this stomach vat system is "rumen," and such cud-chewing mammals are called ruminants.

The ruminant system must be reckoned as one of the best devices mammals have evolved for coping with tough plant food. Most of today's successful big plant-eating mammals are in fact ruminants—all the cattle, sheep, goats, antelope, deer, giraffes, and others. But the gizzard system must be considered superior.

Imagine a twenty-ton *Brontosaurus* equipped with an advanced, avian-style, rock-lined gizzard. A two-hundred-pound ostrich may possess a gizzard four inches across and a pound in weight. A roughly proportionate gastric grinder would provide a twenty-ton brontosaur with a gizzard of approximately one hundred pounds. One hundred pounds of tough muscle contracting a lining of big quartz pebbles could crush up Jurassic vegetation at a rate more than adequate to supply any level of metabolism. A hundred pounds of gizzard muscle weighs more than four times the jaw muscles of a five-ton African elephant. So four elephants, totaling twenty tons, possess less chewing power than the single hypothetical brontosaur.

But what about that tiny head—would a brontosaur be able to engorge enough food to keep a giant gizzard apparatus going at full capacity? That question can be answered by turning to New Zealand, where up until a few centuries ago a giant, long-necked, pinheaded herbivore waddled about the landscape plucking leaves from trees and crushing them with its gizzard. This native New Zealand plant-grinder was the moa—or more precisely, the moa family, a group of flightless species of bird that achieved a weight of half a ton. New Zealand's ecosystems evolved without any native land mammals, so the role of large plant-eater was filled by the evolution of these big ground birds. Unfortunately for modern science, the Polynesian colonists, the Maoris, who arrived in New Zealand about A.D. 1300, found the moas tasty and easy to kill, so moas were extinct before Western civilization could meet them alive.

But moas created a sensation when they first turned up as fossils in New Zealand bogs and stream gravels. European zoologists already knew ostriches well, because they had been circus favorites from the time of the Caesars. But no one suspected that a plant-eating bird as large as a small buffalo could have existed. In 1838, Sir Richard Owen, Queen Victoria's favorite anatomist, received a packet from New Zealand containing a curious bone fragment the size of an ox's femur. Owen was such an accomplished comparative anatomist that he instantly recognized the fragment as from a bird—a bird possessing a body five times heavier than any previously known. With a courage few other young scientists might display, Owen publicly announced his discovery.

Six-foot Maori hunter and the great moa, *Dinornis*

Based on this single fragment, he deduced the existence of huge birds rivaling the mammals in size. Owen's name for the extinct bird was an emphatic superlative: *Dinornis maximus* "enormous terror bird."

Owen's announcement met with skepticism, but his judgment was vindicated by more and better discoveries in New Zealand: partial hind limbs, vertebrae, and then astonishingly complete skeletons found standing upright, buried in quicksandlike deposits. Owen's deductions, based on one thigh fragment, were on target—the greatest species of moa were twelve feet tall and must have weighed a half a ton. Other species varied down to pony size. Moa anatomy was full of surprises for biomechanical anatomists. The wings were nearly totally absent—unlike ostriches, moas retained not even a tiny feathered remnant. And the moa's head was tiny—a twelve-foot-high moa carried a skull no bigger than a poodle's.

Moas are delightful objects of study in their own right, but their importance in the present discussion lies in the pinheaded configuration of their head and neck. At a distance, moas would have appeared as microcephalic as any brontosaur, with the tiny moa skull perched atop a very long, gracefully tapering neck. And moas were without doubt herbivores; their beaks were constructed like those of living leaf-eating species. Unassailable evidence for their food preferences subsequently came from skeletons found in bogs and caves, where the stomach contents from the giant birds' last meal were mummified with the bones. These fossil meals consisted entirely of shredded leaves.

Moas, of course, were birds, and birds are brontosaur nieces (descendants of brontosaur relatives). The moas therefore present an unparalleled opportunity to study how evolution equipped a long-necked, pinheaded giant plant-eater with a high avian metabolism. And how did moas keep their body furnaces stoked with sufficient fodder? Gizzards. Moa adaptations are known in great detail, because specimens have been found with eggs (some containing unhatched moa chicks), with pieces of skin, feathers, footprints, and gizzard linings. Stones found within moa skeletons often represent types of rock found nowhere else in the entombing sediment. Moa bone-and-gizzard sites have been surveyed for miles around in the search for the original source of the gizzard stones.

It is often found that the birds must have traveled as much as ten miles to acquire pebbles of the desired consistency. Such careful selection of gizzard stones implies that the moas were driven to seek the hardest rocks for their gizzards, even if long searches were required. Why expend so much effort looking for hard pebbles? The high polish of moa gizzard stones provides the explanation. Unlike turkey and alligator gizzard stones, which are often pitted and dull, moa stones glisten with a fine patina which can be achieved only by constant rubbing in very hard grit. We must infer that the high polish was acquired in the course of day-to-day gastric function. The constant grinding of the hard pebbles proves that the moa's mill had to operate nearly continuously to pulverize the great masses of leaves and twigs needed to meet its metabolic needs.

Moas prove that a pinheaded brontosaur could process enough fodder to support a high metabolism *if* the dinosaur possessed a similar powerful food mill, equipped with very hard rocks. Can unequivocal evidence for dinosaur gizzards be found? Yes. A few years ago in the Victoria Museum in Rhodesia (now Zimbabwe), its curator, Mike Reath, showed me dinosaur treasures excavated from the red Forest Sandstone of the Zambesi Valley. One splendid skeleton belonged to a small, long-necked pinheaded dinosaur, *Massospondylus*. Nestled within its ribcage was a neat cluster of rounded, finely polished pebbles of a rock quite unknown elsewhere in the sandstone. The curator related how a careful search had demonstrated that such pebbles could not be found any closer to the skeleton's quarry than fifteen miles. The conclusion he drew was the same as the New Zealanders had drawn for the moas. This very primitive Zimbabwean dinosaur was finicky about the quality of its gizzard stones and would go out of its way to find only the most resistant pieces. *Massospondylus* is a critical case in point for the study of gizzard functions because it is closely related to the direct ancestors of the great brontosaurs. If *Massospondylus,* a brontosaur uncle on the dinosaur family tree, was equipped with a gizzard like a moa, it wouldn't be surprising to find *Brontosaurus* itself so equipped.

Perfectly preserved gizzard linings are perplexingly rare as fossils. The Zimbabwean *Massospondylus* is unique, to my knowledge, in preserving the gizzard lining from this species. Many other specimens have been excavated from the forest sandstone, but none show gizzards so well preserved. Why did the gastric rock disappear before fossilization? Probably the gizzard rocks fell out of some carcasses as they were ripped apart by scavengers. Other dinosaur bodies probably became bloated after death, floated awhile in floodwaters, then dropped their gizzard contents when the stomach burst from the gases of decomposition. Some dinosaurs may have belched up their gizzard lining while in their death throes. Whatever the causes, it's unusual to find good fossil gizzards even in those species that certainly had them in life. A good example of how rare fossil gizzards are is provided by crocodiles and alligators. Although crocodilians of all sorts possess muscular, rock-lined gizzards, only a tiny fraction of the hundreds of good fossil croc specimens preserve the gastric mill in recognizable form. So the rarity of fossil gizzards in more recent species is worth remembering when the gastric functions of the more ancient dinosaurs are discussed. Even if dozens of skeletons without gizzard stones are found, it is not certain the species in question lacked stones in life.

I don't know of a single brontosaur skeleton that shows a perfect pile of polished gizzard pebbles. However, I have seen a half-dozen brontosaur bodies in the field where smoothly rounded pebbles were scattered through and around the ribs. Could these be the gizzard's contents, a bit displaced after death? I'm firmly convinced they are. In each of these six cases the skeleton lay on an ancient floodplain surface and the bones had been buried by fine-grained sediment. In each case large, polished pebbles could be found only near the brontosaur bones, nowhere else for hundreds of yards in every direction. Streams can and do polish hard pebbles to the same high patina found on the pebbles around the brontosaur skeletons. But the geological circumstances surrounding these six specimens absolutely rule out action by any stream. The size of the grains of sediment is too fine to indicate anything but the gentle slosh of mud-rich water flowing over the floodplain. There is no way water could have moved those polished stones across the plain to the brontosaur sites. If water didn't move the pebbles, then they must have traveled to the site inside the dinosaur's stomach, to be deposited when the great beast breathed its last and collapsed on the fern-covered meadow.

Other experienced dinosaur diggers have told me that they

too have found irregular patches of polished stones littering brontosaur gravesites. And there is a further clue: the phantom stomachs. Years ago, when the pioneering brontosaur hunters rode on horseback across badlands etched in the Morrison Formation, they noticed piles of polished rocks lying isolated on the outcrops, nowhere near any dinosaur skeleton. I've seen them too, not only in the Jurassic strata at Como but in the overlying sediment layers deposited during the early days of the Cretaceous. Old-timers and rock hounds sometimes call these polished-pebble heaps "dinosaur belches." Chickens sometimes cough up their gizzard lining when it gets worn, so that they can restock with fresh rocks. So maybe the twenty-ton brontosaurs did the same. A seventy-footer might feel a bit out of sorts. It would stop feeding, an involuntary convulsion would ripple through its gizzard and forestomach, then through the long neck, until out from the *Brontosaurus*'s scaly lips would drop a bushelful of outworn rocks. If such a scenario of dinobelches has any truth to it, geologists would have to take the rock-transporting function of brontosaurs very seriously. Four or five brontosaur species coexisted at any one place during this period, together with two multi-ton species of stegosaur. Such numbers imply that rock-carrying gizzards were potentially quite abundant all over the landscape. If each big dinoherbivore were equipped with an outsized rock grinder, and they all regurgitated a couple of times each year, then the Morrison landscapes would have been the passive recipient of an endless series of pebble showers from belching *Stegosaurus, Brontosaurus, Camarasaurus,* and many others.

I was naturally skeptical the first time I heard this belch-a-bushel theory from a wizened old Utah rock hound at a shop in the Eden Valley of Wyoming. After all, he also was selling fossil algae as "dinopoop." However, the phenomenon of patches of pebble found without any related bones demands an explanation. These masses of alien pebbles had to be dumped on the floodplains by some agency to be subsequently covered by a blanket of fine mud. There is no evidence that the current of water carrying the mud blanket was strong enough to roll these big polished pebbles over the meadows. Furthermore, no known hydraulic mechanism could concentrate the pebbles in heaps. Dinobelches? Could be.

An alternative theory derives from the experience of paleontologists in New Zealand. Diggers of moa skeletons found not only cases of gizzard stones within the ribcages, but also stone piles in "ghost" skeletons—bones almost totally destroyed by soil water and erosion. For their gizzards, moas preferred types of rock that were highly resistant to geochemical decay. Pebbles rich in quartz suffer little harm though buried for millennia in stagnant, soggy soil. But bones will soften as the bone mineral (calcium phosphate) dissolves and the bone's connective tissue rots. New Zealand geologists found perplexing piles of pebbles scattered over the countryside—pebbles too big and too concentrated in location to be the product of any natural processes. And sometimes these pebbles are foreigners, rounded fragments with a crystalline composition completely out of place where the piles of pebbles are found. Only some unusual agent could have transported these masses of pebbles. I am convinced *some* of the New Zealand pebble masses are either moa belches or phantom moa stomachs—from carcasses where bones have rotted leaving only the gizzard stones. Almost certainly therefore *some* of the Jurassic and Early Cretaceous piles of pebbles are ghost dinosaur stomachs.

The certainty of gizzards in dinosaurs has been with us for a long time. A tiny skeleton complete with gizzard stones in perfect order was found in Mongolia and was announced in popular and scientific publications in the 1920s. Gizzards in dinosaurs would discredit the time-honored orthodoxy which preaches that brontosaurs *had* to be sluggish. We may even entertain the notion of warm-blooded brontosaurs as a viable possibility.

By themselves, brontosaur gizzards don't indicate how much or what these dinosaurs ate each day; other lines of evidence must be employed to explore these questions. But brontosaur gizzards and teeth together indicate what brontosaurs did not eat. They didn't eat soft, mushy vegetation. Birds that subsist entirely on soft fruits don't possess muscular gizzards and don't use hard pebbles for their gizzard linings. Soft, watery food requires only a short, simply constructed gut—with just enough contractile force to squeeze out all the juices.

Brontosaur teeth, moreover, confirm the heretical idea that they ate a tough vegetable diet. If the brontosaurs dined only on soft water plants, then very little wear would be found on their teeth. But in fact the teeth of *Camarasaurus, Brachiosaurus,* and their kin manifest very severe wear, which could only have been produced by tough or gritty food. Like the dental battery of other dinosaurs, the teeth of these brontosaurs were continuously renewed. As one tooth wore out, it was pushed out of its socket by a new tooth growing from beneath. So the wear on a single *Camarasaurus* tooth represents the abrasion not from an entire brontosaur lifetime but from a much shorter period of use—perhaps a year or less.

Most shed *Camarasaurus* teeth are scalloped out on their front and back edges by wear against some tough food. Such wear is especially impressive considering the large size of the tooth's edge. *Camarasaurus* teeth are very big—up to an inch and a half across. Each camarasaur front tooth is something like a thick wooden spatula in shape and is coated by a thick and roughened layer of enamel. What sort of food could wear the broad grooves in such teeth? Twigs and branches with tough bark, or big, palmlike fronds from the cycadeoid trees which flourished all through the Jurassic Period.

Diplodocus among the conifer needles. Conifers dominated the forest canopies all through the Jurassic. There were conifers with tight-packed, pineconelike needles (*Brachyphyllum*—upper left), and spirally arranged sharp needles (*Pagiophyllum*—right), and very long, pointed spear-shaped needles (*Podozamites*—lower right and in the *Diplodocus* mouth).

Camarasaurus and its kin represent the thick-toothed brontosaurs. The slender-legged *Diplodocus,* on the other hand, was a
member of the pencil-toothed family. It was the *Diplodocus*'s mouth
that inspired much of the orthodox view of pinheaded, weak-
toothed, and therefore sluggish brontosaurs. Compared to *Camarasaurus, Diplodocus* really does seem poorly supplied with dental
equipment. *Diplodocus*'s teeth are limited to the very front of its
jaws; there are no teeth whatever in the posterior position where
most mammals have molars. Each *Diplodocus* tooth is very thin, and
all the teeth in the row are packed closely together in an arrangement much like a miniature log palisade from an old-time Western fort. *Diplodocus* did, almost certainly, employ gizzard stones.
In the field I have seen two specimens together with scattered
polished rocks. However, *Diplodocus* must have been a much more
careful eater than the *Camarasaurus*. Its teeth do, however, show
severe wear; usually the tips of the crowns are beveled from
grinding against some resistant food items. So at least its food wasn't
simply soft leaves and mush.

Consideration of the *Diplodocus* is complicated by a published
account of the "stomach contents" of one specimen. Allegedly, this
carcass contained the remains of a last supper preserved in the area
of the forestomach. The menu was strange: clamshells, bits of wood,
and bone fragments. Perhaps, as the authors of the published report argued, *Diplodocus* was a scavenger picking through leftover
hunks of meat, shellfish, and whatever else.

I have learned that this "junk-food *Diplodocus*" is a hoax. Utah
geologists who know firsthand of the discovery have informed me
that the skeleton in question was a badly shattered brontosaur
preserved in a stream-bed sandstone. The alleged last supper was
not found within an undisturbed ribcage but was located very generally in the area of the torso. So the bits of clam and bone were
most likely placed there by the regular process of stream currents.

Patterns of tooth-wear in another *Diplodocus* specimen suggested to one scientist that the pencil-toothed brontosaurs were
specialized clam-eaters. When the upper and lower jaws were closed
in this *Diplodocus*'s skull, the worn tips of its upper teeth didn't fit
against the worn tips of the lower. To wear the teeth in this fashion, the *Diplodocus* must have been biting wedged-shaped objects,
and since clamshells are wedged in shape, it is not totally impos-

sible that a clam-nibbling habit caused the wear. But there are many other possibilities. If the *Diplodocus* used its teeth to strip leaves from conifer branches, for example, then the gritty bark could equally have beveled the tips of the teeth.

An important clue to the feeding habits of the pencil-toothed brontosaurs is to be found in their peculiar head–neck posture. In most living reptiles the neck is more or less horizontal, and the head stretches straight forward from the end of the neck. Birds, on the other hand, have a posture like our own—the head is held horizontally but the neck is vertical and joins to the back of the skull from beneath. In birds and people, the neck joints must accommodate this erect posture, so the joint surfaces of the skull face downward, not backward, as they do in lizards and crocodilians. *Diplodocus*'s head–neck joint was very much like ours: the back of the skull faced downward, so the joint with the neck permitted the snout to be horizontal when the neck was held upright. The thick-toothed brontosaurs also had a deflected skull–to–neck joint, but not nearly to the degree found in *Diplodocus*.

In living species, the position of the head relative to the neck is often determined by the animal's feeding habits. Hence the horizontal head and the vertical neck in *Diplodocus* imply that its neck was held nearly vertically during feeding. Since the neck is very long, *Diplodocus* must have been feeding at very high levels—twenty or thirty feet above the ground. Not many clams are found living at such heights. More likely the *Diplodocus* searched the upper reaches of Jurassic trees for select vegetarian morsels. With its sharply tapered snout, *Diplodocus* could probe deeply in among the branches, choosing its menu with more care and delicacy than the big-toothed *Camarasaurus* or *Brachiosaurus* could.

Diplodocus's head–neck anatomy simply contradicts those traditional restorations of the beast portrayed as feeding exclusively on ground level with its long neck outstretched. Why evolve a twenty-foot neck at all if feeding was done exclusively on the ground? *Diplodocus* had short front legs, so a six-foot neck would have sufficed quite nicely for ground feeding. Ostriches are long-necked ground feeders, but they have very different problems—they are very long-legged and require their long neck just to reach the ground.

The most troublesome part of a *Diplodocus*'s head is not its

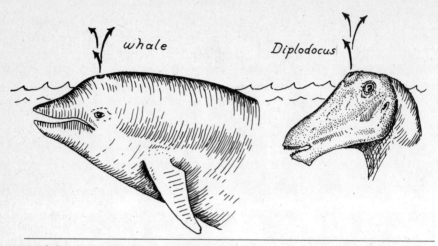

Diplodocus nostrils were in a whale-type position—on the forehead between the eyes.

teeth but its whalelike nostrils. Most air-breathing vertebrates have their nostrils at the tip of the snout. Air is drawn in through the nostrils, passes through a tube in the snout, and is then drawn downward through a hole in the roof of the mouth into the windpipe. The windpipe lies just behind the base of the tongue. But whales do it differently. Whale nostrils—their blowholes—are located way back on the skull right above the eyes. When a whale exhales after a deep dive, a geyser of humid air is blown nearly directly upward from its forehead. (Sperm whales have a long tube running through their fleshy snout from the blowhole in the skull, so, rather exceptionally, sperm whales blow from the front of the snout.) Nostrils in the whale position seem an obvious advantage for a swimming air-breather. The typical whale can inhale and exhale from its blowhole without danger of ramming water into its nostrils. And nostrils at the tip of the snout would be more vulnerable to the rush of water caused when the head plunges back below the ocean's surface.

Diplodocus had nostrils in the whale position—just in front of and above the eyes. If you are inclined to believe the water-living theory, the interpretation of *Diplodocus*'s nostrils is obvious: The

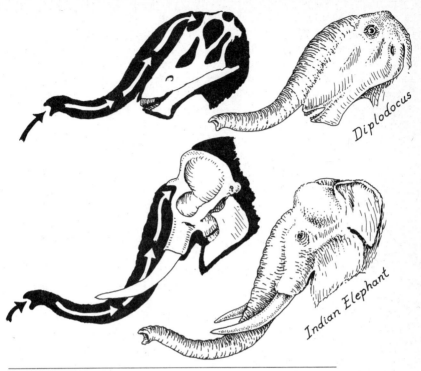

Did *Diplodocus* have an elephant-style trunk? Modern elephants have bony nostrils located in the forehead position.

beast used its skull as a combination snorkel-periscope to simultaneously breathe and look around while only the forehead was exposed above the level of the water.

An alternative explanation is however possible. There is one type of forehead structure found among living species that matches the *Diplodocus*'s—the foreheads of mammals with trunks. Elephants have nostrils located exactly in the *Diplodocus* position, between the eyes on the forehead. Tapirs—short-legged relatives of horses—possess nostrils located halfway between the elephant position and the usual mammal location at the end of the snout. Tapirs have trunks of moderate length. A trunk is actually a highly modified set of upper lip muscles that surround the fleshy nostrils and wrap around to form a mobile muscular tube. Usually the fleshy nostril—the hole in the skin through which the breath passes—is located in the flesh that more or less directly covers the bony nostril hole in the skull. But in a trunk the fleshy nostril is carried at the end of the mobile tube. In fossil mammal skulls, a trunk can be hypothesized if the bony nostril is located in the elephant or tapir position, and the skull bones around the nostril show attachment sites for the modified lip muscles.

I find the similarity between a *Diplodocus*'s forehead and an elephant's thoroughly unsettling. Could *Diplodocus* have been a dinosaur equipped with a proboscis? A horrendously heterodox thought, but not a new one. The possibility of trunked dinosaurs has been raised in paleontological journals on and off for half a century. There are all sorts of evolutionary problems generated by this theory. First of all, to produce a trunk, evolution requires a start with a set of muscular lips. Nearly all mammals possess a complex set of lip and face muscles, so evolving a trunk from any given mammal ancestor poses no great difficulty. But reptiles possess hardly any lips at all. Lizards have thin muscular bands running along the inner edges of their lips—just enough muscular tissue to flare the lips a bit to bare the teeth. But lizards don't have enough lip muscle to pucker, suck, flare the nostrils, or wiggle the nose. Crocs are even more lipless. The muscular lip band found in lizards is gone entirely in crocs, which have only a thin, scaly layer of skin over the gums. These thin croc lips are so tightly connected to the jaw and skull bones that they can't move at all. The thin band of the lizard lip hangs down enough to hide the teeth when the mouth is closed. But the croc lips hide nothing; its upper teeth are visible sticking down out of the gums even when its mouth is closed. Crocs have achieved the ultimate tight-lipped condition. The croc jawbone curves upward at the rear, which accounts for a smile the animal seems to have frozen on its face. In point of fact, crocs can't smile at all.

What sort of lips did dinosaurs have? Primitive brontosaur relatives, like *Massospondylus,* possessed bony gums just as modern lizards do. On the fossil gumline along the outer edge of the upper and lower jaw there is a gently beveled edge which must have been the attachment site for thin, muscular, lizard-style lips.

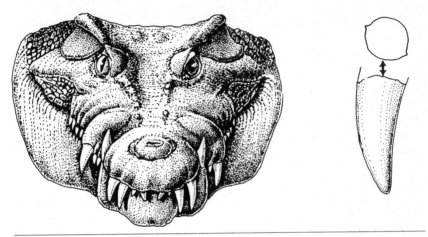

Crocodile lips. Croc facial skin is thin and tightly fixed to the skull bones, so there are no movable lips along the gum line. Tooth shape shown at right.

Lips require blood for nutrition and nerve fibers to carry sensory information to the brain. *Massopondylus* shows a series of holes in the jawbones precisely where the lips would lie in life. Through these holes passed the requisite blood vessels and nerve tracts. An identical pattern of holes can be found in the jawbones of living lizard species. So *Massospondylus,* the brontosaur uncle, was equipped with a little bit of lip, and primitive dinosaurs of the Triassic Period were all similarly lizard-lipped. The predatory dinosaurs, *Allosaurus, Ceratosaurus,* and the tyrannosaurs, retained this lizard-lipped condition into the later periods. Therefore *Tyrannosaurus* can be restored accurately with a sneer on its face or in the act of baring its teeth.

It is not totally impossible that evolution could convert the lips of *Massospondylus* into a big complex system of elephantlike face muscles, complete with proboscis. If *Diplodocus* really walked around with a trunk hanging from its forehead, some evidence of big proboscis muscles attaching to the skull bones near the edges of the bony nostril would have to be found. I can't find any such marks. But the *Diplodocus*'s lips were definitely different from those of lizards—the gum lines along its jawbones were not beveled, and the holes for blood vessels and nerves did not make an evenly spaced row like the one in lizards. *Diplodocus*'s lips were different from those of crocs, too. In the tight-lipped crocs, the skull bone beneath the thin scaly lip tissue is pitted and grooved so that the horny skin can attach very firmly to this roughened bone surface. *Diplodocus*'s jawbones were quite smooth compared to crocs'.

It's very unclear how *Diplodocus*'s lips were attached to its skull, but the possibility of a proboscis must be explored by more anatomical research. Alternative explanations for the locations of *Diplodocus*'s nostrils should also be explored—were they perhaps adaptations for tooting and honking? Primitive dinosaurs close to the brontosaurs, such as *Massospondylus,* may have snorted. The bony nostril hole is quite capacious in these early species and must have housed a series of pockets and compartments structured from skin, cartilage, and nasal lining. Compartments of soft, nonbony tissue in horses' skulls amplify their snorts and whinnies. Perhaps *Massospondylus*'s nostrils performed the same functions. Stout-toothed brontosaurs like *Camarasaurus* and *Brachiosaurus* had truly gigantic bony nostrils, so large that the eye sockets appear small by comparison. The bony nostril hole on each side of the skull is so enlarged in these species that only a tiny strip of bone separates the right from the left orifice. In life, these gigantic apertures were filled by some form of enlarged nasal device. The nasal organ was sufficiently large to spread over the snout bones, because the mark left by the soft-tissue nose can be seen on the top surface of the fossil snout. What important biological function required such a huge nose? The *Camarasaurus*'s bony nostril vaguely resembles a

Tyrannosaur lips. *Tyrannosaurus*, like *Massospondylus* and other primitive dinosaurs, had a lizard-style lip band along the gum line. A row of bony lip holes in the skull and jaws shows how blood vessels and nerves reached the movable lips.

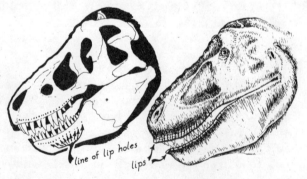

line of lip holes

lips

tapir's, and therefore the possibility of a short proboscis cannot be dismissed. Or the nostril hole may have housed resonating chambers to provide its owner's voice with a rich and varied timbre.

If brontosaurs in general possessed nasal adaptations for bellowing out Jurassic songs, perhaps, just perhaps, *Diplodocus*'s forehead nostrils were part of its nasal symphony. *Diplodocus* had small nostrils but probably evolved from an ancestor with gigantic bony nostrils like those of *Camarasaurus*. If *Diplodocus*'s ancestors had been nose-honkers, roofing over the narial tissue would have altered the tone—probably making it brassier.

Brontosaur faces and noses are still full of mystery. Right now, fossil tongues are the exciting topic in paleontology. The tongue itself doesn't fossilize, but the tongue bones in the throat do—and tongue muscles leave their marks where they attach to skull and jaws. But too few researchers are studying lips and their evolution. Someday someone will win his or her place in the history of science by solving the mystery of brontosaur noses and lips. When I'm asked by students what they should study, I always reply, "Think lips."

Diplodocus nostrils as nose flutes. Primitive brontosaurs—like *Camarasaurus* and *Brachiosaurus*—had huge bony nostrils that must have been covered with a fleshy chamber. The big chamber may have been used to amplify sound. In *Diplodocus* the nasal chamber is roofed over by the snout bones, so that the sound produced would have been brassier than that of a camarasaur or brachiosaur.

7
THE CASE OF THE DUCKBILL'S HAND

On the fourth floor of the American Museum of Natural History in New York, in a square glass case, stands the nearly complete, mummified carcass of a twenty-foot duckbill dinosaur. That mummy is famous worldwide for its hands.

In the early days of paleontology, the scientists of Europe and America expected that if dinosaur skin impressions were ever found, they would reveal a scaly hide. Up to 1900 only a few small patches of skin marks had been recovered from European dinosaurs and none from American. The Late Cretaceous delta beds of the American West changed all that. Expeditions from the American Museum to the Red Deer River in Alberta uncovered complete duckbill skeletons, including enormous patches of skin impressed in the sandstone around the ribcage, tail, and neck. The skin's substance itself was of course not preserved—it had rotted away after the duckbill had been entombed by a sudden influx of sand. But sand and mud had infiltrated the duckbill's body cavities while the dried hide still separated the animal's insides from the surrounding sediment. Consequently, when the sandstone was carefully chipped away, the surface of the vanished skin acted as a separation layer, so the stone faithfully recorded the living skin's texture. The American Museum's technicians—probably the best in the world at the time—erected twenty-foot slabs of sandstone in their exhibit hall, displaying the Alberta duckbill skeletons as

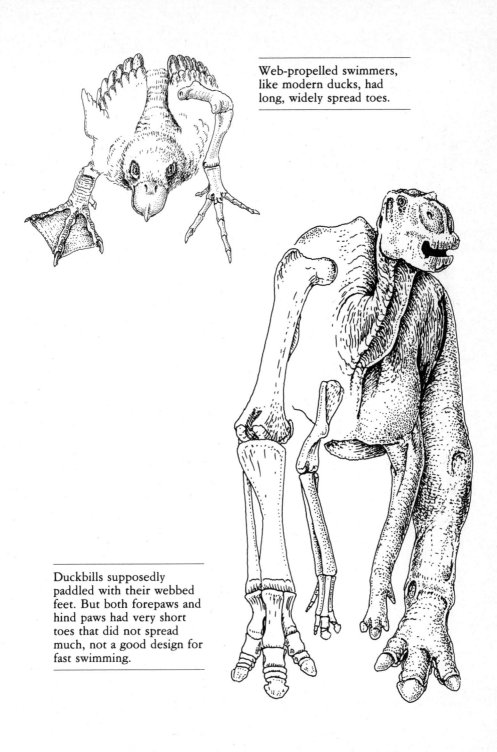

Web-propelled swimmers, like modern ducks, had long, widely spread toes.

Duckbills supposedly paddled with their webbed feet. But both forepaws and hind paws had very short toes that did not spread much, not a good design for fast swimming.

they had lain, half-embedded in rock, still partially clothed in their skin texture.

Even more spectacular mummies subsequently came from Wyoming's Lance Creek beds. The Alberta skeletons lay on their sides, their skin impressions flattened by the overlying rock bodies—producing a two-dimensional appearance. The Wyoming mummies however were nearly entire bodies, preserved lying on their backs, their chests expanded as if in a last gasp for breath, both fore and hind feet extended outward in an agonized pose. Impressions of skin covered each carcass on all sides. So lifelike do these three-dimensional mummies look, it is easily possible to imagine driving across a dried-up lake and happening upon just such a carcass, victim of a drought, sprawled with its collapsed belly pointing upward, legs contorted into unnatural angles by the contraction of sun-dried skin. Something about these contortions of death, so eloquently preserved in sand, drives home the message: This twenty-foot carcass was once alive, with full, pulsating muscles filling out the now cadaverous torso.

Duckbill dinosaur skulls and skeletons had entered the annals of science during the 1880s. The broad, ducklike muzzles suggested in a vague way some sort of mud-grubbing habits. But when the hands of the recently arrived three-dimensional mummy were finally cleaned, they caused a sensation. The skin impressions continued down the wrist, and between the duckbill's fingers. The conclusion was obvious—the duckbill's feet were webbed! The concept of aquatic duckbills, up till then an ill-defined theory, crystallized into a solid scientific "fact"—duckbills had webbed feet, duckbills swam. From that moment on, nearly every popular and scientific account of duckbill dinosaurs portrayed them paddling through lakes and rivers. The exhibits in the New York Museum depicted crested duckbills rushing into the water, with huge waddling strides; the caption read: "Escape to the Swamps." And once the aquatic theory became "fact," all the quirks of the duckbill's anatomy were forced into supporting the notion of an aquatic mode of life. The hollow crests of some duckbills were even hypothesized as air reservoirs adapted for prolonged diving.

A more careful consideration of the duckbill mummy's webbed paddle raises important heretical doubts about its aquatic role. If three-ton duckbills paddled at fair speed through the swamp waters of their Cretaceous delta home, they would require a paddle of considerable area. In other words, the fingers supporting the web would have to be long and spread out. Present-day ducks are effective paddlers—with their hind legs, of course—and their hind toes are exceptionally long and widespread, so that the paddling web is relatively huge compared to the bird's body size. Beavers are among the best of modern mammalian web-propelled paddlers, and their toes feature a similar wide-spreading arrangement. Otters, muskrats, and bullfrogs follow the same rule. But duckbill dinosaurs don't. The forepaws of the wide-billed *Edmontosaurus* or of the crested *Parasaurolophus* are the exact opposite of what could be expected in a specialized paddle. Those duckbills' front toe bones are short and the three main fingers are carried closely together with hardly any spread at all to the overall hand pattern. A very strange arrangement indeed if these dinosaurs were as fond of swimming as orthodoxy says.

So the duckbill's forepaw was manifestly inadequate for effective propulsion in water; but the orthodox theory maintains a second line of defense, the hind feet. Duckbills concentrated almost all their limb power in their huge hind legs, which were twice the thickness of their forelimbs. If evolution wanted to design a swimming duckbill, it would be logical to make the hind paw, not the forepaw, the main underwater propulsive organ. Therefore the toes of the hind feet should be long and spreading. But, once again, they are not. The duckbills' hind toes are among the very shortest ever evolved in the Dinosauria—much shorter, for example, than those of the meat-eating tyrannosaurs or the plant-eating horned dinosaurs.

Duckbills trace their lineage back to a gazelle-dinosaur, a small, long-legged, fast runner like *Dryosaurus* from Como. *Dryosaurus* possessed relatively longer toes than its duckbill descendants and thus might have paddled better. But according to the orthodox theory, *Dryosaurus* was strictly terrestrial in its habits, a confirmed landlubber, while the shorter-toed duckbills were supposedly committed to a watery life style. This paradox demands further scrutiny. Evolutionary processes are supposed to alter adaptations so that their possessors become better fitted to their new environments. But duckbills became progressively shorter-toed and therefore progressively worse adapted for paddling, the very habit they were supposedly evolving toward.

The argument for the swimming duckbill presents a third, ap-

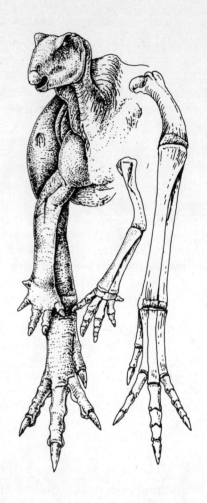

Duckbill ancestors had long toes.
Dryosaurus and *Laosaurus* (shown
here) had much longer hind toes
than did the more advanced
duckbills.

parently very strong point—the flattened tail. Crocodiles and Nile
monitor lizards are excellent swimmers, which employ the sculling
strokes of their deep, flat-sided tails to propel themselves through
tropical lakes and rivers. Deep, flat-sided tails are the characteristic locomotor equipment of other reptilian swimmers, too—the sea
snakes of Indonesia and the extinct giant sea lizards of the Cretaceous. When the first complete duckbill skeletons turned up in the
1880s, one striking peculiarity was immediately noticed: the unprecedented depth of the tail vertebrae. The bony spikes rising

from the upper vertebral bodies were of great height, and the lower vertebral spines, called chevrons, were nearly as long. And so it was proved: duckbills swam with sculling movements of their deep caudal organ. And nearly all the textbooks repeat this conclusion today, with complete assurance. But this hypothesis is rendered highly dubious by every important detail of the anatomy.

First, there's the problem of muscle power to support the alleged swimming stroke. Swinging the flat-sided tail back and forth against the water's resistance would require great muscular power. Strong tail-scullers today have thick muscles at the base of the tail and great spines of bone, called transverse processes, beneath them, sticking out sideways from the vertebral bodies to provide strong sites for the attachment of the muscles. Wide transverse vertebral processes are found in some dinosaurs—the armored ankylosaurs possessed outstandingly wide tail bases and must have had great power in the sideways swing of their tails. As the ankylosaur's tail ended in a bony war club, it's not surprising that the muscles at the base of that tail were provided with strong attachments. The duckbill's ancestors were also fairly strong in the rump; the *Dryosaurus* clan show good-sized transverse processes on the first ten tail vertebrae, counting from the hips back. Since evolution supposedly made duckbills better swimmers than their ancestors, we should expect to find duckbills outfitted with very wide transverse bony spines. We find no such thing. *Edmontosaurus,* the wide-mouthed duckbill, had a tail base of only moderate width, narrower relative to its body size than that of the ancestral dryosaurs. It would have required a massively muscled rump to send *Edmontosaurus* sculling through the Cretaceous bayous, and its modest tail base was manifestly inadequate for that.

There are even more serious problems. The most specialized duckbills were the hollow-crested group, *Corythosaurus, Parasaurolophus,* and their kin. They are inevitably portrayed as water-lovers with prodigious natatory prowess. But their transverse tail spines were absolutely puny—short, thin, weakly braced prongs of bone, which could have supported only an atrophied set of tail-flexing muscles. Thus hollow-crested duckbills, allegedly among the strongest swimmers in all dinosaurdom, were actually weak-rumped, puny-tailed creatures incapable of the powerful contractions required of a fast-swimming sculler.

Further problems concerning the caudal configurations are raised by the geometry of the upper tail spines. The duckbill's upper spines are indeed very long. The tails of living crocodilians also feature long spines. But the geometrical arrangement is totally different. Crocodile tail spines rise almost straight up, from the segments of the backbone. So do those of the Nile monitor lizards. Most land-living lizards—the iguanas, for example—also possess tall tail spines, but in them the bony spikes slant strongly backward. The difference between vertical and slanted spines is explained by

---

Why duckbills couldn't swim well. A good tail swimmer—like a modern crocodile—has a special arrangement of bony prongs in the backbone. Wide prongs that stick sideways attach to the powerful tail muscles. And tall vertical prongs give sinuous flexibility. Duckbill dinosaur tails were wrong on both counts. Their sideways prongs (transverse processes) were too short to support big muscles. And the upper prongs (neural spines) were too short and slanted too far to the rear.

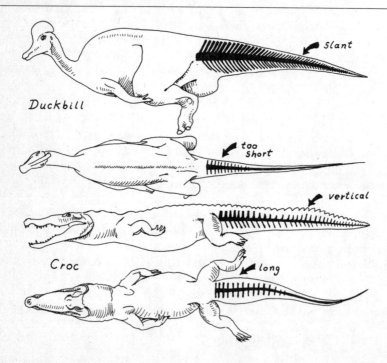

ligament action. A layer of tough ligament runs along the midline of the tail, and each spine is embedded into it. When the spines slant backward, this ligament is stretched when the tail is bent sideways. Such stretching impels the tail back in the direction from which it came, so the land lizard can flex its tail back and forth quickly as it runs. Underwater, much slower, stronger flexing would be required, and the elastic-rebound effect produced by slanted spines would be useless. So specialized tails of swimmers have vertical spines.

Duckbills were supposedly croc-style swimmers, moving by strong, easy, side-to-side flexures of their tail. Therefore, the optimal design would feature vertical tail spines. But duckbill spines all slanted strongly backward, exactly as in land-living lizards, not in swimmers.

Another problem in the duckbill's swimming equipment lies in the profile of the tail. The deepest part of the croc's tail is close to the end, because the end swings through a wider arc than does the base in moving side to side. Thus the tail is deepest where it can do the most good in pushing against the water. All powerful tail-scullers have such deep tail ends. But duckbill tails were deepest at the hips and become progressively narrower from top-to-bottom toward the tip—another caudal feature nearly totally maladapted for its alleged primary function.

An argument very eloquently expressed in 1964 by John Ostrom of Yale administers the coup de grâce to the theory that duckbills swam. Any sort of tail-propelled swimming requires a smooth ripple of tail flexure from the hip out to the tip of the tail, a sort of muscular sine wave that pushes against the water's resistance to propel the animal forward. Even with the handicap of a weak rump and their maladapted shape, duckbills could have swum at least at slow speed if they could undulate their tails. The only anatomical feature that would have entirely prevented a dinosaur from swimming would be a tail corset, a stiff latticework of bone which would hold the entire tail assembly together as one stiff immobile mass. Tail corsets evolved in a wide range of dinosaurs, including some meat-eaters, some plant-eaters, and the long-tailed flying dinosaurs (pterosaurs). No modern reptile has a tail corset—all the lizards, turtles, and crocs can wiggle their tails freely, and even the weak-tailed lizards (desert horned toads, for example) can employ their caudal undulations to swim when the animal is forced to. But when the skeletons of duckbill tails were found, their single most startling feature was a basketwork of long, stiff bony rods crisscrossing over the backbone from the mid-chest all the way to near the tail's end. A perfect tail corset. Each rod consisted of dense bone up to half an inch thick and was attached at one end to the bones of the vertebral column by a short stiff ligament. Dryosaurs had a tail corset, too, as did horned dinosaurs, but the ultimate in caudal basketwork was developed in the duckbills.

The duckbill's tail corset evolved for an obvious mechanical purpose: to keep its backbone stiff and immobile from a point just behind its shoulders all the way down to the hindmost tail section. Even the most devout believers in swimming duckbills are forced to admit that this bony latticework would make for an unusually unsupple spine, the very reverse of what is necessary for swimming with smooth, horizontal undulations.

The supposedly definitive monograph on duckbills came out in 1942. Its two authors (one was the senior professor of paleontology at Yale) had to engage in quite a twisted form of logic to explain away the problem of the duckbill's stiff backside. They admitted the bony system of rods must have evolved to maintain the backbone rigid for perambulations on land when the duckbills chose to walk about on terra firma. But perhaps, the authors argued, the tendons were a little loose so that a small degree of side-to-side movement was possible in the tail. But they ignored a critical problem. The duckbill's ancestors had been land livers, with moderately strong tail corsets, and the duckbills themselves *increased* the stiffness of the corset. The monographers of '42 failed to explain why duckbills would evolve in the wrong direction—why the duckbill family had stiffer, not more supple, tails than their ancestors.

The sum of evolutionary evidence is thoroughly damning. In nearly every modification of the evolutionary process made in the duckbills as they developed from their dryosaur ancestors, the duckbills suffered a diminution of their swimming potential. Their fore- and hind paws became shorter and more compact, not longer and more widely spread. Their tails got weaker and stiffer. Far from being the best, the duckbills must have been the clumsiest and

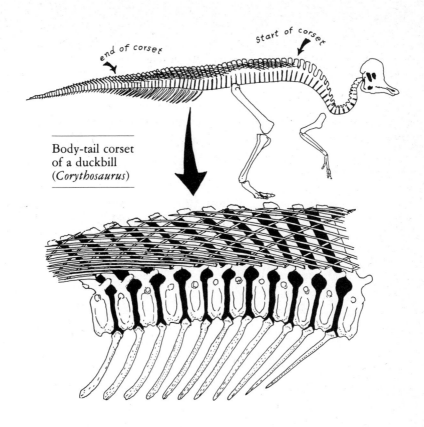

Body-tail corset
of a duckbill
(*Corythosaurus*)

end of corset

start of corset

slowest swimmers in all the Dinosauria. If pressed, they probably could paddle slowly from one riverbank to another. The central theme of their bodily evolution was indeed specialized—orthodox theory was right on that point—but the direction of specialization was landward. These dinosaurs were specialized for a totally terrestrial existence.

Every so often some paleontologists attempt placing some other major dinosaur group in the water. A young Canadian would have Alberta horned dinosaurs wading through the sluggish backwaters of the Judith Delta. But his evidence derived from the dubious notion that the horned dinosaurs spent most of their lives in the water because their fossils are found buried in river-channel sandstone. American buffalo often are found buried in river sands where their bodies came to rest after a flood, yet the buffalo is hardly an aquatic creature. One quick way to calculate how well a dinosaur would cope with swampy terrain is to calculate the area of its hind foot available to support the downward thrust of its hind leg. Anyone who has tramped around as many bogs and swamps as I have knows that feet get stuck not when you stand still, but when you step too forcefully and drive your leg down into the muck. The faster you walk, the more downward thrust is applied to the sole of the foot. Hence to move speedily over mud or soft earth, devices such as snowshoes must be used to expand the foot's area. Hippos follow this pattern: they have wide-spreading toes relative to the power of their legs, a contrast to the small, short-toed feet of the rhino. Since all dinosaurs had stronger hind limbs than fore, the largest thrust would be exerted by their thigh muscles. So the thickness of the thigh bone (femur) works as a useful gauge of the

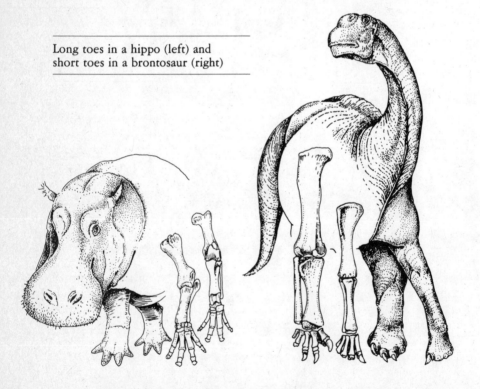

Long toes in a hippo (left) and
short toes in a brontosaur (right)

force applied to the sole of the feet. From fossil footprints, the area of a dinosaur's foot can be calculated and then compared with the cross section of the femur.

This exercise in quantitative paleopodiatry produces distinctly counterorthodox conclusions. All the popular books list brontosaurs and duckbills at the top of the list in the preference for swamps. But brontosaurs and duckbills had among the smallest feet in area relative to the size of their thigh. If these giants had tried to spend their lives paddling around marshy terrain, they'd have found themselves stuck in the mud with genuinely maladaptive frequency.

Stegosaurs and ankylosaurs were also compact of foot. But the horned dinosaurs had much bigger feet per pound of thrust from the thigh. A few dinosaurs were especially large-footed—some little horned dinosaurs, the primitive anchisaurs (brontosaur ancestors), the dryosaurs, and most of the meat-eaters, both large and small. Strangely enough, it's these dryosaurs and meat-eaters that are supposedly least adapted for soft swampy soils according to orthodox dinosaur ecology. But like so much else in traditional dinolore, the standard story about feet and mud is not accurate. Museum exhibits teach that brontosaurs and duckbills escaped their predatory enemies by wading out into the marshes where meat-eaters feared to tread. But if an allosaur pursued a brontosaur or a tyrannosaur pursued a duckbill, it would be the ponderous pads of the plant-eaters that would mire first into the mud to hold their hapless owners fast as the killer descended.

After all this calculation of tail mechanics and foot areas has been done, that duckbill mummy's hand, that webbed forefoot raised forever skyward in its fourth-floor glass case at the American Museum of Natural History still requires an explanation. If it was not for swimming, then what was it for? A dead camel I observed in the Transvaal might solve this mystery. While I was in South Africa studying the ancient mammal-like reptiles of the Permian Period, a colleague from Johannesburg Museum took me for a weekend outing to one of their famous parks. There were camels—though the species isn't native to South Africa. They had been imported for use in crossing desert regions and are popular exotic displays in the outdoor parks. Camels have thick cushiony pads under their toes, and these pads spread out under their own-

The mummy's forepaw

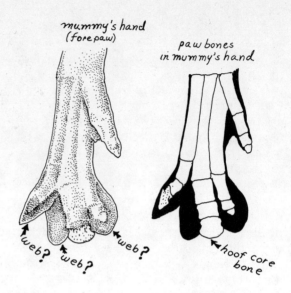

er's foot as they walk. In the course of our visit I happened upon a dead camel in an unkempt corner of the park. It lay like a desiccated mummy, all its natural juices evaporated by the hot Transvaal sun. The camel's mummification is not uncommon in such dry climates. Beneath its outstretched feet, its cushions, plump and elastic in life, were now dried-out bags of skin, which had flattened against the dusty soil surface. A spark of recognition shot through my brain. If camels were extinct and this carcass were found covered by flood-borne sand, wouldn't paleontologists conclude that the camel had webbed toes? The flattened skin of its paws created the perfect imitation of a web.

The skin of the duckbill's paws was not marked by calluses the way camels' paw skin is. But the way the duckbill mummies are preserved permits the hypothesis that in life those flattened hands were in reality plump, rounded cushions of connective tissue—elastic shock absorbers for the impact of the ground on the wrist when the animal moved fast over hard ground. Duckbill forepaws were so narrow and compact that a paw cushion would do invaluable service by lessening the load of impact within the joints of the toes. Fossil duckbill trackways, just now being excavated in

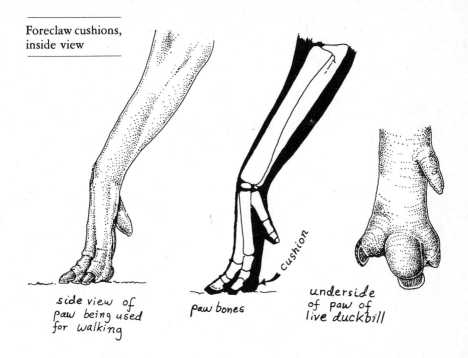

Foreclaw cushions, inside view

side view of paw being used for walking

paw bones

cushion

underside of paw of live duckbill

Canada, suggest that indeed this line of reasoning may be correct. The forepaw impressions resemble smooth crescents, as though the individual toes were all imbedded within a single, insulating mitten. There is definitely no sign of a spreading ducklike web.

This may be the true solution to the century-old mystery of the mummy's hand. That brown withered paw may have misled four generations of paleontologists into believing in a series of nonexistent adaptations for swimming. The mummy's hand, when alive and full of healthy tissue, may have worn a shock-absorbing glove, an earth-mitten entirely designed for walking on dry ground.

8

DINOSAURS AT TABLE

Orthodox paleontologists insist most of their dinosaurs ate mush. They condemn both of the great tribes of plant-eaters—the brontosaurs and the duckbills—to a way of life at the water's edge, forced to eat nothing but soft water plants. In its own way, this theory epitomizes the traditional view of most dinosaurs as swampland creatures, virtual dead ends in evolution's race to develop lively, active species. In 1915, William Diller Matthew, a very respected mammal paleontologist, wrote a highly influential book, *Climate and Evolution,* which argues that evolution bogs down in the soggy lowlands. Matthew believed that only on the high, dry soil of plains and plateaus did evolutionary forces create the most vigorous, most advanced creatures. There's a lot of truth in Matthew's thesis. It has been ascertained, for example, that water-loving turtles and crocodiles evolve most slowly, changing so little on average through geological time that a single genus can be followed for thirty million years or more. So the orthodox concept of a mush diet is consistent with the overall theory of sluggish dinosaurs: soft, plant food was all they required for their sluggish metabolic needs, and the consequent swampy habitats limited dinosaurs to slow rates of evolution.

There may be some ground for believing the brontosaurs ate such soft foods. If the possibility of gizzard stones is ignored, the brontosaurs' dentition does seem little equipped to deal with meals

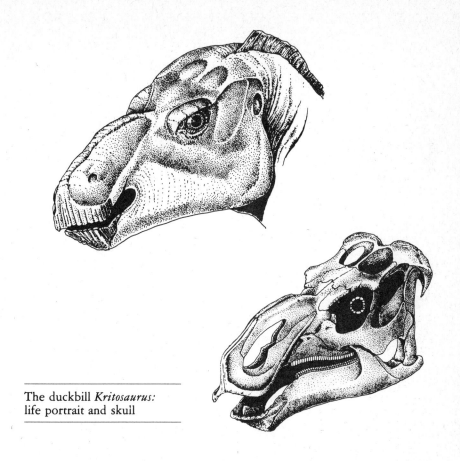

The duckbill *Kritosaurus:*
life portrait and skull

of tougher plants. But there are no grounds whatsoever for believing it of duckbills. The mouth of a duckbill dinosaur contained
one of the most efficient cranial Cuisinarts in land-vertebrate history. Duckbill teeth and jaws were incomparable grinders, designed to cope with foods right inside the duckbill's oral
compartment.

The myth of mushy foods for duckbills began with a single
error by one of the great pioneering American dinosaur hunters.
Edward Drinker Cope discovered a fragmentary duckbill jaw in
1885. His specimen had cracks running through the row of teeth,
so that individual teeth fell out of the fossil jaw when he examined it. Cope mistakenly assumed this condition was natural and

jumped to the conclusion that a duckbill's teeth would break off whenever the beast tried to chew tough food. This error should have been corrected by 1895, when complete skulls and jaws revealed that duckbill teeth were firmly packed together and no one tooth could possibly fall out before it was totally worn down. Even then, whenever a worn tooth dropped out, a new tooth already stood beneath it ready to take over chewing duties. Duckbills apparently never ran out of teeth. No one has ever discovered a senile duckbill mouth; not one specimen exists with all its teeth either

The head of *Edmontosaurus,* a duckbill. Life portrait at top, skull in the center, and skull cut through the tooth rows at the bottom.

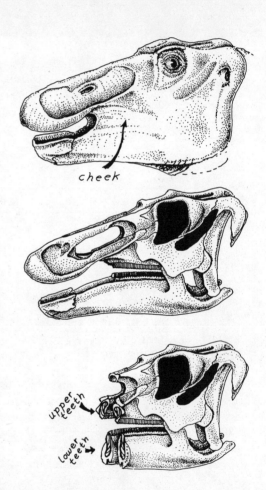

cheek

upper teeth

lower teeth

worn out or fallen out. To all appearances, from the day they hatched out of the egg to their last breath, the duckbills enjoyed the use of healthy dental machinery, continually renewed by young teeth growing in to replace the old.

Not only were the duckbills' teeth never-ending, their arrangement was designed especially for powerful grinding. At any one moment many rows of young teeth were growing into the mouth, providing the animal with grinding surfaces made up of hundreds of closely packed teeth. Each tooth was built up from two different biological materials: a thick layer of very hard enamel and a central core of softer dentine. Since many rows of teeth were packed together in each jaw, and all the rows together participated in chewing action, the chewing surface was a mosaic of enamel ridges and dentine. Enamel ridges always protruded a little higher than the dentine cores, because the enameled parts of the teeth got worn down a bit more slowly than the softer cores. This arrangement was very effective. No matter how hard the duckbill

How duckbill teeth work

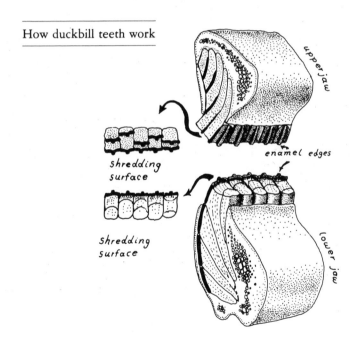

chewed or how hard its food was, the enamel stuck up further than the dentine, young teeth kept replacing the old, and the duckbill maintained a grinding surface that worked much like a self-sharpening vegetable grater.

Although Professor Marsh of Yale clearly illustrated the real qualities of the duckbills' chewing equipment in 1896, most paleontologists retained the mistaken theory and ignored the obvious adaptations for tough food. It required yet another Yale professor to set matters straight. In 1961, John Ostrom published his heretical interpretation of duckbills. He defined them as land creatures and emphasized the mechanical–ecological implications of their dental Cuisinart. He pointed out that the teeth of duckbills had a pattern that virtually necessitated tough food. Their characteristic bills were also consistent with a tough-food diet, despite a superficial resemblance to the bill of modern water-feeding ducks. Way back in the 1880s, Cope had already found fossil remnants of the horny edge that had lined the bony beak of duckbill dinosaurs while alive. This horny edge was sharp and deep from top to bottom, more like the edge of a cookie-cutter than the soft, sensitive rim of a mud-dabbling duck. After Cope's initial discovery, other horny fossils turned up, making it clear that all duckbills possessed deep, sharp-cutting edges along the entire upper and lower beak. Such sharp edges were obviously for cropping tough plants—not for grazing on mush. So soft-beaked ducks were never good analogues for duckbill dinosaurs, but modern tortoises are; the tortoise's beak is tall and sharp-edged, and constantly used to cut through tough blades and stems.

If duckbill dinosaurs were truly efficient shredders of tough fodder, they would also have required good tongue–cheek coordination. Consider what it takes to chew something as recalcitrant as a piece of celery—your tongue contributes by moving the fibrous lump between palate and teeth. Your cheeks play their role by retaining the mass of celery and preventing it from slipping. Tongue-in-cheek skill is characteristic of the best shredders among today's Mammalia—horses, cows, elephants, rabbits, kangaroos. All these herbivores possess large, active tongues and strongly muscled cheeks. Incidentally, that lump of food while being chewed in the mouth has been dignified with a technical scientific label: "bolus."

All of today's Reptilia are cheekless. Their open mouthline extends all the way back to the joint of the jaw before the ears. There is no skin to hold any food being chewed. Consequently, herbivorous reptiles—tortoises and iguana lizards, for example— are sloppy eaters; when their jaws slice off a piece of leaf, the part sticking out of the mouth simply falls to the ground. Each time they chew, they lose nearly half their mouthful, quite a wasteful business. Primitive meat-eating dinosaurs had similar wide-open mouthlines.

Traditionally, duckbill dinosaurs have been portrayed as cheekless, with the mouthline running from chin to ear like a lizard's. A dissenting voice was raised by Yale Professor Richard Swann Lull (Yale's tradition of duckbills seems to have been consistently heterodox). In 1942, Lull restored duckbills with cow-style cheeks walling the sides of the oral space. But most of Lull's colleagues rejected the idea because everyone knew dinosaurs were reptiles, and reptiles, by definition, didn't have cheeks. Such objections were specious. No living reptile has cheeks. But no living reptile has grinding teeth anything remotely resembling those of a duckbill. If the duckbills could have evolved such unreptilian teeth, why couldn't they have evolved unreptilian cheeks?

The final Yale duckbill–cheek conclusion was joined in the late 1960s. Peter Galton, an English paleontologist resident at Yale as a research associate, reinvestigated the question of the duckbill's oral tissue. He concluded that Lull's reconstruction of cheeky duckbills was almost certainly correct. All duckbills had deep recesses in their skull and jaw bones running parallel with their mouthlines above and below where their teeth came together. This recessed zone resembled the deep hollowed-out areas found in the jaws of gophers, chipmunks, and other rodents which have capacious cheeks for holding food while they chew. A slightly roughened ridge often marks the top and bottom of the duckbill recess, and some sort of skin or muscle or both must have attached to it. Peter Galton drew diagrams of the cheek–pouch recesses in modern species such as pigs, horses, elephants, and rodents, which demonstrated how duckbill pouches must have been as well developed as any of these.

What, then, did duckbills eat? Considering their prodigious dental powers, the flip answer might be "anything they wanted."

But in terms of serious theory, those powers expand the boundary conditions of any hypothesis about their diet very widely. The duckbills might have masticated extremely tough leaves, stems, twigs, pinecones, even roots and tubers. Some clues as to their actual dining habits can be gleaned from their very curious body posture. All duckbills had much longer and stronger rear than forelegs and probably moved semibipedally, striding on their hind legs and using their forepaws only to touch down lightly for balance. Old restorations showed duckbills standing in a tripodal posture, their hind legs and tail supporting their weight, with their back and neck nearly vertical. Such a posture would have permitted the duckbills to feed high in the pine trees of their habitat. Yet that upright body posture was wrongly conceived. The build of the duckbill was clearly designed for low, near-the-ground feeding, not for tree-browsing. If duckbills had specialized in high-level feeding, they would have had shoulders and necks designed for reaching upward. But that is not the case. Instead, in the region of the shoulder their backbone bends permanently downward. This sharp flexure locates the base of the neck and the head on a very low anatomical level. The downward bend in the chest area is so marked that even when a duckbill raised its neck as far as it could go, the head was still below the level of the topmost point of the shoulder.

Some mammals today exhibit this same downward curve of the backbone. In the American buffalo, for example, the line of the vertebral column curves sharply downward as it passes from shoulder to neck. Thus they must always hold their heads low, with muzzles close to the ground. As the song says, buffalo roam where the deer and the antelope play, but deer and buffalo represent divergent tactics for eating plants. Deer can carry their heads much higher than can buffalo, and can reach up into the trees to nibble on twigs, leaves, and bark. Buffalo stick to ground level and use their strong, wide snouts to pull up tough grass deer cannot deal with. Clearly, the duckbills were more like buffalo than like deer. And the entire tribe of duckbills must have spent most of their time feeding at or near the ground.

These considerations dramatically narrow the boundary conditions for any hypothesis about their diet. The duckbills' preferred food must have been low-growing herbs or shrubs (grasses had not yet evolved in Cretaceous times). These boundaries still

Permanent downflex in the American buffalo and a duckbill (dashed line shows line of backbone)

allow for a wide selection of Mesozoic roughage and greens: horsetails, ground pine, ferns, low tree ferns, seedling evergreens (pines, cypress, etc.), cycads and other tough-frond types, low-growing palms, magnolialike shrubs, and so on.

It's probably barking up the wrong herb to try to find the one duckbill food. Duckbills were so varied in snout design that it's unlikely all species fed on the same plant stuffs. Today the antelope family demonstrates how snouts can be custom-tailored to fit each species' method of feeding. Cape buffalo (cows and buffalo are members of the antelope family) have very wide muzzles, fine for biting a wide swath through the sward but much too clumsy for picking out individual succulent tidbits. Royal antelope have slender snouts which they can use to pick and choose. Among the duckbills, *Edmontosaurus* had a huge, blunt muzzle and must have cropped wide batches of leaves with each bite. Duckbills with hollow head crests, *Lambeosaurus* and its kin, adopted a totally different approach; their muzzles were narrow, and allowed them to poke around for a more discriminating bite.

Everywhere on the Late Cretaceous deltas the duckbills' constant companions were the great horned dinosaurs. Side by side, three-horned *Triceratops* and wide-mouthed *Edmontosaurus* cropped the greenery. Duckbills and horned dinosaurs were distant cousins—both had beaks and traced their ancestry to the same ancient little dinosaur of the Triassic Period. But what an extraordinary contrast in bioengineering these two beaked clans displayed. The *Triceratops*'s snout was a mammoth set of pincers, with a sharply edged upper and lower beak, narrow from side to side, and covered with horn. Such deep, powerful beaks must have given those horned giants the power to slash and cut long, tough fronds and branches—fodder probably too coarse even for the wide beak of duckbills.

After he had completed his unorthodox treatment of duckbills, John Ostrom, in 1963, attacked the problem of *Triceratops*'s diet. To maximize their biting strength, jaw muscles require three biomechanical properties: muscles must be thick, they must be long, and they must have great leverage. Great leverage can be developed by designing the muscles so that their line of pulls is located far from the jaw joint. John Ostrom showed that *Triceratops* and its horned relatives evolved high bony cranks on their lower jaws to move the line of muscle pull up and thereby increase leverage.

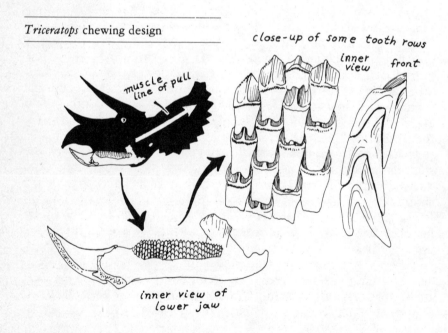

Triceratops chewing design

close-up of some tooth rows

inner view   front

muscle line of pull

inner view of lower jaw

The evolution of duckbills remodeled the lines of muscle pull in the same way, but the horned dinosaurs went far beyond them when it came to muscle thickness and length. All primitive dinosaur heads had some basic design problems that caused difficulties when evolution tried to enlarge the biting muscles. Most of the jaw muscles in the skull were housed inside of bony compartments located behind the eye sockets, so the outer bony walls of the skull tended to limit possible muscle size. If the muscles grew too big, they would bulge out during a strong bite with a force that would burst open the skull bones—certainly maladaptive.

But *Triceratops*'s ancestors required even larger jaw muscles because those dinosaurs were locked into an evolutionary path leading to diets of ever tougher, thicker foods. They needed an escape from the limitations imposed by the architecture of the older skulls—and they found one. On the top of the primitive dinosaur's head, behind the eye socket, were a pair of holes, covered in life by a tough membrane. Holes like these evolved many times, probably because the stresses of chewing were concentrated along a certain few trajectories in the head bones. The most effective way to construct a head was thus to evolve thick bone where stresses were great and holes were stresses were minimal. Large upper-rear head holes (formally called "temporal fenestrae" in anatomical parlance) gave the horned dinosaurs their escape route to freedom for the design of their jaw muscles. As the head evolved, the rear rim of the hole grew upward and backward, forming a gigantic frill. Since the jaw muscle was attached to the membrane that covered the rear rim, as the rim grew backward, so did the muscle, and the entire mass of the jaw muscle could enlarge to unprecedented proportions. Marks on the *Triceratops*'s frill illustrate how far the muscles had enlarged in both length and width: On a big skull, the distance across the mass of muscles is often three feet and the maximum tract of muscle fiber often three and a half feet in length. These muscles must have delivered an astounding bite in life, with a force greater by far than any other land herbivore in life's history.

Triceratops could use this prodigious biting power either for nipping branches at the beak end or for cutting up the fodder into smaller chunks with its teeth. Horned dinosaurs had cheek pouches and could employ tongue–cheek coordination to keep chopping

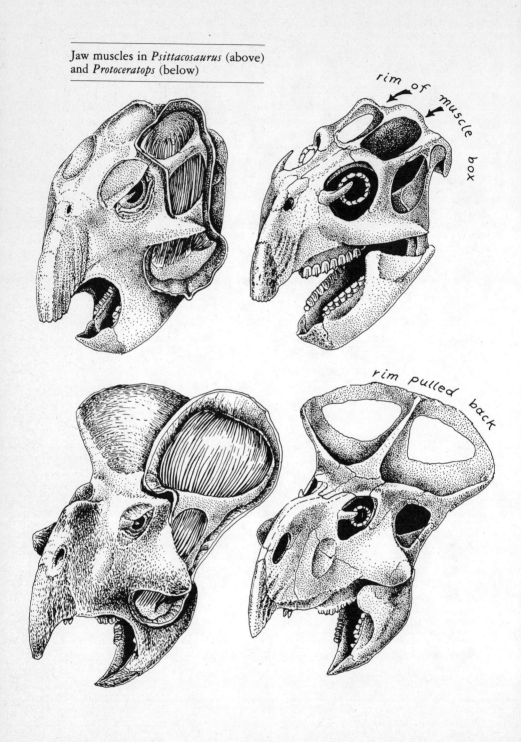

Jaw muscles in *Psittacosaurus* (above) and *Protoceratops* (below)

rim of muscle box

rim pulled back

the bolus into ever finer slices. The geometry of their teeth resembled the duckbills'—several rows of teeth were packed tightly together so that enameled ridges provided a self-sharpening cutting mosaic. But the horned dinosaurs' teeth were arranged to provide more of a vertical slicing action and less fine shredding than those of the duckbills.

What did *Triceratops* eat? Ostrom suggested cycadeoid fronds, probably a good guess. Cycadeoids were plants with large fronds, their leaves resembling the cycads popular today in Florida as decorator shrubs. Cycadeoids were so common that the Cretaceous is known as the Age of Cycads. Both cycads and cycadeoids had fronds two, three, even four feet long, characterized by especially strong fibers and prickly pointed leaflets, so that cutting such leaves was a nasty business. But a rich source of protein and calories lay in those Cretaceous fronds, awaiting any beast that could evolve the proper chewing armament. Horned dinosaurs were late arrivals; they didn't make their evolutionary debut until halfway through the Cretaceous. But once they got going, they developed with explosive success, proliferating species by the dozen, the result of their mechanical prowess in chopping the previously inaccessible fronds.

Beaked dinosaurs featured another adaptive device in their plant-eating repertoire, an extra-long digestive tract for soaking and fermenting stubborn plant tissue. Paleontologists usually dismiss any theorizing about the soft parts of dinosaurs. Stomachs rot, intestines decay . . . both disappear without a trace in the fossil. Ergo, all speculation about gastrointestinal tracts in the Dinosauria is futile. This is a serious problem because obviously there's no hope of understanding the dinosaurs' approaches to plant eating without at least some knowledge of their innards. Some skepticism about the study of digestive systems is justified—only very rarely do sediments preserve direct evidence of inner architecture. Digestive structures possess neither bones nor other hard tissue, so the only way their outlines can be preserved is on rare occasions when mud fills the stomach and intestines before the tissue rots (a few Coal Age amphibians were preserved that way, but no dinosaurs).

Christine Janis, a fellow graduate student at Harvard in the mid-seventies, was the first to excite my interest in the digestive organs. She pointed out that teeth and jaws tell only half the story. Nearly all plant-eaters have fermenting vats, enlarged chambers where food sits and soaks while microbes attack it with powerful enzymes. Janis stressed how enormous is the variation in location chosen by natural selection for the fermenting site. Ruminants— the deer–cattle–antelope family—chose a forward site and re-modeled their stomach into a complex multi-chambered rumen where the bolus is soaked by enzymes. Since the rumen is located in the forward stomach compartment, a deer, antelope, or buffalo can crop leaves, wad them up into a bolus, pass it down for pre-softening, then pass it back up to the teeth for a thorough chew after the leaves have been softened. Forward locations offer sub-stantial advantages—teeth are saved from unnecessary wear when all food is pre-soaked and softened. Horses, rhinos, and ele-phants, on the other hand, chose a rearward location, a pocket evolved far back in the intestine or colon. Rearward location has one major disadvantage—the bolus can't make any sort of return to the mouth. But since the rear of the body cavity is spacious, its advantage is that rearward fermenting vats can be huge.

A dinosaur's fossilized ribcage can reveal a great deal about the organs it housed—information largely ignored until recently. For one thing, how big the dinosaur's digestive chambers were can be gauged by the size of the ribcage. Orthodoxy maintains that many dinosaurs were too weak-toothed to eat tough plants; but large digestive tracts could compensate for weak teeth. Janis made a point often ignored by bone-and-teeth paleontologists: The bet-ter the enzyme soak given to food, the fewer the teeth needed to deal with any specific food texture. A good case in point: Today's herbivorous lizards usually have relatively small, weak teeth and until recently had the reputation of being inefficient plant-eaters, but recent experiments show that some lizards carry out very ef-fective rearward fermentation in their extra-long intestines. Giant ground birds—rheas and ostriches—have tiny heads and no teeth whatever, yet these birds successfully employ rearward fermenta-tion on a large scale.

The precise details of any dinosaur's plumbing cannot be de-termined, but the overall body contours outlined by the ribcage and hips do show how large the entire digestive apparatus was and in what locations. Humans don't have ribs in their abdominal sector—their ribs end at the posterior edge of the lung compartment. But dinosaurs had long ribs attached to every segment of the backbone from chest to hip, so the cross section of the digestive tract is preserved by the skeletal architecture. Brontosaurs clearly had short, deep, crowded gastric tracts, because the abdominal ribcage was compact front to back, and the ribs over the belly arched widely outward from the backbone like barrel staves. In general configuration the brontosaur's intestines followed the proportions of modern elephants. And, just as in elephants, the front edge of their hip bone (ilium) was flared outward to support their wide belly. Elephants are big rearward fermenters, and brontosaurs must have been so too. Some brontosaurs had larger digestive organs than others; *Brontosaurus* itself had a very short torso from front to rear and must have had a less voluminous intestinal apparatus than *Brachiosaurus* with its long torso. Equipped with both gizzard stones and rearward fermenting vats, brontosaurs could have tackled really tough plant food.

Early beaked dinosaurs were bipeds, using hind legs alone for their fast locomotion. That presented a special design problem: They had to evolve a large digestive tract without upsetting the balance necessary for bipedal walking. In most primitive dinosaurs, the digestive tract ended where it butted against the wide pubic bones, which formed a rear bulkhead for the entire abdominal cavity. Located above the pubic bones was a narrow passage through the hips, through which must have passed all the animal's outlets—colon, urinary tube, and birth canal. This pubic bulkhead was an obstacle to redesigning the intestines: in order to lengthen the gastrointestinal tract, the pubic bulkhead had to be pushed backward and with it the entire pelvis. So a longer digestive system implied a longer, heavier body cavity before the hips—a shape hard to balance. Primitive beaked dinosaurs were forced to face this design problem squarely from the very beginning of their evolutionary development, because nearly all the earliest species were very long-limbed and lively, built for ultrafast bipedal running. Fast locomotion placed special strain on the back, and a long, heavy stomach in front of the hips would be difficult to support. Evolution was thus pulling in two opposed directions—shorter torsos provided better balance for running, but longer torsos were required by the need to deal with the problem of digesting leaves.

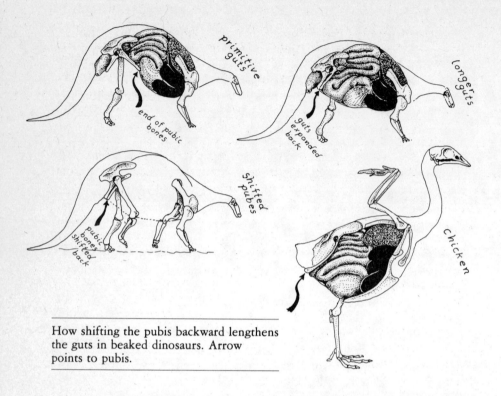

primitive guts

end of pubic bones

long guts

guts expanded back

shifted pubes

pubic bones shifted back

chicken

How shifting the pubis backward lengthens the guts in beaked dinosaurs. Arrow points to pubis.

Escape from this biomechanical dilemma was engineered by a clever bit of anatomical sleight-of-hand. Peter Galton was the first to work out the details in his Ph.D. thesis in 1966. If beaked dinosaurs managed to shift their digestive system beneath the hip bones, its length could be increased without relocating the hip joint. That effect was obtained by bending the pubic bones backward from the point where they already attached to the other hip bones, so that the intestines could lengthen while the hip joint remained as it was. In their new position, the pubic bones slanted downward and backward instead of straight down as before. Beaked dinosaurs finally shifted the lower end of the pubic bones so far back that they were far behind the hip socket. In this remodeled position the thick coils of intestines and the colon continued without interruption backward from the belly, to below the hip socket, and all the way to the base of the tail.

The solution was elegant. The digestive system could be lengthened and bipedal balance improved at the same time, for more intestinal tubing located to the rear of the hip joint clearly helped to balance the weight of the body before it. Although the beaked dinosaur classes are totally extinct nowadays, a quite vivid picture of their intestinal arrangements can be obtained from a chicken, turkey, or any modern bird. In birds, the gastrointestinal tract passes right under the hips all the way to the end of the pubis below the base of the tail. And birds obtain the same advantages for balance and strength as did the dinosaurs. Birds—like early beaked dinosaurs—are bipeds when they walk, and their intestinal arrangement allows them easy balance on their hindlegs. In the air, all the force of the wingbeat passes through the upper shoulder joint, and the design of the intestines lets birds have very short, very strongly braced torsos to anchor the stresses and strains of flying.

Some beaked dinosaurs had every possible plant-digesting device—the parrot dinosaurs, for example, had (1) strong, deep beaks, (2) closely packed teeth, (3) large jaw muscles, (4) a forestomach gastric mill with stomach stones, and (5) a long intestine. These parrot dinosaurs were the ultimate in dietary adaptation. They were, however, exceptions to the rule. Most advanced beaked dinosaurs heavily developed one type of digestive device or another, not all at once. Duckbills possessed cranial Cuisinarts par excellence. They had developed the best teeth for shredding leaves and twigs into tiny bits. Large horned dinosaurs were the long-slicers, the best at cutting tough vegetables into digestible chunks. Yet neither duckbills nor horned dinosaurs have ever been found with gastroliths, even though dozens of good skeletons have been hewn out of the Late Cretaceous rocks. Probably these strong-jawed dinosaurs substituted the power of their teeth for stone power in the stomach. And the size of their digestive system varied considerably in these species: wide-beaked duckbills had broad tail bones that supported enlarged colons and intestinal appendices; hollow-crested duckbills and horned dinosaurs were less enlarged in the rump.

The successful development of rearward fermenting vats provides the solution to one of the biggest puzzles about plant-eating dinosaurs—the weak-gummed giants. Although brontosaurs, duckbills, and horned dinosaurs were all apparently well provided with grinders of one anatomical sort or another (teeth or gizzard stones or both), two groups of big plant-eating, beaked dinosaurs seem to have been totally unprepared for grinding plants: the armored ankylosaurs and the dome-headed dinosaurs. Both groups seem to have followed the wrong evolutionary path—their teeth are smaller, weaker, and less tightly packed than in the ancestral early beaked dinosaurs. Ankylosaur and domehead had tiny crowns on their teeth, and the teeth were loosely spaced in the huge bony jaws. Traditional paleontologists looked at those mouths and concluded they couldn't have chewed anything tough with teeth like those. These dinosaurs must have had restricted diets of soft food and therefore low metabolic rates. As usual with such interpretations, they immediately arouse suspicion. After all, ankylosaurs and domeheads were advanced dinosaurs bristling with specialized neck, skull, limbs, vertebrae, and armor. So the very logical question is: Were there any equally advanced gastrointestinal adaptations that might have compensated for this chewing apparatus? There was indeed: giant afterburners.

No American museum presently displays an entire ankylosaur or domehead, even though a dozen good skeletons lie in storage drawers in the United States and Canada. The effort to do so would be incomparably rewarding for both professionals and the public, if for nothing other than the opportunity of at last viewing one of the most formidable gastrointestinal systems in the Dinosauria. From the side, the ankylosaurs and domeheads present a tubby appearance—deep ribs arching out from the chest and belly. Looking from above straight down on the ribs and hips, the entire hind region from belly to tail was enormously expanded, nearly beyond the anatomically credible. Ribs became longer from mid-torso to hips, until the rearmost ribs arched out so far that the afterbelly must have been wider from side to side than it was deep from top to bottom. This extra-wide fermenting compartment continued beneath the upper hip bones (the ilia), where the normally narrow pelvic architecture was transformed into an immensely broad horizontal roof. The ensemble was a dinosaurian body broadened to twice the usual width through the compartments housing the intestines and colon.

No other dinosaur's gastrointestinal system was nearly so en-

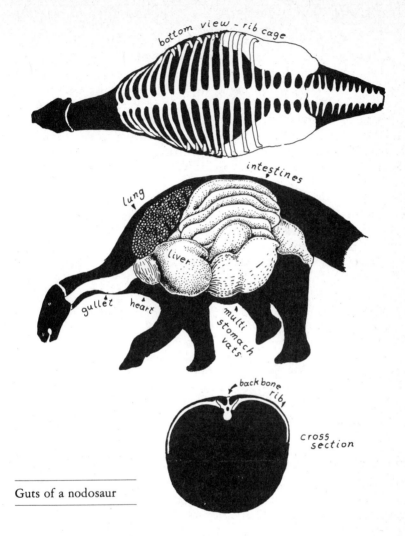

bottom view ~ rib cage

intestines

lung

liver

gullet heart

multi stomach vats

back bone

rib

cross section

Guts of a nodosaur

larged relative to the body mass. No mammal or bird possesses comparable skeletal architecture today. The exact layout of stomach, intestines, and colon in ankylosaurs will never be certainly known. It is certain however that every leafy bolus received an extraordinarily thorough biochemical treatment in a long series of enzyme baths and fermentation vats. The ankylosaur's teeth were indeed weak, but its beak was strong and sharp-edged. So the ankylosaur began the process of feeding by stuffing broad mouthfuls of leaves into its capacious cheeks. Then with its simple row of teeth it cut the longer leaves and stems a few times, and wadded up the pieces between its cheeks and tongue into a coarse bolus. The entire ball passed down to the superenlarged gastrointestinal chambers. Now, the coarsely chopped wad was broken down by successive biochemical assaults. The huge compartment for the colon at the base of the tail provided room for the enormous afterburner, so a final posterior appendix exposed the fodder to one last digestive procedure.

The ankylosaur's rearward digestive system with its special afterburner surely was big enough to make up for its weak teeth. Even quite tough vegetation could have been handled in large volumes. My colleague and friend Ken Carpenter has evidence indicating some species had gizzard stones as well. Could gastrochemical treatment have supplied the ankylosaurs with enough food energy to be warm-blooded? Absolutely—at least the boundary conditions from the dietary perspective must include this possibility. And at the very least the ankylosaurs too are rescued from the category of soft-food-eating, low-energy semi-invalids.

Besides the major families of herbivorous dinosaurs discussed so far, there were a dozen smaller groups all outfitted with plant-eating equipment of the sort already described for the major families. *Iguanodon* was a relative of the duckbill. It won international fame as the first dinosaur made known to science, when it was dug from road-gravel quarries in Sussex, England, in 1822. The iguanodont's adaptations were styled after the duckbill's—closely packed chopping shredding teeth (although iguanodont's weren't as complex as duckbill's). Dryosaurs must have been very selective eaters, using their narrow muzzles to crop carefully chosen fodder. The fabrosaurs, the most primitive beaked dinosaurs, were bipeds with small, loosely packed teeth like those of the much later ankylosaurs.

Altogether, each dinosaur dynasty, from Early Jurassic to Late Cretaceous, was equipped for a comprehensive attack upon foliage, buds, bark, tubers, and fruit. Not one plant-eating dinosaur has been found to subsist on aquatic plant mush. Every herbivorous clan could have harvested land plants at rates and quantities sufficient for high metabolism.

9

WHEN DINOSAURS INVENTED FLOWERS

Darwin and his followers regarded the ecological drama as a complex, choreographed struggle among competitors, predator, and prey. "Nature red in tooth and claw" expressed the violent aspect of natural selection, the killing and bloody rending of flesh by predators' fangs, the maiming of sexual rivals during the vicious combats between dominant males during the mating season. But Darwin was clever and observant; for all the violence of nature, he knew that most evolutionary dramas were played to a subtler script, the day-to-day interaction between the antelope and the grass, the squirrel and the acorn. Plants and plant-eaters coevolved. And plants aren't the passive partners in the chain of terrestrial life. Hence today's Pop Ecology movement is quite wrong in believing that plants are happy to fill their role as fodder for herbivores in a harmonious and perfectly balanced ecosystem. A birch tree doesn't feel cosmic fulfillment when a moose munches its leaves; the tree species, in fact, evolves to fight the moose, to keep the animal's munching lips away from vulnerable young leaves and twigs. In the final analysis, the merciless hand of natural selection will favor the birch genes that make the tree less and less palatable to the moose in generation after generation. No plant species could survive for long by offering itself as unprotected fodder.

Plants evolve all sorts of devices to foil plant-eaters: They

The pygmy dinosaur, *Nanosaurus*, in the Late Jurassic underbrush. *Nanosaurus* was a four-feet-long omnivore and a very primitive beaked dinosaur. The understory plants are: a gingko (upper left), two cycadeoids (on both sides, with diamond-sculpture trunks and big fronds), a fern (lower left), and ground pine (foreground creepers).

poison them with deadly alkaloids; they keep them away with thorns and spines; they render plant tissue unchewable by incorporating rock-hard phytoliths into the plant cells or by toughening plant fibers with cellulose; they avoid being eaten by producing new leaves in early spring when plant-eating populations are low. Of course the plant-eaters fight back. The evolution of herbivores leads inexorably to better teeth for crushing the toughest leaves, to more complex digestive systems where enzymes can detoxify plant poisons, to taller shoulders and longer necks to reach higher into the trees, or to lower heads and square muzzles perfect for cropping ground-hugging leaves.

The warfare between plants and herbivores began on land 400 million years ago, when the first algae colonized the bare ground during the Silurian Period and the herbivorous arthropods evolved to follow them. Vertebrate plant-eaters on land appeared much later, during the last epochs of the Coal Age, 270 million years ago. Dinosaurs captured the herbivorous niches on land during the Triassic, 200 million years ago, and subsequently maintained their dominance through the entire Jurassic and Cretaceous. But how did dinosaurs co-evolve in relation to the plants of their world? Dinosaurs held the roles of large land herbivores for longer than any other vertebrate group, so there must have been a rich history of adaptive attack and counterattack between plant-eater and plant. Moreover, herbivorous dinosaurs suffered several episodes of extinction and adaptive revolution that must also have been reflected in contemporary plant systems. And there was a momentous development in the plants during the Mesozoic, for the Jurassic and Cretaceous witnessed the single greatest event in the evolution of the modern system of plants—a turning point that must have changed the life of every plant-munching dinosaur—the appearance of the flowering plants.

Today flowering plants, known collectively as angiosperms, are by far the most numerous of land foilage, literally thousands of species, including nearly all the plants that feed mankind and our mammalian relatives. So numerous are angiosperms that to the average person, the term "plant" is synonymous with "flowering plant." Oaks, birches, maples, and all the other broad-leafed trees are angiosperms, as are nearly all the berry-producing bushes and shrubs. Palms, grasses, sedges, and dandelions also belong to the angiosperms, as do tulips and all the other species with showy flowers: squash, beans, coconuts, lilies-of-the-valley, peaches, apples, oranges, rhubarb, tomatoes, cucumbers, onions, garlic, potatoes, scallions, leeks, lettuce, spinach, broccoli, and thousands more. All angiosperms are members of one natural group, descended from a common ancestor that first appeared at the midpoint of the dinosaurs' reign.

The anatomy of the angiosperms is the key to their success.

They have distinctively complex reproductive organs—flower and fruit—and most woody species additionally possess highly advanced conduction tubes in their roots, stems, and leaves, which give them enormous advantages over other plants. Angiosperms use their brightly petaled flowers to attract animal pollinators (insects, bats, birds), and many use large fruit containing tough seeds to attract animals as agents of dispersal. (Some modern angiosperms are wind-pollinated, but this is an evolutionary reversal. The earliest flowering plants probably exploited animal vectors exclusively.) Different flower shapes attract different species of insects, bats, and birds, and thus each angiosperm creates the opportunity of spreading its pollen efficiently without the wholesale waste inevitable in pollination by wind. The same is true for angiosperm seeds and fruit, which are far more diverse and distinctive than those of non-angiosperms.

So overwhelming is the advantage of the angiosperms today that non-angiosperms are forced to play subordinate roles in the flora of most areas. Today, the most conspicuous non-angiosperms are conifers, cycads, ferns, ground pine, and horsetails. None of these non-angiosperms produce flowers, and most rely upon the wind to spread their spores, pollen, and seeds. Conifers—the needle-leafed trees—are important in temperate forests, but they are outnumbered by angiosperms ten to one on a worldwide average. Cycads with their spiny fronds are always a tiny minority in every flora. Ferns, ground pine, and horsetails, very ancient relics of Coal Age flora, make important contributions to the forest undergrowth and to swampy herbiage. But these living Coal Age fossils are outnumbered thirty to one by angiosperm species in nearly all habitats.

How did flowering plants begin to win this unchallenged hegemony? Whatever the story, dinosaurs must have had a hand in it because the earliest angiosperms sprouted up in a landscape dominated by dinosaur plant-eaters. And they remained the major outside factor for plants all through the first forty million years of the angiosperms' evolution. But, for no apparent reason, modern science has ignored the dinosaurs' role in plant evolution nearly completely. Paleobotanists theorize about new insect groups which might have co-evolved with the flowers in Late Cretaceous times. Mammal paleontologists assert that Cretaceous mammals, no mat-

Iguanodon browses among the broadleaf saplings. Flowering plants began their spectacular evolutionary career during the Early Cretaceous, when big-beaked dinosaurs like *Iguanodon* fed close to the ground. Early angiosperm leaves included some sassafraslike species (upper left), the broadly rounded *Proteaephyllum* (lower left and in *Iguanodon*'s mouth), and the oaklike *Vitiphyllum* (right).

ter how tiny and unimportant, made a major impact on the evolution of angiosperm fruits, nuts, and leaves. But hardly anyone has argued for the interaction of Cretaceous dinosaurs with the plants that fed them—an extraordinary oversight, considering the dinosaurs were the only herbivores large enough to gobble an entire flowering shrub in one gulp or strong enough to push an angiosperm tree so as to get at the tender young leaves at the top.

The consistent neglect of the dinosaurs' potential role in the evolution of plants is one of the most pernicious examples of the orthodoxy that relegates the dinosaurs to what amounts to an evolutionary sideshow, a menagerie of irrelevant dead ends that can be ignored so far as any large implications are concerned. Today, large herbivores can change the structure of the flora overnight. Rhinos and elephants can level acacia groves and rapidly crop down thickets, converting dense African bushland into open woodland. In the early nineteenth century, the American buffalo kept pushing back the boundary between prairie and forest by its intensive grazing on seedlings. Surely four-ton nodosaurs and three-ton iguanodonts did the same in the Early Cretaceous system.

Another bias also works against herbivorous dinosaurs, however. Paleobotanists are a bit chauvinistic about their objects of study. They tend to regard plants as the movers and shakers in evolution, and the plant-eaters are consigned to the role of reactors and followers. As one paleobotanist expressed it, "The sun gives energy to plants, and plants give energy to the animals. Therefore, the plants evolve and the animals must co-evolve." Stated thus, the assertion is understandable, but it's misleading. Co-evolution works both ways. When plant-eating dinosaurs evolved more effective teeth or fermenting chambers, the plant species had to adjust to the new weaponry or die. Whichever evolved faster, plant or animal, had the evolutionary initiative. And plant-eating dinosaurs evolved fast, faster than the plants. On average, a species of dinosaur endured two or three million years before becoming extinct and being replaced by a new species. That's a brisk rate of evolutionary turnover, as fast as the mammals'. Such rapid replacement of old adaptive models by new ones guaranteed that the dinosaur plant-eaters were always coming up with novel ways to bite, chew, ferment, and digest plant tissue. Mesozoic plants, on the other hand, usually evolved more slowly—the average species of plant lasted eight million years before being replaced by a new one. Since the turnover wasn't as fast, the plants must have been lagging behind the dinosaurs in the evolutionary race.

Herbivorous dinosaurs in fact were the fastest-evolving part of the entire Mesozoic land ecosystem, even faster at adaptive remodeling than their meat-eating relatives. *Tyrannosaurus rex,* the fifty-foot-long Cretaceous killer with seven-inch teeth, was really just a sophisticated variation on the basic predator plan first evolved a hundred million years earlier in the Late Triassic. Bone by bone, *Tyrannosaurus rex* was fundamentally little different from its ancient Triassic ancestors. But the Cretaceous plant-eaters—three-horned *Triceratops,* club-tailed *Ankylosaurus,* broad-beaked *Edmontosaurus*—carried skull and jaw developments totally unknown in the Triassic.

To follow the pattern of co-evolution between dinosaurs and plants, the major turning points in the development of each must be defined, then laid side by side. Among the herbivorous dinosaurs, three grand periods of development are clearly marked:

I. *The Age of Anchisaurs.* The Late Triassic and Earliest Jurassic, when the long-necked anchisaurs ruled. Anchisaurs were primitive, crude plant-eaters by Cretaceous standards. They had simple, iguanalike teeth, suitable for soft leaves only, and their digestive system wasn't much expanded.

II. *The Age of the High Feeders* (stegosaurs and brontosaurs). The Mid and Late Jurassic, when the spike-tailed stegosaurs joined the gigantic *Diplodocus, Brachiosaurus,* and *Brontosaurus.*

III. *The Age of the Low Feeders.* The Cretaceous, when all the terrestrial habitats were overrun by big beaked dinosaurs which fed close to the ground. Each of these types had its own unique approach to cropping the foliage, so each must have made a distinctive impact on the co-evolutionary history of plants.

Orthodox dinosaurology has muddied the conceptual waters here by relegating the herbivorous dinosaurs to the swamps, where they are supposed to have gummed nothing but water plants. Now that we have a corrected view of them as dry-land herbivores, it is possible to begin reconstructing a much more accurate context for the Mesozoic evolution of the flowering plants. The first clue to the interaction of dinosaurs and angiosperms can be found in the timing of extinctions. Flowering plants first appeared in the Early Cretaceous just *after* the extinctions which occurred at the end of the second grand period (the age of stegosaurs and brontosaurs), and as the replacements for the third grand period (the age of the low feeders) were taking place. This sequence is highly suggestive. When the coalition of stegosaurs and brontosaurs died out at the end of the Jurassic, the plant-eating dinosaurs changed so profoundly that the rules of co-evolution must have been reset. Could this dramatic shift from Jurassic-type to Cretaceous-type dinosaurs have opened the way for flowering plants? It's an exciting hy-

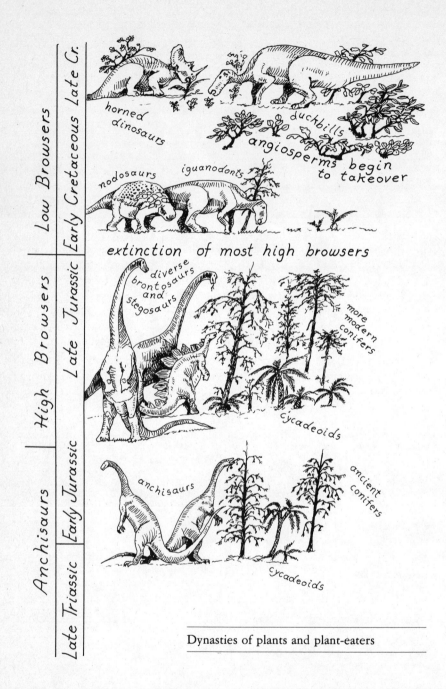

horned dinosaurs

duckbills

angiosperms begin to takeover

nodosaurs

iguanodonts

extinction of most high browsers

diverse brontosaurs and stegosaurs

more modern conifers

cycadeoids

anchisaurs

ancient conifers

cycadeoids

Late Triassic | Early Jurassic | Late Jurassic | Early Cretaceous | Late Cr.

Anchisaurs | High Browsers | Low Browsers

Dynasties of plants and plant-eaters

pothesis—that the revolution in dinosaur plant-eaters caused the single most far-reaching development in the kingdom of land plants. And the evidence suggesting such a cause-and-effect relationship is very good.

To understand how the dinosaurs' plant eating changed and how dinosaurs may have invented flowers, *Stegosaurus* and its strange adaptations for eating plants must be understood. Here again orthodoxy has obfuscated some obvious truths about dinosaur biomechanics. *Stegosaurus* is often portrayed as something of a misfit, a quadruped endowed with two sets of mismatched legs—the front pair too short and the hind too long. Stegosaur skeletons mounted in museums pose the beast with a clumsy, shuffling gait, its hips towering above its low-slung shoulders and its nose nearly at ground level. *Stegosaurus* and its close kin were the only common large, beaked dinosaurs in the Late Jurassic. Therefore its feeding habits must have had a major influence on the evolution of plants. The orthodox restorations depict *Stegosaurus* as an ungainly low cropper, plucking plants from within a few feet of the ground. That is the precise inverse of the truth—*Stegosaurus* wasn't a badly designed low feeder, it was a superbly designed high feeder.

The point the usual reconstructions miss completely is that the plan of the stegosaur's body was carefully balanced so the beast could rear up on its hind legs and tail to feed upright, in a tripodal stance. The very tall bony spines at the hips become understandable only in that position. Tall vertebral spines supplied great leverage to the back's muscles and ligaments, so that the entire weight of the trunk could be supported by the hindquarters. The brilliant English biophysicist D'Arcy Thompson pointed out the mechanical significance of tall spines at the hip as early as 1924. He demonstrated that *Stegosaurus*'s spine-and-ligament construction worked much like a single-span suspension bridge. Such bridges are based on a tall central tower (equivalent to the stegosaur's hind legs). The span's vertical steel supports are shorter and shorter the further they are from that tower for the sake of easier leverage (just as the stegosaur's vertebral spines become shorter the further they are fore and aft from the hips). Just as the thick cables of the suspension bridge based on the central tower hoist the weight of the span, so did the thick ligaments of the stegosaur's vertebral column based on the hips hoist the weight of its body. Modern cranes

Diracodon laticeps   q.13 Como

Diracodon laticeps, a three-ton Late Jurassic stegosaur, tilting up its body with a push of a hind foot

and derricks work on the same principle—the jib works like the Stegosaur's backbone, the cables work like its back muscles.

All the details of *Stegosaurus's* construction meshed perfectly to produce a body machine that could swing up easily from a four-footed stance into a hind-legs-plus-tail posture. Its shortened front legs reduced the dead weight forward, so the animal could hoist its front end up with less effort. The base of the stegosaur's tail was provided with huge bony flanges to anchor immensely strong tail muscles designed to brace its body against the ground. The stegosaur's ancestors had had stiff, brittle tails containing long bony rods, like those of the duckbills, which held the tail and back rigid. Clearly the stegosaurs had evolved away from the ancestral stiff-tailedness and had acquired strong, supple tail joints all the way from hip to tail tip, so the rear half of the tail could be flexed flush against the ground. A masterful final touch completed the stego-saur's tripodal equipment. In most dinosaurs, the lower bony prongs at the tail's end (chevrons) were usually simple, straight spines. But on stegosaurs those bones were expanded into a shape like the runners on a sled, to provide a superior brace for the body's weight.

Tall vertebral spines give back muscles and
ligaments leverage for raising the body—just like a
crane's jib is braced by a cable. Stegosaurs had
much larger back-raising leverage than does an
elephant of the same weight.

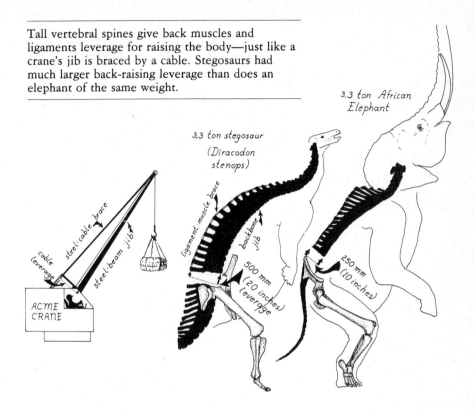

Live stegosaurs might have been slow when moving on all
fours, but the excellence of their design would appear when they
walked into a grove of trees that appealed to their taste. With a
slight upward push from the front paws, the entire body would
pivot upward from the hips, raising the head up to twelve feet above
the ground—and now the stegosaur could poke its snout into high
shrubs and the low canopy of trees, selecting the choicest leaves
and branches.

As an undergraduate at Yale, I published several papers attacking the orthodox theories and arguing for the tripodal habits
I have described here. I believed I had arrived at some quite new,
revolutionary ideas, until I discovered some papers actually published in the last century. I had in fact merely resurrected a view

carefully worked out some eighty years earlier. Professor Marsh had dug the first stegosaur skeleton in 1878, and many descriptions published in the 1880s and 1890s portrayed the animals rearing on their hind legs and tail, exactly as their bony anatomy indicated. Somehow this totally logical interpretation was lost in the 1920s; in its place orthodoxy substituted the nonsensical view of stegosaurs as ill-designed quadrupeds.

Stegosaurus was not the only animal whose bodily configuration was adaptively modeled for high feeding. The brontosaurian dinosaurs, *Diplodocus* and *Brontosaurus,* evolved exactly the same set of tripodal characteristics: short backs and short forelimbs to lessen the weight of the trunk; tall, "suspension-bridge" vertebral spines at the hips; huge and supple tail bones; gigantic muscles at the base of the tail; and sledlike lower tail bones as final support for the body's weight. When a *Diplodocus* raised itself into a tripodal posture, its feeding height was phenomenal by today's standards—its nose would have reached to forty feet or more above the ground. *Brontosaurus*'s reach would have been a little less, but *Barosaurus,* a close relative of *Diplodocus,* could have reached to forty-five or fifty feet.

A modern giraffe at its full height can reach only up to eighteen feet. Just as was the case with stegosaurs, these tripodal adaptations of the diplodocines were clearly understood by the pioneering American paleontologists; Elmer Riggs of the Chicago Museum wrote a detailed explanation in 1904. But, as with the stegosaurs, the orthodoxy of the 1920s forgot all about this work, so that between 1930 and 1960 the standard view likewise maintained that *Diplodocus* was a quadruped with maladaptedly short front legs.

Every single one of the giant Late Jurassic dinosaurs was in reality a high browser of some sort. *Brachiosaurus*'s tail was too weak for a tripodal posture, but it compensated with its spectacularly long neck, so that even on all fours it could reach up forty feet. *Camarasaurus* was the shortest-necked brontosaur found at Como. It could nevertheless stretch up to twenty-five feet with its forelegs on the ground, and much higher if it assumed a tripodal stance. Never before nor since the Late Jurassic has the world witnessed such a profusion of high-feeding plant-eaters. This was nothing less than a unique epoch in the history of herbivorous

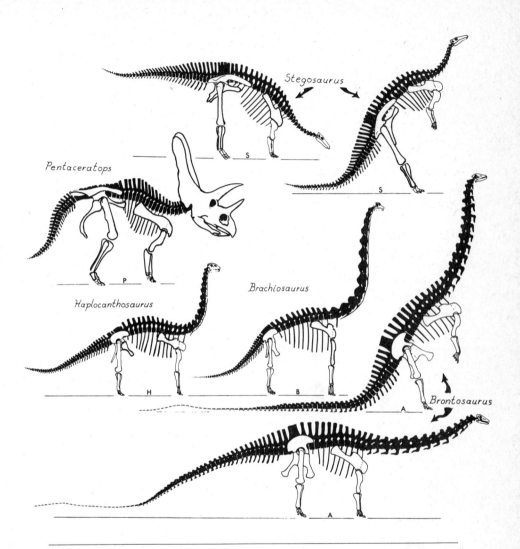

Stegosaurus

Pentaceratops

Haplocanthosaurus

Brachiosaurus

Brontosaurus

Some plant-eating dinosaurs were designed for rearing up on hind legs and tail. *Brontosaurus* and *Stegosaurus* had tall vertebral spines over the hips, so that the back muscles and ligaments had strong leverage for raising the body. Other big herbivores had shorter vertebral spines and must have stayed on all fours. (Each of these skeletons is drawn to the same length, hip to shoulder socket, to show the proportions of the backbone for easy comparison.)

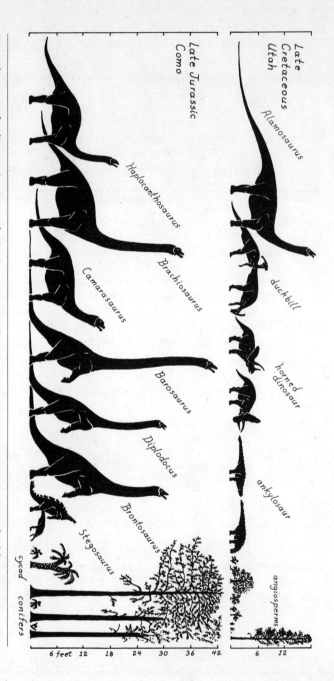

Late
Cretaceous
Utah

Alamosaurus

duckbill

horned
dinosaur

ankylosaur

angiosperms

Late Jurassic
Como

Haplocanthosaurus

Brachiosaurus

Camarasaurus

Barosaurus

Diplodocus

Brontosaurus

Stegosaurus

cycad

conifers

6 feet 12   18   24   30   36   42   6   12

How dinosaur feeding style changed. At Como in the Late Jurassic, there were six long-necked brontosaurs plus a stegosaur. The stegosaur and three brontosaurs could rear up and feed tripodally (*Barosaurus*, *Diplodocus*, and *Brontosaurus*). But in Late Cretaceous habitats in Utah, there was only one kind of brontosaur—*Alamosaurus*— and low-feeding beaked dinosaurs dominated the plant-eating role.

habits. No plant or leaf was safe from a dinosaur's mouth unless it stood over fifty feet above the ground! Since no flowering plants of any sort existed yet, the principal trees were broad-needle conifers and tall, spiny-fronded cycadeoids. These plants had to be able to protect at least some of their growth through a combination of alkaloid poisons, heavy oils, or spiny branches. As conifers and cycads grow slowly compared to many angiosperms, Late Jurassic trees had to guard themselves carefully to obviate the destruction of entire breeding stands.

During these Late Jurassic times, one danger was minimal. Since few dinosaurs were specialized for feeding at ground level, conifer seedlings and other sprouts wouldn't suffer the same degree of cropping we find at work in the modern ecosystem, where cattle, horses, sheep, goats, and many other mammals pluck their plant food from the surface of the soil.

When the Late Jurassic curtain fell and the Early Cretaceous began, most of the old, established high croppers died out. The mysterious hand of worldwide disaster swept across the continents, killing off whole families of herbivorous species and thinning the ranks of the surviving clans. Stegosaurs disappeared forever. Hardly a species survived into the Cretaceous. *Diplodocus, Brontosaurus,* and nearly all the other tripodal brontosaurs died out too. Some of *Brachiosaurus*'s relatives survived, but that was an exception to the rule. This tremendous disaster destroyed the high-browsing system permanently; it never recovered. And no Cretaceous dinosaur evolved high-browsing adaptations to replace *Stegosaurus.* A few new brontosaurs evolved—*Alamosaurus,* for instance—but they never came close in number to the glorious days of the Late Jurassic.

As the dust settled after these extinctions, the opportunists— evolutionary carpetbaggers—started to move into the devastated ecosystem. New herbivorous groups blossomed into clusters of new species. And nearly all these new Cretaceous herbivores were committed to cropping near ground level. Among the earliest of the newcomers were the big, spike-shouldered nodosaurs. Iguanodons also evolved early in the Cretaceous, reaching weights of two to three tons and developing broad muzzles for close cropping. Parrot dinosaurs emerged shortly after, as did the queer-looking domeheads, both groups with their backbones curved to bring their snout close to the ground. More and more low feeders appeared as the Cretaceous continued—the wide-snouted ankylosaurs, the large and small horned dinosaurs, the myriads of duckbills.

This was an unprecedented alteration in the nature of plant eating. In place of the Late Jurassic's tall browsers, the Cretaceous concentrated on munching close to the ground. Low shrubs and seedlings now faced a threat from herbivores magnified many times compared to the conditions prevalent during the Jurassic. In this ecological context the very first flowering plants appeared on the earth's surface. How did the Ur-angiosperms react to all this munching close to the ground? One ideal adaptive strategy for a

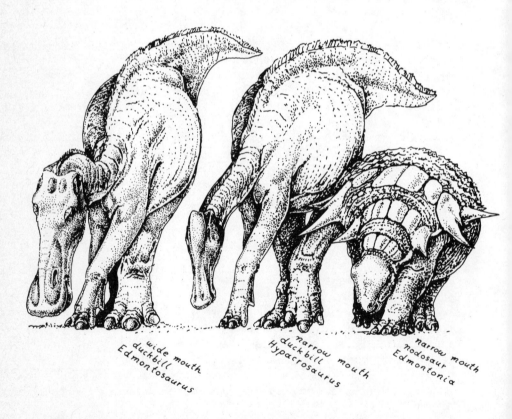

wide mouth
duckbill
Edmontosaurus

narrow mouth
duckbill
Hypacrosaurus

narrow mouth
nodosaur
Edmontonia

low woody shrub or seedling would be to grow as fast as possible to achieve a height where the low browsers could not threaten it. Early angiosperms could do that—their basic reproductive equipment gave them a fast-growing edge over many contemporaneous plants. A second approach would be a "dicey" strategy—scatter sufficient seeds onto an overgrazed patch of bare soil to produce a clump of shrubs quickly, and thus spread another generation of seeds before the herbivores returned and once again mowed everything to the ground. Early angiosperms could do that too— their seeds and flowers allowed them to spread more quickly than conifers or other nonflowering plants.

The low-level cropping brigade of the Cretaceous. On a typical Late Cretaceous meadow in Alberta, all the big plant-eaters were specialized for feeding close to ground level. Broad-snouted duckbills bit off wide mouthfuls of vegetation; delicate-snouted parksosaurs nipped precision bites of selected leaves. And the other dinosaurs added their cropping activity to the medley of ground-floor herbivory.

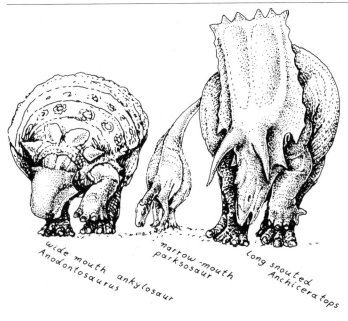

wide mouth
Anodontosaurus ankylosaur

narrow-mouth
parksosaur

long snouted
Anchiceratops

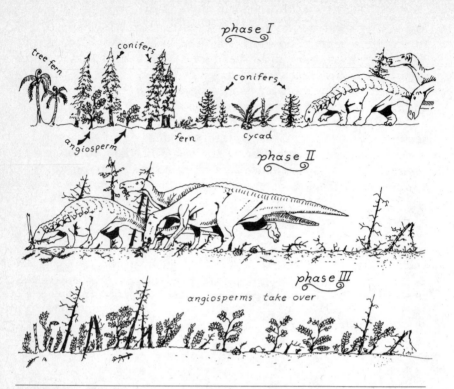

How Cretaceous dinosaurs invented flowers. Nodosaurs, iguanodonts, and the other plant-eaters that fed near the ground threatened meadows with overgrazing. Nonflowering plants (conifers, cycads, ferns) couldn't regenerate as quickly as angiosperms and they couldn't recolonize as rapidly. So after a heavy raid by the plant-eaters, flowering plants took over.

Obviously intense low cropping placed a premium on any and all plant adaptations for fast spreading, fast growing, and fast reproduction. And early angiosperms performed exactly these biological functions especially well. From this, a quite plausible scenario emerges: Low-feeding Cretaceous dinosaurs opened the way for the initial waves of angiosperms. Conifers, cycadeoids, and other non-angiosperms were probably far less adaptive for handling the assaults of new Cretaceous herbivores. Anywhere the plant-eaters thinned out the conifer groves and cycadeoid thickets, an opportunity for species of flowering plants to win a foothold in the habitat was created. Early angiosperms were probably cropped just as severely as their neighbors, but their basic adaptations permitted them to continue growing and reproducing in the face of the intense mowing action of the dinosaurs.

Footholds are crucial for major adaptive revolutions. Any new group finds it difficult to break into an ecosystem already full of old, established groups. Conifers were highly adaptive old-timers in Early Cretaceous times—the conifers had begun indeed long before the first dinosaur evolved in the Triassic. Cycadeoids were old-timers too, already diverse in the Early Jurassic epoch. When angiosperms first evolved, they were confronted with meadows, woodlands, and forests full of long-lived, highly refined plants. Early angiosperms possessed adaptations that were new and potentially revolutionary. But they needed some edge—an opening for breaking out of the confines imposed by the older groups to start proliferating species. Herbivorous dinosaurs gave them that initial break.

Flowering plants and low-feeding beaked dinosaurs must have co-evolved in a mutually beneficial way. As more and more new kinds of Cretaceous beaked dinosaurs entered the system, more and more angiosperm families evolved. From the meek beginning of a few Early Cretaceous species, the angiosperms grew into a mighty clan by the Late Cretaceous, boasting more species than conifers and cycadeoids combined. When the Cretaceous ended, massive extinctions again swept through the terrestrial habitats. Cycadeoids disappeared, but angiosperms not only survived, they increased their share of the species count in epoch after epoch down to the present day.

Modern angiosperms no longer depend upon intense low feeding for their advantage over conifers and ferns. They now contain hundreds of specializations for every habitat from tall climax forests to swamps and deserts, windswept mountain meadows and bare rock faces. But in those first critical years of their evolutionary history, the angiosperms were struggling newcomers. Square snouts and pincerlike beaks helped the flowering plants beat the floral competition and establish the angiosperms as the fastest-evolving plant group. There was of course no plan in this. *Stego-*

saurus didn't die out purposely to permit its low-cropping cousins to take over and make the world fit for flowers. Wide-mouthed ankylosaurs didn't plan to munch down the competition. It was all serendipitous. Nonetheless, because of the way they suffered extinction and then rebuilt their herbivorous groups, the dinosaurs played a central role in one of the grandest dramas of the flora. In their way, dinosaurs invented flowers. Without them, perhaps our modern world would yet be as dull green and monotonous as was the Jurassic flora.

PART 3

DEFENSE, LOCOMOTION, AND THE CASE FOR WARM-BLOODED DINOSAURS

10

THE TEUTONIC DIPLODOCUS: A LESSON IN GAIT AND CARRIAGE

Anyone who doesn't believe that God was looking out for America at the turn of the century should look at the dates of key American victories in war and science. On July 4, 1898, Admiral Sampson announced a "Birthday present for the Nation," the complete victory of the fleet over the Spanish squadron in Cuba. (Commodore Schley actually won the battle. Sampson was away conferring with generals—but claimed the credit.) On July 4, 1899, Arthur Coggeshall found *Diplodocus carnegiei*. On July 4, 1900, Elmer S. Riggs, hunting dinosaurs for Marshall Field's Chicago museum, found the first-known *Brachiosaurus,* king of the brontosaurs, a giant that dwarfed even *Brontosaurus excelsus.*

American museums were erecting dinosaur skeletons as fast as American shipyards erected new steel battleships to protect the fledgling star-spangled empire. Europe viewed both developments with mixed admiration and alarm. For a century Old World scientists had been digging and studying dinosaurs, but no one had found Jurassic giants nearly as complete as the ones that tumbled out of almost two dozen American quarries, starting in 1878. When the first *Diplodocus,* named for the American millionaire Andrew Carnegie who funded its discovery, arose on its metal scaffolding in Pittsburgh, John Bell Hatcher, in charge of the operation, directed the placement of the thigh bone into the hip socket very carefully. He drew upon the anatomical expertise of Marsh, Cope,

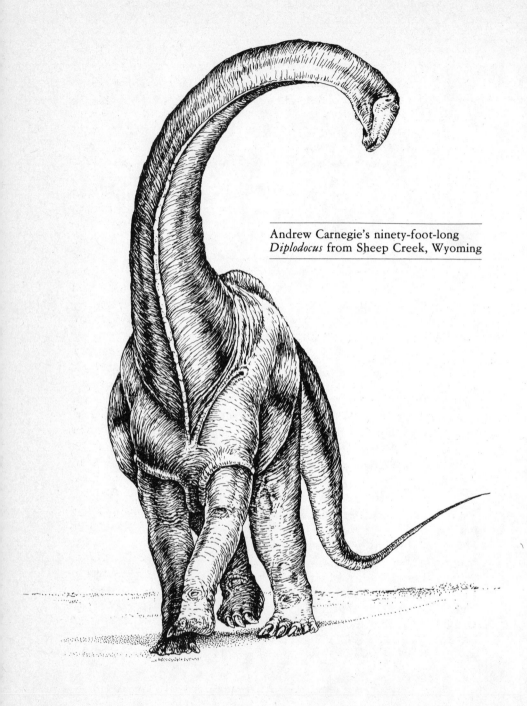

Andrew Carnegie's ninety-foot-long
Diplodocus from Sheep Creek, Wyoming

and other Americans who had engaged in two decades of intense study. Hatcher decreed a vertical stride, with the thigh bone swinging fore and aft directly under the hips, for *Diplodocus carnegiei*. And the other American museums consequently agreed. Riggs's *Brontosaurus* went up that way in Chicago, as did Osborn's in New York. These Americans were convinced that the great dinosaurs strode through their Mesozoic world with the upright gait and carriage that are characteristic of the biggest African elephants today.

America, which had had to import all its scientific apparatus, and had sent its scholars to England and Germany for doctorates only decades earlier, now began to export scientific wealth.

On one of his frequent visits in British high society, Andrew Carnegie met with Edward, Prince of Wales, the future Edward VII. Aware of Carnegie's enthusiasm for the exploration of dinosaur sites in Wyoming as an aspect of his new-found passion for public service, the Prince of Wales suggested Carnegie might be pleased to have his people find another *Diplodocus* for the British Museum, which had no complete specimen.

Back in America, William J. Holland, director of the Pittsburgh Museum, was aghast at Carnegie's request. New quarries not already being worked by other new American museums were extremely difficult to find, and it would be impossible to guarantee quick delivery. Holland proposed a complete plaster replica instead. So, in due course, the Pittsburgh technicians assembled beautifully accurate plaster casts of *Diplodocus carnegiei* and shipped them to London complete with instructions for assembly. A characteristic American approach—the prefabricated, instant dinosaur kit. Soon Carnegie was besieged by envoys from Berlin, Vienna, and St. Petersburg for matching gifts of a *Diplodocus*. And Andrew was delighted to comply. Within a few years nearly every major European capital had its own prefab *Diplodocus*.

Once assembled, most of these European *Diplodocus* replicas were posed in mid-stride, with the same high-hipped posture prescribed by Hatcher and the other American experts. This reconstruction bore important implications for dinosaur biology, precisely because no living species of reptile walked that way. Most present-day lizards scuttle over the ground with their thighs sticking out sideways. Their hind limb strokes back and forth in horizontal arcs, parallel to the ground. Hence, these modern lacertilians are decidedly low-slung in posture, with a ground-hugging configuration that allows them only slight clearance. Crocodilians and chameleon lizards are a bit more upright in carriage, raising their bellies higher off the trackway. Yet their elbows and knees still stick out sideways more than those of an elephant, rhino, or most other large mammals. Endowing the *Diplodocus* with elephant-style posture was therefore a clear statement that the biomechanics of dinosaurs were unlike those of any living reptile. *Diplodocus*'s posture as envisioned by Hatcher, Osborn, and Riggs was equal to the most advanced mammalian adaptive machinery.

Enter the Germans. No culture had a more illustrious nineteenth-century tradition of paleontological scholarship. A German, Hermann von Meyer, had first recognized the unity of all the great Mesozoic creatures we now call dinosaurs. And German anatomists were acknowledged worldwide as the best in laboratory dissections and microscopy. In the early 1900s, Germany was a new and ambitious nation, and it was perhaps to be expected that a certain chauvinism should manifest itself in many different areas, including the scientific. It was not surprising that German paleontologists didn't immediately accept the conclusions about the posture of dinosaurs advocated by the Americans. What was surprising was the condescending tone the Germans resorted to when they published their scathing criticism of Carnegie's *Diplodocus*. The Germans insisted that the Americans had missed the point when they put elephant's legs on a dinosaur. *Diplodocus* was a genuine reptile, the elephant a genuine mammal, and nature did not mix the two. Tornier, the dinosaur expert in Berlin, described his version of the corrected *Diplodocus*: it was portrayed in a slinking pose, with the thigh and arm sticking out sideways, and the belly close to the ground. The Berlin school proclaimed this was a proper reptilian posture for a proper reptilian body.

The Americans did not concur. They had a well-earned national reputation for a hard-headed approach to the functions of machinery. Hatcher had handled scores of fossil hip joints. He knew that *Diplodocus*'s thigh had a cylindrical surface at the joint that faced predominantly upward and forward. *Diplodocus*'s hip bone contained a deep socket at the joint, whose surface correspondingly faced mostly downward and backward. Put the thigh bone into the

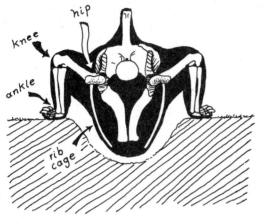

knee

hip

ankle

rib cage

Diplodocus in a rut. Carnegie Museum's Professor Holland poked fun at the German reconstruction of *Diplodocus* with its outspread knees. Holland remarked that the poor German *Diplodocus* would have had to find giant ruts to run in because its rib cage was so deep. (The diagram shows a cross-section view at the hip joint.)

How Carnegie's men mounted *Diplodocus*. The Americans put together the hind limb the right way—with the thigh bone in a tall, erect posture.

hip socket, and only one correct fit was possible: the hind leg stood vertically, with the knee facing directly forward. The *Diplodocus*'s knee did not sprawl sideways like that of a "genuine lizard."

American scholars, one after another, rebutted the arguments of the Germans. A key point all too often forgotten today was made in the course of this debate. Dinosaur biology cannot be reconstructed by assuming these beasts were merely "good reptiles."

By 1920, the Great Trans-Atlantic War of the posture of *Diplodocus* was over. The Americans clearly had the better of it. Expeditions from Berlin had found fabulous *Brachiosaurus* graveyards in German East Africa before World War I, and as the postwar research was published, the German scholars' restoration of *Brachiosaurus* became like Carnegie's, with a nearly vertical, elephant-like posture. Incontrovertible evidence for the correctness of the Pittsburgh-style hip joint came during the 1930s and 1940s, when brontosaur footprints were found in abundance impressed into the Cretaceous limestone of Texas. The right and left hind prints proved to have been made very close to the trackway centerline, and therefore without question the two thighs had swung in great vertical arcs close to one another under the animal, exactly as Hatcher had reconstructed them. Indeed, all of the dinosaur fossil trackways without exception showed the very same narrow-tracked gait, evidencing no splaying out of the knees. No dinosaur splayed its knees or toed out its feet. Duckbill dinosaurs actually toed in their hind feet, like enormous pigeons.

Sad to relate, some modern reconstructions done in the 1960s and 1970s still portray dinosaurs everting their knees and planting their hind feet down, wide-set, with right and left hind paws spread far out to the side—like enormous lizards. An entire series of postcards in the British Museum reproduces paintings with such essentially dislocated dinosaur hips and ankles. Andrew Carnegie and Prince Edward knew better in 1906.

As a student at Yale in the sixties, I observed that nearly everyone was restoring dinosaur hips with a mammal-type posture and narrow trackway. But the forelimbs were a different matter entirely. In the Great Hall of the Peabody Museum, trailing behind the nobel strides of Marsh's *Brontosaurus excelsus,* stood a finely preserved horned dinosaur, *Centrosaurus.* Professor Richard Swann Lull had mounted it in 1929. He had been with Osborn at Como and was highly regarded. He gave the centrosaur tall, erect hind limbs, but the elbows were mounted lizard-style, sticking out sideways, and the upper arms paralleled the ground. I marveled at this curious combination, which looked like two totally different locomotor apparatuses welded together at mid-torso. Furthermore, Lull had also published a monograph on the horned dinosaur in which he employed this mismatched front end as an argument for a slow and plodding gait in these animals. But Marsh had already seen it otherwise as early as 1896. He had published drawings of horned dinosaurs with fully erect carriage in both forequarters and hind. Marsh's *Triceratops* imparted a light-footed air to the huge three-horned beast, as though it were about to go trotting off the page, head down, to charge its mortal enemy, *Tyrannosaurus*. Lull completely reversed Marsh's ideas about these forelimbs.

But why had views on the forelimbs of dinosaurs changed? I could find no good reason. Smithsonian scientists endowed their *Triceratops,* a skeleton excavated by Hatcher, with bowed-out elbows because, they said, the elbow's huge "funny bone" would be of use only in a sprawling posture. This made no sense to me. The "funny bone," properly called the olecranon, is a projection on the lower arm bone (ulna) to provide the elbow-opening muscles leverage for their work. A large olecranon implies the elbow joint can be opened with great force. Now, turtles, crocs, and lizards possess splayed-out elbows, but they all have short olecranons. The rhino's forelimbs stride vertically, and its olecranon is big. *Triceratops* possessed a large, rhino-style funny bone. Why wasn't *Triceratops* accorded a rhino-style posture?

Al Romer, the greatest dinosaur anatomist alive in the 1960s, explained the anomaly by arguing that evolution had worked faster on the hind legs than the fore. All primitive dinosaurs, as Romer told the story, had been two-legged bipeds, standing on hind limbs only. Thus when some families later dropped back down onto all fours, they didn't bother to rearrange their elbows to match their knees. This account was in all the textbooks, but I could not accept it. Yale had a program—the Scholar of the House—in which an undergraduate could dedicate one full year exclusively to research. For my project, I focused on the problem of the alignment of the forelimbs of dinosaurs.

It was a shoestring operation. My parents gave me a tiny handwound movie camera, and with Yale funds I bought the alligators and lizards that I kept in the museum basement. I'd lie on my stomach watching a lizard or 'gator walk around on an old rug, and built up a library of motion-analysis film of a dozen species. These filmings had their dangers. No three-foot 'gator was afraid of a Yale undergraduate, and I had to remember to move smartly when an open mouth filled the viewfinder. Usually I was in time to avoid having my nose bitten.

Alligators are dinosaur uncles—relatives of the direct ances-

Sprawling Front End

Yale's Centrosaurus

Fully Erect Rear End

Fully Erect Rhino

Semi-erect Gator

Sprawling Lizard

angle of upper arm bone

The riddle of the mismatched legs. Living species have a range of postures—most lizards sprawl, crocodilians have a more upright, semierect stance, and most big mammals have a fully erect carriage. But orthodox dinosaurs—like Yale's *Centrosaurus*—had front ends that didn't align with the rear ends.

tors of early dinosaurs—and as such they should be living representatives of the ancestral dinosaurs' forelimb arrangement. I was therefore surprised at the upright arrangement of the alligators' forelimbs. They kept their elbows close to their sides without spreading nearly as much as most lizards do. So I labeled this posture "semi-erect," to set it apart from the "fully erect" posture of rhinos and other large modern mammals and from the "sprawling gait" of the ground-hugging lizards.

Alligators sprawled at the elbow much less than Professor Lull's *Centrosaurus,* and yet the horned dinosaur was supposed to be a much more advanced evolutionary design than the 'gator. Something was deeply wrong here. Why would an advanced dinosaur exhibit a more sprawled posture than its more primitive relative? I needed evidence from the shoulder-bone structure which I could use to evaluate dinosaur forequarters. Two pieces of evidence came immediately to hand: First, the shoulder socket's shape. An elephant or rhino's shoulder socket is shaped like an oval saucer. It is a hollowed-out joint surface, elongated fore to aft, which faces downward and backward to fit over the top of the upper arm bone. But lizards and crocs, whose elbows sprawl, have a saddle-shaped shoulder joint, concave from bottom to top and convex from the inside out. This saddle-shaped notch lets the upper arm swing out and back and twist around like an axle, a complicated set of movements required by the sprawling and semi-erect gaits. Now, what kind of shoulders did dinosaurs have?

I spent a year digging into museum drawers, and covering myself with dust while I diagrammed the shoulder sockets of the Dinosauria. Almost all had rhino-type joints. When properly mounted, dinosaur shoulder joints were concave sockets facing downward and backward. Markings on the bones showed clearly that the joint didn't curve around to face sideways as it did in 'gators or lizards. Professor Lull's *Centrosaurus* had a misaligned front end, as did the mounts of most other horned dinosaurs.

The second piece of evidence reinforced the first. Crocodilians and chameleon lizards had a semi-erect gait, and when I measured their shoulder joints oriented to a side view, I found that both of these reptiles displayed a joint which slanted so that it faced slightly downward as well as outward and the upper edge of the joint overhung the lower edge. Fully sprawling lizards didn't exhibit a trace of this downward slant. On the other hand, dinosaurs

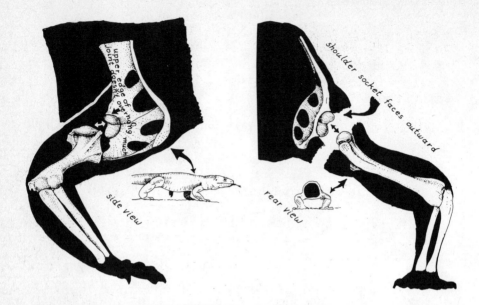

Shoulders designed for sprawling. Lizard shoulder joints are doubly curved notches shaped like a saddle, and the normal walking posture is with the elbows stuck far out to the side. (The upper-arm bone—the humerus—is pulled out of the socket a bit in the diagram to show the fit.)

Horned dinosaur shoulders were designed for fully upright posture. The upper edge of the shoulder socket overhung the lower edge a great deal, even more than in crocodilians. And, viewed from the rear, the shoulder socket faced mostly downward, not outward.

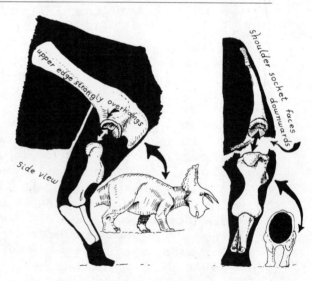

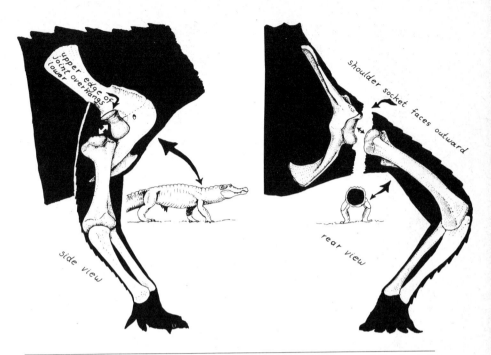

Shoulders for a more upright gait. Alligator shoulder joints are saddle-shaped but face more strongly downward than do those of lizards, and so the gator can hold its body higher off the ground.

all manifested very strong downward slants, so that their entire shoulder socket had been reoriented from the primitive arrangement. This strong downward orientation meant that the dinosaur's upper arm could swing fore to aft in an upright stride. And the upward force of this limb's stroke would be braced against the downward-facing shoulder socket.

Finally, there was the acid test of fossil footprints. Quadrupedal dinosaur footprints aren't as common as those left by bipedal types, but each and every set of four-legged footprints showed forepaws working on a very narrow track. *Triceratops* and the rest of the four-legged Dinosauria did not splay their forelimbs. Marsh had been right in the 1890s, Lull wrong in the 1930s.

Lull's own account of why he mounted the *Centrosaurus* with wide-set forepaws was quite surprising. Lull wrote that he had carefully studied the fossil footprints of big quadrupeds found in Canada as his guides for posture. Charles Sternberg had published illustrations of those prints in 1930, several years before Lull mounted his sprawl-elbowed beast. But Sternberg's diagrams showed right and left forepaws quite close to the centerline, and not spread widely apart. Lull simply ignored this, because he was so convinced, *a priori,* about splayed forelimbs that the obvious facts simply didn't register, as they still don't for some. Several large quadrupedal skeletons have been erected in various museums during the last decade, and some still faithfully cling to the traditional stance with the widely splayed forepaws, despite the publication of dozens of footprint diagrams.

I was pretty proud of myself when I finished my undergrad thesis on posture evolution. I published a couple of articles arguing that the dinosaurian fully erect gait was superior to the sprawling gait because erect posture didn't waste as much muscular effort. It seemed like a logical idea, and Al Romer had used it way back in the 1920s. For example, if you do push-ups on the floor, you can put your arms in the lizard-style posture by bending your elbows at right angles and holding your body halfway off the floor. In this position, you feel very uncomfortable strain in your arm muscles. If you hold your arms straight up and down, in a fully erect

Footprints don't lie. All dinosaur tracks show that the forepaws were put down right under the body with only a little space between the line of march of the left and the right set of prints. But many museum reconstructions still show dinosaurs with widespread forepaws that would have left a sprawling-style trackway. (This drawing is from a model in the National Museum of Canada.)

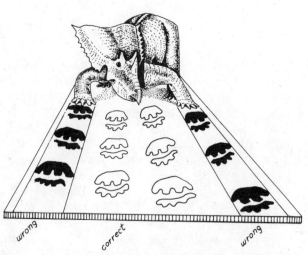

posture, you can keep your body off the floor with less effort.

When I got to Harvard, I had fun chatting with Romer about how my theories agreed with his. But then I got my comeuppance. As part of my Ph.D. work, I had to run lizards on miniature treadmills inside micro-environmental chambers to measure just how hard they had to breathe to run at different speeds. (Hot, boring work for me and the lizards—each run was thirty minutes and I needed twenty runs per lizard.) When the results came ticking out of the oxygen analyzer, I was devastated—and my theory was totally deflated. My sprawling lizards were *more* efficient than fully erect mammals and birds. All the lizards used *less* energy to run at any given speed than did birds or mammals of the same size. As the old laboratory saying goes "The theorist proposes, Nature disposes."

I trotted into Romer's office the next day and sadly announced, "Our theory is dead." Then I plopped the computer printout on his desk. Romer scrutinized it. Then with a twinkle in his eye and a mock inquisitorial tone in his voice he said, "Your data are probably correct. But they must be suppressed. Our beautiful theory has got to be preserved." I felt better. If Romer could chuckle, so could I.

So what advantage is the fully erect gait? Probably it allows for much higher speeds even if efficiency is sacrificed. Having a

Correct stance. Here's the proper reconstruction of a horned dinosaur (genus *Chasmosaurus*) made to fit the fossil trackways.

vertical limb stroke means that you can exert more of a thrust downward onto the ground with your paws. And the speediest gaits require such thrust to propel the body when all feet are airborne.

When I finally arrived at Harvard in 1972, I was still interested in the gait of dinosaurs. All the anatomical footprint evidence vindicated Marsh's light-footed and lively postural restorations of the 1890s. The forelimbs of dinosaurs were aligned quite perfectly to match with the stride of the hind limbs. I now asked myself, "How fast might the big dinosaurs have been?" Most twentieth-century paleontologists had been willing to concede lively locomotion to the small, long-legged ostrich dinosaurs and to the smaller predators, but the big two-ton-plus species were always reconstructed as slow shufflers. But large mammals can gallop. While in South Africa I observed three-ton white rhino bulls at a full gallop with all four huge feet off the ground simultaneously in mid-stride. In fact, rhinos can accelerate and turn faster than horses, though in the stretch a horse can outdistance the short-winded rhinos. Perhaps big quadrupedal dinosaurs could also quick-start off into their own clomping high-speed charge.

A useful piece of evidence about the speed of dinosaurs can be extracted from the angles in their joints. Seen from the side, a running rhino always exhibits greater flexure at the elbow, knee, hip, and shoulder than does an elephant. Elephants run straight-legged, thigh lined up with shank and upper arm with lower arm, so their legs look rather like mobile Doric columns. Rhinos run with a more bent-legged stride and are consequently faster than elephants—top speeds are thirty-five miles per hour for the rhino, twenty-two for the elephant. The rhino owes its greater velocity precisely to the bounce it gets from the stretching tendons at its joints each time its feet plant down. Flexing joints provide more of this bounce, and all the big mammals that gallop are so jointed. Elephants can never get all their feet off the ground simultaneously, even at top speed, and their fastest gait can best be labeled a running walk. If we could compare the angles in dinosaur joints to those in these living mammals, we would have an important clue to the bounciness of their gait and hence their speed.

Brontosaurus has a reputation for being a relatively slow dinosaur, and here orthodoxy is correct—all the brontosaurs had rather straight elephantine legs that didn't flex very much and must

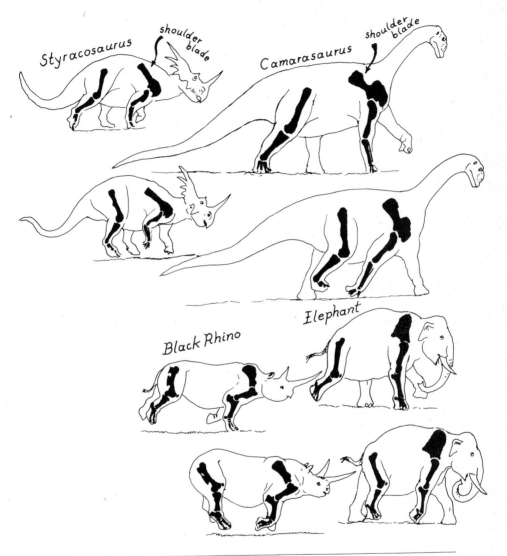

Swinging shoulders and bouncing knee joints. Big modern gallopers—like rhinos—have more flexure at their joints than do elephants. Brontosaurs, such as *Camarasaurus,* had little flexure and must have run like elephants. But horned dinosaurs had much more bend in each joint and must have been more rhinolike in gait. Both brontosaurs and horned dinosaurs had very long shoulder blades.

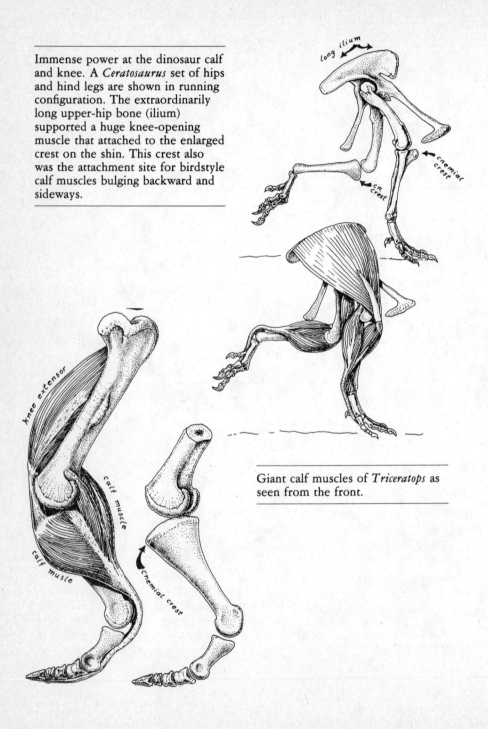

Immense power at the dinosaur calf and knee. A *Ceratosaurus* set of hips and hind legs are shown in running configuration. The extraordinarily long upper-hip bone (ilium) supported a huge knee-opening muscle that attached to the enlarged crest on the shin. This crest also was the attachment site for birdstyle calf muscles bulging backward and sideways.

Giant calf muscles of *Triceratops* as seen from the front.

long ilium

cnemial crest

cn. crest

knee extensor

calf muscle

calf musle

cnemial crest

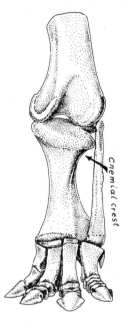

cnemial crest

have limited them to a running walk. But the bipeds and the quadrupedal horned dinosaurs display much sharper joint flexures and probably bounced quite a bit as their thick tendons stretched out and snapped back with each stride. How strong, then, was the bouncing stroke of such a limb? Big gallopers today possess strong knee muscles that attach to the kneecap and shank in such a way that the knee joint opens and closes under tremendous muscular power. A bony ridge, the cnemial (silent *c* here: "nee-mee-al") crest, marks the point of attachment for the knee tendons, and one can directly gauge the muscle power of a knee from the size of a cnemial crest. Elephants, turtles, and salamanders are all slowpokes in their body-size classes and all have puny knee muscles and low cnemial crests on the shank bone. Rhinos have big cnemial crests, as do other large-bodied gallopers, such as water buffalo, giraffe, bison, and gaur. Big crests would also mean big calf muscles.

All dinosaurs had bigger cnemial crests than do elephants, even those groups with relatively straight hind legs—the giant horned dinosaurs, stegosaurs, and brontosaurs. When these systems of oversized knee muscles contracted, the power exerted on the hind paw would have had no equal today. The biggest meat-eater, three-ton-plus *Tyrannosaurus,* had an absolutely huge cnemial crest, even by dinosaurian standards. At full speed, a bull *Tyrannosaurus* could easily have overhauled a galloping white rhino—at speeds above forty miles per hour, for sure. The consistent pattern of huge cnemial crests is documentary evidence of super-powerful knees and calves that gave fast top speeds in most big dinosaurs.

A quite different approach to the question of dinosaur speed is provided by calculating the maximum strength of the bone shafts of the limbs. Legs do break in nature, and evolution usually outfits a species with bone shafts strong enough to withstand the highest strains imposed when muscles contract. Rhinos have relatively stout, thick-shafted legs. Elephants feature a more spindly design. To measure the shaft strength of dinosaur limbs, I constructed scale models in clay of the life appearance of various species. I then calculated the live weight by measuring the volume of the model (most land animals are a little less dense than water, so live weight is about 95 percent of the body's volume in water). Brontosaurs and stegosaurs were somewhat thin-thighed, and in cross section their bones are about as thick as we would expect in an elephant of similar size. But *Triceratops, Tyrannosaurus,* and the other predators were much more massively shafted, far stronger in girth of bone, and these dinosaurs could exert positively prodigious force through their limbs without fear of fracture.

Tyrannosaurus moving at forty-five miles per hour is a horrendously heretical concept, and when I began to publish reconstructions of galloping dinosaurs, the shrill voice of outraged orthodoxy rose to deafening heights. The advocates of slow dinosaurs had two strong arguments. They pointed out that the dinosaurs' joint surfaces usually weren't smooth and polished as are those in mammals, but were roughened and pitted. Those pits held cartilages. It was therefore alleged that dinosaurs had too much gristle in their knees to stand the strain of fast trots and gallops. But this argument is flawed.

In point of fact, cartilage is excellent biological material for absorbing shocks—better than dense, brittle bone, because cartilage will compress under load, building up hydrostatic pressure in its fluid-filled micropores and springing back when load is released. Adult mammals and birds have only a thin film of cartilage over their joint surfaces, but their young often possess thicker pads of cartilage, which fill pits in their bones like the pits found in dinosaurs'. And adolescent animals usually display greater locomotor vigor than adults, not less. The pitted limb bones of dinosaurs would be no handicap to high speeds.

The other argument against galloping concerns the question of long shanks versus short shanks. Many fast mammals have long shank bones in comparison to the length of their thighs and even more elongated ankle bones (called metatarsals). Gazelles and most other fast-running antelope show bones and shanks that are very long relative to the thigh. Ostriches are fast runners and also have long shanks and ankles and short thighs. According to the traditional theory of shanks, to estimate the top speed of an extinct creature, one measures the length of shank + ankle and divides by length of thigh. If the resulting number is over 1.5, the animal is moderately fast; if over 2, the animal is in the gazelle category. Very few dinosaurs possessed shanks and ankles as long and thin as a gazelle's or an ostrich's, and most large dinosaurs had a shank + ankle ÷ thigh index of 1 or perhaps a little higher. Therefore, it's been concluded that short-shanked dinosaurs were limited to low-gear locomotion. *Triceratops* had a quite stubby ankle index and was therefore allegedly incapable of any fast movement at all. But all horned dinosaurs had shanks that were actually much, *much* longer than a rhino of the same weight. These dinosaurs only *seem* to have relatively short ankles and shanks because the thigh is much larger than a rhino's.

Triceratops was indeed shorter in the shank than a modern rhino is, but that doesn't prove *Triceratops* couldn't run as fast or faster. *Triceratops* had tremendously strong limb bones, and that strength must have evolved to withstand great forces. The unbelievers who scoff at the notion of a galloping *Triceratops* will have to explain why dinosaurs evolved such strong, thickly shafted limbs if they were going to do no exercise more strenuous than a shuffle through the swamps.

A third argument has occasionally been advanced against the notion of fast speeds in quadrupedal dinosaurs. Mammals today use their shoulder blades as arm extenders, swinging each long blade fore and aft with every stride. Dinosaurs supposedly possessed rigid shoulder blades that had to remain in place against the ribcage. If this theory of the stiff shoulder is correct, *Triceratops* would have had considerable trouble locomoting because its forelimbs were much shorter than its hind limbs. If both fore- and hind limbs were working at full stride, the rear end would move faster than the front end and the five-ton monster would have the option either of turning circles or of flipping over altogether—a most maladaptive model of locomotion!

Working on my undergraduate thesis, I had toyed with the hypothesis that the dinosaurs' shoulder blades might have swung across the ribcage, but I was unable to build a reliable support for

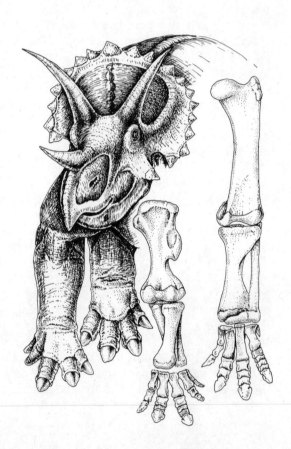

such heterodox mechanics. Later, at graduate school, I met a fellow student, Jane Petersen, who had just completed a thesis about the shoulders of chameleons. She proved that chameleons could swing their long shoulder blades fore and aft more freely than other lizards, because the chameleon's blade was not locked onto the chest by a bulky collarbone. This impressed me because I had already noticed that chameleons were the only lizards that looked like dinosaurs in the shoulders. Both dinosaurs and chameleons have very long, slender shoulder bones that work completely free of restraint from the collarbone, which anchored the shoulder blades

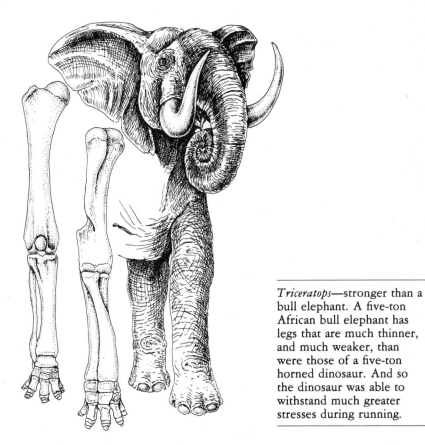

Triceratops—stronger than a bull elephant. A five-ton African bull elephant has legs that are much thinner, and much weaker, than were those of a five-ton horned dinosaur. And so the dinosaur was able to withstand much greater stresses during running.

in all most primitive reptiles. Chameleons evolved from some "normal" lizard ancestor that possessed a thick, stiff collarbone which held the shoulder blade in place. But chameleons shed that collarbone along their evolutionary path to provide themselves with more participation from their shoulders in the strokes of their forelimbs. Dinosaur evolution must have been the same—dinosaurs experienced the same reduction of the collarbone and must have developed a similar free-swinging shoulder. And the big quadrupedal dinosaurs evolved the longest shoulder blades of any vertebrate, past or present. As its yard-long shoulder swung alongside

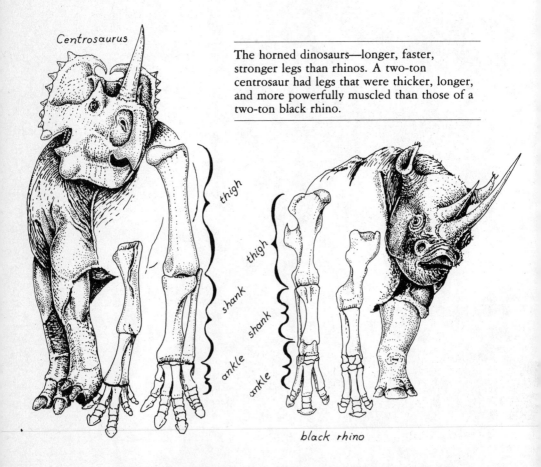

Centrosaurus

The horned dinosaurs—longer, faster, stronger legs than rhinos. A two-ton centrosaur had legs that were thicker, longer, and more powerfully muscled than those of a two-ton black rhino.

black rhino

Triceratops's ribcage, the extra length added to its forelimb must have given the animal a grand propulsive boost. Both fore- and hind limbs were consistently designed for fast, maneuverable movement.

Such outlandish heterodoxy proves doubly sweet when supported by independent confirmation. Fossil footprints are the only direct evidence left by locomoting dinosaurs, so a set of tracks left by some speeding *Tyrannosaurus* would provide dramatic confir-

Swinging shoulder blades—a modern horse, a modern chameleon, and the three-ton horned dinosaur *Centrosaurus*

BELOW: Collarbone prevented shoulder-blade swinging. Primitive dinosaur ancestors—like this Early Triassic *Chasmatosaurus*—couldn't use their shoulder blades for long fore and aft swings because the collarbone held the shoulder blade tightly against the sides of the chest and the breastbone.

collarbone (clavicle)

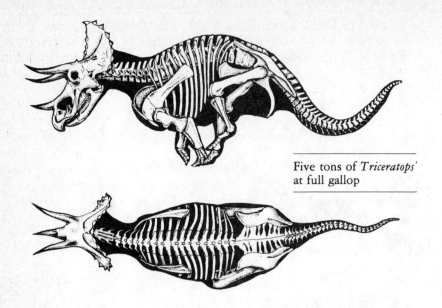

Five tons of *Triceratops* at full gallop

mation. The English biologist McNeil Alexander has worked out a clever formula for computing speed from trackways: all that is necessary is the length of stride and the toe-to-hip measurement. When first applied to some samples of dinosaur prints, the formula yielded low speeds—two to four miles per hour. Some commentators immediately jumped to the conclusion that this conclusively proved the theory of slow dinosaurs. That is nonsense. Most fossil trackways represent slow cruising speeds, not top speed, because all species spend most of their time moving along in an unhurried fashion. Bursts of maximum velocity erupt only rarely, when a predator charges or a plant-eater scampers for its life. Most tracks left by gazelles and rhinos today are made at a slow speed when these animals are feeding or going to or from water holes. Rhinos don't live their entire lives at thirty-five miles per hour; a trackway that caught one of these rare moments when the rhino was galloping full tilt would be a most extraordinary find. Trackways from big quadrupedal dinosaurs are rare—there exist only four sites with good brontosaur tracks—so the sample is far too poor to argue any case about *top* speed.

Bipedal dinosaurs are represented by more tracks—hundreds altogether—so a few tracks might conceivably capture a moment of high speed. And a few two-legged trackways do provide such proof. Several medium-sized, fifty-pound to half-a-ton bipedal predators have left long-striding tracks which compute to speeds of twenty, thirty, or even forty miles per hour.

Narrow tracks, swinging shoulders, stout-shafted limbs that bounced at every stroke—all these bits of modern evidence agree with the lively restorations drawn for Marsh and Cope way back in the 1890s. Cope had a painting made of *Dryptosaurus,* showing a pair of the giant meat-eaters excavated from the phosphate mines of New Jersey. Cope's dryptosaurs were portrayed in violent locomotor exercise. One was flung on its back, hind legs lashing out in claw-tipped defensive strokes; the other was painted in mid-leap, its great hind legs having propelled its body far above the ground. A good painting, far more faithful to the real structure of dinosaur locomotion than the shuffling reconstructions popular in most orthodox textbooks today. Speed and vigor were the way of the dinosaurs, multi-ton monsters able and ready to break into a fast-paced charge whenever necessary. The Mesozoic was life in the behemoth fast lane.

11

MESOZOIC ARMS RACE

Humans are one of the least armored products of evolution. Perhaps our own defenseless hide renders the apparently bizarre armor plate sported by three great clans of beaked dinosaurs—the Stegosauria, the Ankylosauria, and the horned dinosaurs—especially fascinating. The story of these armored dinosaurs is a drama out of the Mesozoic arms race, the co-evolutionary link between ever deadlier meat-eaters and ever more formidably protected prey.

Stegosaur tails were without question one of the most dangerous weapons ever evolved by a plant-eating animal. At the extreme end of the stegosaur's long tail sprouted a fearsome war club, composed of four or eight sharply pointed spokes between two and three feet long. Extra-thick connective tissue in the skin anchored the bases of these bony spikes so that the points extended outward, and upward, and backward. And pits left by blood vessels on these spikes show that they were sheathed by a very thick horn cover in life, much like the outer sheath of longhorn cattle today. Horn constituted the ideal sheathing material for such sharply pointed weapons because it is more flexible and less brittle than bone and thus can be honed to a much sharper point.

To drive all those pointed tail spikes deep into the body of its adversaries, *Stegosaurus* required a tail of great power and flexibility, and both qualities were in abundant supply. To acquire flexibility in the tail, the stegosaurs' evolution had to dispose of a major feature of their ancestry, the system of stiff tendons. Most beaked dinosaurs featured a latticework of bony tendons running down either side of their backbones from torso to tail. And all the earliest, most primitive beaked dinosaurs possessed such equipment. As has already been discussed, this latticework—best seen in duckbills and horned dinosaurs—would have provided an advantage for supporting the body weight without muscular effort. But such bony tendons would have stiffened the stegosaur's tail

The big-plate stegosaur *Diracodon* battles a *Ceratosaurus*

too much for easy swinging. Evolution therefore eliminated the system of tendons and the stegosaurs were the only beaked dinosaurs to do away with bony tendons entirely. But merely eliminating bony tendons wouldn't have been enough to render the stegosaur's tail optimally dangerous. Since the spikes stood at the tail's extreme tip, the bones of the tail had to be both strong and flexible all the way to the end. In most dinosaurs the tail joints grew progressively stiffer toward the end, but not so in stegosaurs. The joints between the successive segments of the tail gave its entire length from rump to tip enough suppleness to flex in a graceful S-shaped curve, and the vertebrae were much stronger than usual near the end.

To achieve the muscular strength necessary to swing its club, the stegosaur evolved enlarged shelves of bone for anchoring its muscles (similar shelves had evolved in the big-tailed brontosaurs, such as *Diplodocus*). A twenty-foot-long stegosaur would have had more strength in its tail muscles than a large modern elephant has in one of its hind legs. And when the mighty tail muscles contracted, the stegosaur's caudal club swung with irresistible authority.

The eight-spiked *Stegosaurus ungulatus*

Stegosaurs had need of such a war club because they faced predators nearly as large as elephants. *Allosaurus* and *Ceratosaurus,* the two most common Late Jurassic flesh-eaters, both grew to lengths of thirty feet and more and would have weighed between one and two tons. Even larger was *Epanterias* (possibly a very large species of *Allosaurus*), a forty-five-foot predator that must have reached four tons, six times heavier than a large lion. If such huge flesh-eaters attacked in groups (a tactic widely believed possible), only the most heavily armed plant-eaters could have survived. Imagine the potential of the stegosaur's tail spikes in such a confrontation. If the three- to four-foot-long spikes were driven full force into the chest or belly of even the largest predator, the result would have been devastating. Not even *Epanterias* would have survived a direct hit.

But to fight well, *Stegosaurus* would have had to maneuver quickly, pivoting about to keep its tail club facing the attacker. *Allosaurus* and *Ceratosaurus* were long-legged and nimble-footed, and could have danced around the stegosaur in order to lunge in for bites at the vulnerable neck or shoulders. How could evolution equip the stegosaur with the necessary maneuverability to employ its tail club to best advantage? The solution was found in its unique body proportions and its short but thickly muscled forelegs. Stegosaurs appear ungainly at first sight—their hind leg was much longer than the fore, the hips much taller than the shoulder. The combination of a heavy rump and tail with short forelimbs placed the point of balance of the stegosaur's body just forward of the hips, so that the beast could easily have pivoted around by pushing sideways with its forepaw.

The muscles employed to push sideways with the arms are known as the deltoids. In most dinosaurs the deltoids were moderately strong but not unusually so. But stegosaurs possessed prize-winning deltoids, and the site where they attached to the upper arms (humerus) was gigantic, larger than in any other vertebrate. Obviously then, when threatened by a predator, the stegosaur shifted its weight back onto its hind feet, then pushed with its fore-feet, to rotate right or left in order to keep its deadly tail facing the foe. Its huge deltoids provided sufficient power for pivoting its entire body mass with ease.

Stegosaurus is, however, best known not for its war club, but

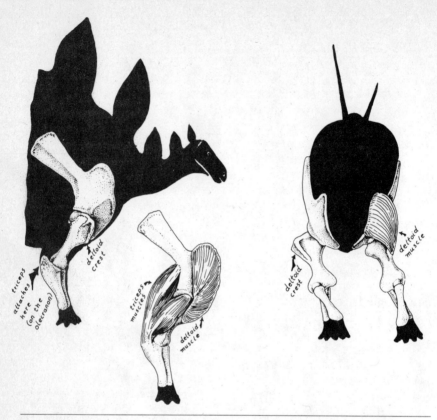

Stegosaur muscles for quick turns. The deltoid muscle group had a huge sideways-facing crest on the upper arm (humerus) so that stegosaurs could push their bodies to one side or another. Powerful triceps muscles running from shoulder blade to elbow gave the stegosaur a forward-lunge capacity.

for the spectacular triangles of bone that rose up to four feet above its backbone. Though tall and broad, they were thin in section and, like the tail spikes, were sheathed in life by an outer layer of horn. Roughened zones along the bases reveal that these bony plates were embedded in the skin along the top of the spine. Most restorations show these plates sticking straight up from the back. But that is a most puzzling orientation for them. What could have been the bioengineering purpose of these strange triangles? Some paleon-

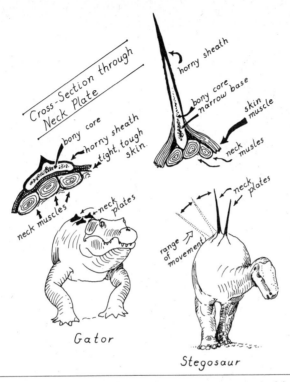

How stegosaurs flapped their plates. Stegosaur ancestors had bony armor plates shaped like those of gators—the plate base was very wide and firmly embedded in the outer layer of tight skin. But during stegosaur evolution the plate base became very narrow and a sheet of skin muscle attached to the sides of the plate to swing it from side to side.

tologists have suggested that if the stegosaur's plates stood up vertically, they might have offered some defense against bites directed at the backbone. But the stegosaur's spinal cord was already well protected without the plates. It lay deep beneath the very tall vertebral spines and the ligaments, which together constituted a very tough hump over the torso and hips, much like the ridge on a modern razorback hog. Any *Allosaurus* unwise enough to bite into that ridge would have broken off its teeth without inflicting significant damage. Moreover, the largest plates were located over the hips and base of the tail, where the spinal cord was already best protected by vertebral spines. The stegosaur's spinal cord was so well armored by the backbone that the triangular plates really wouldn't have added extra protection. And it appears like a redundant use of bone to place the tallest plates over the strongest segments of the back.

American paleontologists of the last four generations have puzzled over the apparent incomprehensibility of the arrangement of the stegosaur's plates. Several prominent museum scientists even concluded there simply wasn't any mechanical function at all for the plates—they were purely ornamental devices to make the stegosaur look more intimidating to enemies and sexier to potential mates. A sexy look and an intimidating profile are worthy evolutionary results. But in fact if the evolution of the stegosaur's plates is carefully considered, it becomes possible to see how they could have functioned as a very effective addition to defensive armament.

Armor plate was a long-standing characteristic in the Dinosauria. Crocodiles sported armor on their neck, torso, and tail when they first appeared during Triassic times, 220 million years ago. And all living croc species retain a flexible body shield of horn-covered bony plates joined together by sheets of ligaments within the deep skin layer. No croc possesses stegosaur-style triangles, but most crocodilians do display big oval plates of armor on the back of the neck, and these usually have raised, pointed ridges. Similar arrangements of armor protected other dinosaur relatives (the Thecodontia) from the Triassic Period.

The process whereby oval, keeled plates evolved into the upright, thin triangles of stegosaurs was fairly simple: by reducing the bony bases and enlarging the ridges, the oval plates quickly became thin stegosaur triangles. However, the key question about stegosaur armor is why would the animals evolve such weak narrow bases for their plates? A broad base firmly embedded in tough skin would have prevented the armor from bending when a predator struck. But stegosaur plates were so tall, and their bases so narrow, it was most unlikely they could remain stiff or upright. Maybe the stegosaur's approach to armor design was dynamic. Muscles in the skin might have moved the plates so they could point upward or downward, depending on the point of origin of the attack. When erect, the points would have warded off attacks from above; when flexed downward, they could have provided the stegosaur with a very effective flank defense. Held horizontally, the tallest plates would have stuck out sideways three to four feet, so that no predator could approach near the stegosaur's hide. Since the muscles and flanks of the hind limb were especially vulnerable zones, the plates over the rump had to be the largest.

A flexible armor defense would at last explain all the most peculiar features of how stegosaurs were designed. The pitted texture of the plates' basal surface would have provided purchase for some sort of tissue—ligament or muscle—to embrace the plates at the base for about half a foot. Thin bases embedded in the skin would have been necessary to permit the hingelike movement of the plates. The thin, triangular shape conveyed the greatest strength to the pointed tip with the least concomitant amount of weight.

Muscles in the skin have evolved several times in different vertebrate groups, and might easily have evolved in stegosaurs. Skin muscles have evolved in mammals to move hair as when a horse twitches the skin on its back to flip off a fly. (Human evolution took a U-turn here. We had skin muscles once but lost most of them when we evolved our naked skin, and now possess few except on our faces.) Most living reptiles bunch skin muscles thickly around the throat, and they are what slips the neck frill forward on the Australian frilled lizard, for example. The skin muscles of birds are used to control the orientation of their feathers. If skin muscle could evolve to flip frills in lizards and feathers in birds, stegosaurs might have evolved them to flip their armor plates.

Stegosaurus must have been a grand performer under attack— a five-ton ballet dancer with an armor-plated tutu of flipping bony triangles and a swinging war club. Browsing peacefully on the tops of bushes, perched upon its hind legs and tail, its keen eye quickly catches the movement of two huge *Allosaurus,* hunting *au pair,* along the floodplain. The pair of allosaurs stride quickly to the attack, one from either side. The *Stegosaurus* alights defensively on all fours. The first *Allosaurus* darts quickly for the stegosaur's neck. Instantly the intended victim pivots and lowers its armored triangles. The allosaur suffers a cut across its snout for its efforts. The second allosaur lunges at the other flank, but the stegosaur's tail slashes out to meet the charge. With an audible "whoosh," the four-spiked tail barely misses the allosaur. That's too much for the would-be predators. They back off from the stegosaur still capably brandishing its weapons. And off they go to find easier pickings.

Sudden extinction interrupted the evolution of dinosaur armor at the end of the Jurassic when the true stegosaurs died out totally or at least became very rare. But as the Cretaceous Period dawned, new dinosaur dreadnoughts appeared: the nodosaurs. Nodosaurs shared some characteristics with the stegosaurs—their hips were high and their deltoid muscles were strongly developed for sideways maneuver. But on the whole, the nodosaur's approach to defense was much more massive. A complete flexible

Four tons of
charging armor—
the nodosaurian
Edmontonia

pavement of small bony plates armored the entire upper surface of nodosaurs—head, neck, torso, and tail. None of these plates were anything like the tail triangles of the stegosaurs, but some nodosaur species had broad-based plates topped by tall conical spines. Nodosaur armor plating was therefore much more continuous than that of stegosaurs, and their hip bones had expanded into an immense solid roof protecting the upper surface of the abdomen.

Some paleontologists believe nodosaurs defended themselves passively. According to this view, these massively armored monsters employed their carpet of plates much like a mobile bomb shelter. Under attack, they would merely hunker down on the Cretaceous forest floor, legs folded under their body, to wait out the tyrannosaur's attack. But there is ample reason to believe nodosaurs could become actively dangerous antagonists and turn the tables on their attacker. What made some nodosaurs dangerous was their sharp shoulder spikes. Mounted on broad bony bases embedded in the skin of the shoulder stood long, sharply pointed horn cores that curved forward. In life, an outer covering of horn made these shoulder spikes as long and deadly as the tail spikes of stegosaurs had been. Like stegosaurs, nodosaurs had very strong elbow muscles (the triceps muscle group) perfect for quick, forward lunges. Altogether, the forward end of a nodosaur resembled the gigantic, short-legged warhorse of medieval times, coated with armor, and ready to charge with wickedly sharp lances jutting forward from either side of its head.

Wrestlers and short-legged fullbacks know the advantage of short, strong legs—a long-limbed wide receiver can run faster, but the shorter legs provide a greater initial acceleration. The nodosaur's enemies were the very long-limbed tyrannosaurs, fast enough to catch any nodosaur very quickly. But once the combatants were close, the advantage of speed disappeared. A tyrannosaur could stretch downward to snap at the nodosaur's tail or back, but the predator would only succeed in breaking its teeth against the impregnable carapace. The tyrannosaur's only hope would have been to flip the nodosaur in order to attack its unprotected belly. But nodosaurs were very wide across the hips and had a low center of gravity. A fully grown nodosaur would have been as easy to flip as a modern wide-track station wagon. And then the tyrannosaurs would have had to face counterattack. The nodosaur could have

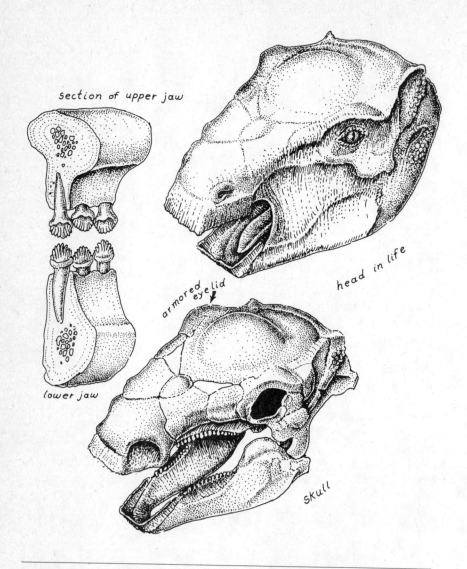

section of upper jaw

lower jaw

armored eyelid

head in life

skull

Weakest teeth in the Dinosauria—the nodosaurs. *Edmontonia* was typical of the strangely constructed nodosaurids. Thick armor plates covered the snout and forehead, and a sharp-edged beak was built around the muzzle, but the teeth were absurdly tiny. Did this dental decrepitude condemn *Edmontonia* to a diet of soft-water plants?

kept pivoting to face its long-limbed attacker, awaiting its opportunity. If the tyrannosaur allowed any opening at all, the nodosaur's powerful elbows and knees could instantly drive its armor-plated body forward. And its murderous shoulder spikes might catch the predator's calf or leg, tripping the tyrannosaur or ripping a nasty wound. Before such a lethal charge, any tyrannosaur might have beaten a well-advised retreat.

Late Cretaceous days were the high point of armor development for the Dinosauria. Nodosaurs of several species stalked the meadows of the Cretaceous deltas of North America. And they were joined by a new family of dreadnoughts, the ankylosaurids. At first glance the ankylosaurids seem less dangerous than the nodosaurs: they were shorter at the hips and weaker in the shoulder, couldn't pivot as quickly, and lacked the lethal shoulder spikes. As compensation, ankylosaurs had better head protection than nodosaurs—overhanging plates protected their eyes and cheeks.

But the tail functioned as the cornerstone of the ankylosaur's defensive tactics. Like the nodosaur's, the root of the ankylosaurs' tail was powerful and supple. Unlike nodosaurs, the last half of the ankylosaurs' tail was stiffened by a series of bony tendons that converted this end into a prolonged handle for a bone-crushing war club. At the very end of the tail, three large masses of bone fused together to create a sort of monster cloverleaf-shaped club head. By contracting its tail muscles, the animal could quickly swing the stiff handle from side to side and powerfully flail its three-leafed club head.

The ankylosaurs' war club was less precise but quicker than the stegosaurs'. A stegosaur's tail joints were supple right down to the very tip of its tail, so the animal could control the movement of the spiked club carefully. Such care was necessary for effective defense because the sharp spikes required accurate aim. Ankylosaurids had less finesse, but the massiveness of their club guaranteed damage no matter how the blow landed.

In nodosaurs, ankylosaurids, and stegosaurs, the dangerous weapons were carried on shoulders or tail. But one group of armored dinosaurs, the boneheads, used their skulls as their principal offensive device. "Greatest Dinosaurian Bonehead!"—touted the label beneath the skull in the New York museum. In the glass case stood *Pachycephalosaurus* ("heavy-headed lizard"), a bonehead indeed, with a two-foot-long skull topped by a dome of solid bone eight inches thick covering the forehead and crown.

Bone-headed dinosaurs at first provoked the same combination of awe and ridicule that originally greeted the stegosaurs. Here was yet another case of outlandish dinosaur construction without any conceivable explanation from body mechanics. Back in the early 1900s, the theory of racial senescence would have served as a respectable hypothesis to explain strange dinosaur adaptations such as bonehead skulls. Flamboyant and senseless crests, plates, and spines were supposedly signs that all Dinosauria had gone senile in their evolutionary old age. Like old ladies wearing out-of-fashion headgear and mismatched gowns and coats, the dinosaurs had allegedly lost their adaptive vigor and had become incapable of evolving anything but nonfunctional ornaments and maladaptive excrescences. As a theory, racial senescence was bankrupt by 1920, but it still pops up here and there in bad books about dinosaurs.

The bodies of boneheads were nothing unusual: a pair of long hind legs, long, stiff tail, short arms, a barrel-shaped body mass to accommodate masses of vegetation. This general configuration wasn't different from that of a host of other bipedal beaked dinosaurs. And even bonehead skulls weren't noticeably deviant in the snout, jaws, and teeth. The strangeness of boneheads was concentrated entirely in the domelike swelling over the top of the braincase. Some species had only a slightly thickened skull roof of otherwise normal construction. But the fully developed boneheads, like *Pachycephalosaurus,* had giant domes that suggested great intellect. As the original discoverers of bonehead dinosaurs were quick to point out, the supposedly brainy appearance was a sham— the brain itself was tiny and occupied only a small volume deep inside the bony dome. Most of that dome was in fact filled with bone cells arranged in a radiating pattern like the fibers in a cross section of grapefruit. Bone cells usually grow in the direction of greatest stress, so the bonehead's pattern of growth is a clue to the head's function. The radiating pattern strongly suggests that the dome was subjected to enormous outside pressures. But of what sort?

Dome-headed dinosaurs can probably best be understood as wearing NFL-style football helmets over their minuscule braincase. Modern football helmets weren't designed for merely pas-

Stegoceras—two individuals ramming

sive protection; they were built so the wearer could ram his head into the unfortunate player opposite him. Old-fashioned leather helmets weren't as good for the head-first block, but coaches soon discovered that the human head could serve as a weapon, so the helmets were redesigned. A domed, impact-resistant helmet was invented to provide the head-ramming linebacker with the ability to strike blows without damaging his own skull. Dome-headed dinosaurs had evolved the physical equivalent of a ramming helmet millennia before, a configuration that strongly suggests a head-first mode of attack.

Peter Galton, a leading expert on beaked dinosaurs (he's the man who studied the cheeks and digestive tracts in duckbills), first worked out the head-butting theory in 1971 and won nearly worldwide acceptance of his basic idea. Galton pointed out that dome-headed dinosaurs had exceptionally strong muscles holding the head at a right angle to the neck, so that the dome would face forward when the beast lowered its head and charged. The animal possessed all the qualities for an optimal butting attack—a bull neck, a low head position in the charge, and a thick skull covering a small brain.

Galton believed sex was the ultimate motivation behind the head-butting behavior. He argued that male domeheads would have banged their crania against one another in ritualized combat, much like bighorn sheep. On this point, I would disagree a bit. A bighorn sheep's horns are wide and flat, so when two males clash, their horns meet across a wide surface. The resulting collision is a true test of strength, because the full force of the sheeps' bodies is delivered. But the rounded shape of bonehead dinosaur domes made a precise head-to-head blow nearly impossible. If two boneheads did clash, their heads would probably have struck only glancing blows. Domed heads, therefore, like football helmets, were probably for butting an adversary in the body, not in the head.

Polish expeditions to Mongolia in the 1960s found a marvelous bunch of bonehead dinosaurs. The Poles wondered whether the head-butting equipment really was for sex-related contests—the head seemed too dangerous a weapon for such encounters. The Polish hypothesis argued that the bonehead was essentially an antipredator weapon. It is difficult to judge between the alternatives. Did the boneheads ram one another, or did they ram meat-eating dinosaurs that threatened attack? Probably both hypotheses are right. Protecting oneself from a predator was always a vital function, and any anatomical device that could be wielded as a ram or club was useful in that connection. But in the evolutionary scheme of things, staying alive means nothing unless one's genes are passed on to the next generation. So, the ability to butt a sexual rival hard in the ribs might provide an edge in the great evolutionary dating game.

Without doubt the most dangerous devices for active defense among the Dinosauria emerged in *Triceratops.* The scene has been portrayed in paintings, drawings, and illustrations hundreds of times, but it remains thrilling. *Tyrannosaurus,* the greatest dinosaur toreador, confronts *Triceratops,* the greatest set of dinosaur horns. No matchup between predator and prey has ever been more dramatic. It's somehow fitting that those two massive antagonists lived out their co-evolutionary belligerence through the very last days of the very last epoch in the Age of Dinosaurs. *Tyrannosaurus* stood over twenty feet tall when fully erect, and a large adult was as heavy as a small elephant—five tons. No predatory dinosaur, no predatory land animal of any sort, had more powerful jaws. Withstanding a *Tyrannosaurus*'s attack required either tanklike armor—the approach taken by *Ankylosaurus*—or most powerful defensive weapons—the approach taken by *Triceratops*.

Triceratops's body was designed for lunging and charging. The torso was very short, the chest broad, the hips wide and strong. Fore- and hind limbs were very thick for the body size—much thicker than an elephant's of the same weight—and the paws were wide and compact. No armor plate encased the hide of *Triceratops,* because its defense was active, head-first, and devastatingly effective. *Triceratops* and its kin carried far and away the largest and heaviest skull ever to evolve on a land creature—six, seven, even eight feet long, up to four feet wide, and of very solid construction. Where the neck muscles attached to the back of the skull, the cranial bones had expanded sideways and upward and were reinforced to support sudden twisting lunges of the great horns located on the brows.

Triceratops's horns were wonderful examples of Mesozoic armature. From the eye socket to tip, the horn cores could reach four feet in length and often had a graceful double curve like the horns of longhorn cattle from the Wild West. When the first *Triceratops* horns were discovered in Colorado in 1880, Professor Marsh thought they had belonged to ancient buffalo. But although *Triceratops* horns were shaped like a buffalo's, they were located on the head in a far more dangerous orientation. Longhorn cattle and buffalo horns face sideways, and their horn thrusts can only be delivered by tossing the head to the left or right. *Triceratops* thrusts could be far more precise. Its horns curved forward and slightly outward over the long snout. Although the head was massive, it was nearly perfectly balanced on the ball-and-socket-type joint between the head and the neck. The heavy snout forward was counterbalanced by the broad head shield extending backward. The entire apparatus was a marvelous combination of delicate musculoskeletal poise and brute power, allowing *Triceratops* to lunge forward at its opponent with the entire set of horn tips.

The neck is a vulnerable point in any vertebrate, and *Triceratops* protected its neck with a flaring collar of bone, fringed by

Two *Triceratops* confront a *Tyrannosaurus*.

Triceratops drumsticks. Muscular power for quick charges came from the huge calf muscles that attached to the inner and outer sides of the big bony crest of the shin (cnemial crest). This same crest also gave the knee-straightening muscles great leverage.

short, sharp, horn-covered spikes. Part of this frill was covered by an extension of the jaw muscles. But the wide periphery of the bony frill was pure armor, covered with tough horny skin. Below each eye and just above the jaw joint was a short horn-covered spike that protected the cheek. This defensive master-machine alive and in action must have been a sight to behold, its eight-foot skull pivoting easily left and right, its neck frill swinging in wide arcs.

Triceratops was not the only giant horned dinosaur found on the Laramie Deltas. It was accompanied by the rare *Torosaurus*—the "bull lizard"—which sported an even longer neck frill. And in New Mexico, during Late Cretaceous days, a splendid long-horned relative of *Torosaurus* walked the floodplain—*Pentaceratops,* the "five-horned face," named for its combination of unusually long cheek horns, brow horns, and nasal horn.

The finest display of horned dinosaur heads anywhere in the world is located in the Cretaceous Hall of the American Museum of Natural History in New York. The horned legions reached their greatest variety during Judithan times, a few million years earlier than the age of *Triceratops.* Judith River beds in Montana and Alberta have been very generous to dinosaurophiles. The New York

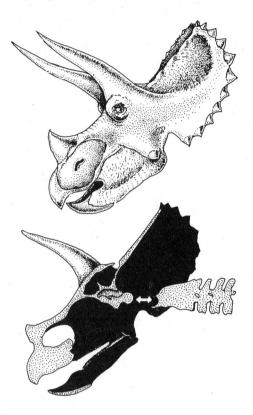

Perfect balance at the *Triceratops* skull pivot. Horned dinosaurs had a ball-and-socket joint connecting the massive head to the neck. Since the joint was placed under the eyes, just at the natural balancing point of the huge head, the neck muscles could toss the head in any direction with great precision of movement. Botton cutaway view shows the ball joint on the skull separated from the socket.

museum displays *Centrosaurus,* a short-frilled variety whose weak brow horns were compensated by an erect and very long horn over the snout. Its close kin *Styracosaurus* possessed the great nose horn plus a magnificent set of curved spikes over the frill, giving its head a monumentally prickly appearance. *Monoclonius* is there too, with its stout nose horn. In general, Judith River horned dinosaurs sort out into two systems of attack. The first includes the genera with huge nose horns and weaker brow horns. These animals probably thrust their powerfully armed snouts straight upward as they tried to gore the softer underparts of a tyrannosaur. The second system includes the *Triceratops*-like configurations—long brow horns curving forward. Such long brow horns are rare among the Judith fauna—but the species *Chasmosaurus kaiseni* is there to represent this second system.

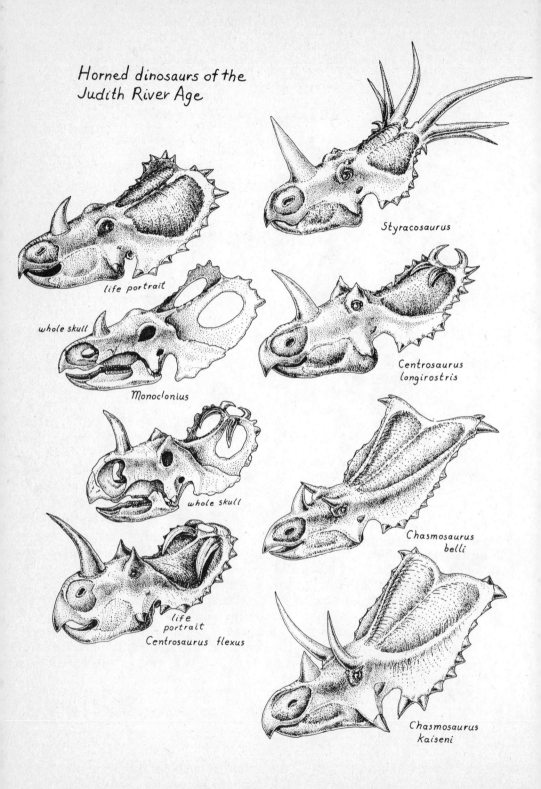

Horned dinosaurs of the Judith River Age

Styracosaurus

life portrait

whole skull

Monoclonius

Centrosaurus longirostris

whole skull

Chasmosaurus belli

life portrait
Centrosaurus flexus

Chasmosaurus kaiseni

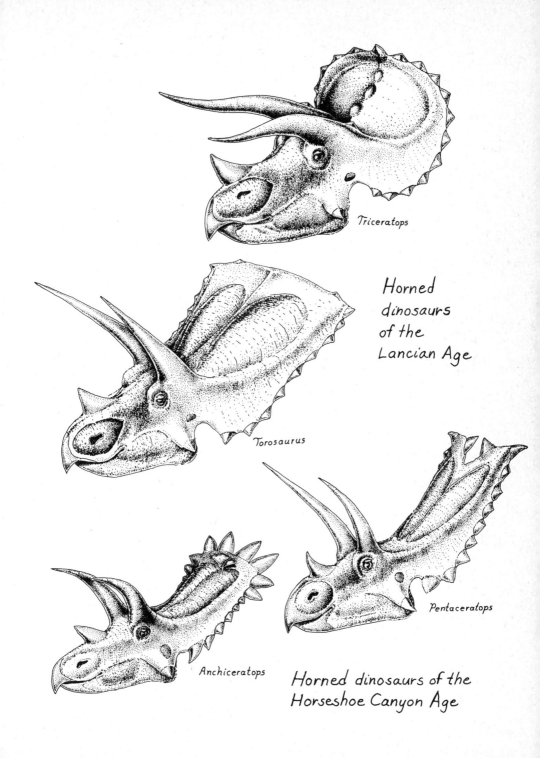

Triceratops

Torosaurus

Anchiceratops

Pentaceratops

Horned dinosaurs of the Lancian Age

Horned dinosaurs of the Horseshoe Canyon Age

In the 1890s, horned dinosaurs confronted science with an evolutionary puzzle: These dinosaurs were so highly evolved for an aggressive defense that paleontologists were at a loss as to how such creatures could have descended from any other kind of beaked Dinosauria. Even the oldest horned dinosaur fossils from North America manifested the very complexly designed snout, horns, frill, and neck muscle attachments in a fully developed state. It was as though the horned dinosaurs had sprung directly from the mind of the Creator.

Today the early fossil record of the horned dinosaurs is still imperfect, but we are two big steps closer to understanding their evolutionary origin, owing to discoveries in the Cretaceous sands of the Gobi Desert. The spectacular discoveries made in the Gobi came about through a colossal error of scientific theory. American scientists in the early 1900s wanted to explore the Mongolian desert because a theory popular at the time maintained that human evolution occurred fastest and most efficiently on a dry, invigorating plateau such as the country of Central Asia—the "Roof of the World." Conversely, tropical lowlands were supposedly evolutionary slums where stagnant water and fetid air suppressed the development of higher life forms. According to this theory, only where air was dry and thin—as on the Asian Plateau—could lively species evolve. These Asian plateaus were *terra incognita* at the time; no thorough scientific surveys of Gobi zoology or paleontology had been made, and no one knew what sort of beasts had evolved there. After the first World War, wealthy Americans contributed funds for a grand American Museum expedition to the Gobi. The trip was billed as the search for the missing link; the key to human evolution was to be found in the windswept desert. Roy Chapman Andrews, naturalist-explorer par excellence, was its leader.

So far as missing links in evolution were concerned, the expedition was a bust. No important protohuman fossils were found. We now know in fact that nearly all the steps in human evolution took place in warm tropical realms, not on high plateaus. But the Gobi expedition uncovered a boundless treasure trove of dinosaurs, whole new families of them. These Gobi Cretaceous dinosaur beds were totally different from the Judith and Laramie Deltas familiar to the American geologists. As we have already noted, Late Cretaceous habitats in America were mostly humid deltas, but

Central Asian habitats of the time were dominated by desert and near-desert conditions. Red Gobi sandstones preserved the sedimentary work of Cretaceous winds that drove sand into dunes around shallow lakes. Over millions of years these Cretaceous dunes coalesced into dune fields, and these fields, in turn, piled upon one another to produce hundred-mile-wide layers of preserving sand. Andrews's field parties found innumerable white skeletons in the red rock—small, chunky-bodied dinosaurs with long hind legs, powerful beaks, and short frills protecting their necks. Andrews's men had found primitive horned dinosaurs, the ancient uncles of *Centrosaurus* and *Triceratops.*

When the scientists in New York unpacked the first crate-loads of dune rock from the Gobi, it was clear a dinosaur missing link had been found. The new dinosaur's name was a tribute to Andrews's leadership: *Protoceratops andrewsi,* "Andrews's ancestral horned-face." *Protoceratops*'s cranial structure was almost perfect as the ancestral state of the large American horned dinosaurs. The basic horned dinosaur design was proclaimed by the deep beak, solidly connected skull bones, and a well-braced neck frill. But *Protoceratops* displayed only the suggestion of horns. In the biggest skulls a roughened bump on the snout must have supported a low horny crest in life. And *Protoceratops*'s legs, hips, and shoulders were delicate compared to the massive strength of the American horned dinosaurs.

Protoceratops and its close relatives must have swarmed over Asia, because their bones and nests of eggs are the commonest dinosaur fossils found in the widespread dune beds. But not one *Protoceratops* has ever been reported from the rich beds of the American Judith and Laramie Deltas. Swampy meadows and broad humid floodplains were evidently not to *Protoceratops*'s liking, though Canada and Montana did play host to relatives in Late Cretaceous times—the genera *Leptoceratops* ("diminutive horn-face") and *Montanoceratops. Leptoceratops* probably was an immigrant from Asia.

Roy Chapman Andrews's team also discovered a second missing link in horned-dinosaur history: an earlier Mongolian family which at last revealed how the horned-dinosaur story began. At first the announcement created little stir—two skeletons and some odd bones from a very small beaked dinosaur out of the Early Cretaceous beds of Inner Mongolia. (One of the skeletons had

Protoceratops: male (left) and female (right)

nicely preserved gizzard stones.) But before long dinosaur anatomists began to discover what they had; these tiny beaked dinosaurs possessed a very deep, parrotlike snout that looked just like that of *Protoceratops*. It was this beak that suggested the name for the little animal—*Psittacosaurus,* "parrot lizard." However, not just the beak, but the rest of this parrot dinosaur's skull began to look suspiciously like that of an ancestor for *Protoceratops*. Parrot dinosaurs protected their cheeks with sharp, crestlike horns, and they showed just the beginnings of a neck frill. Especially striking to the anatomist's eye was the core of the upper beak; it was formed by a bone separate from the rest of the skull, a most unusual trait, found elsewhere only in true horned dinosaurs.

Thanks to these Asian discoveries, and those made in recent decades by Polish, Russian, and Chinese expeditions, we now possess an outline of horned-dinosaur history. Parrot dinosaurs must have been close to the original ancestral stock—they have the extra-long hind legs and short forelimbs so common among primitive beaked dinosaurs of all sorts. In parrot dinosaurs we already find the trend toward an exceptionally strong head with a powerful beak and strong bite. Parrot dinosaurs were leaf-eaters, it's clear from their teeth. But their beak served both as an herbivory and an antipredator device, equally snipping off branches and snapping menacingly at predators that threatened attack. The crest-armored cheek bones would protect the parrot dinosaurs when they lunged to bite at an enemy. From such a beginning it was only a short evolutionary step to *Protoceratops* and *Leptoceratops* with their incipient horns.

The horned-dinosaur story shows how paleontologists can trace the major evolutionary events. Rarely do fossils yield a complete evolutionary sequence from mother to daughter to granddaughter species. Evolution is too bushy to permit such a straightforward story, too full of side branches. As clans evolved, the ever-branching species spread over the continents. Since fossils come from a few small areas, it is impossible to follow every stage of an evolutionary line. But it's possible to make out an overall progression of uncles and nieces, even when the parent–daughter sequences cannot be found. Parrot dinosaurs probably weren't the direct ancestors of *Protoceratops,* since the parrot clan had already branched off in their own unique direction (parrot dinosaurs had evolved

very narrow forepaws with fewer fingers than true horned dinosaurs). And *Protoceratops* probably wasn't the direct ancestor of
Triceratops or *Monoclonius*. *Protoceratops* and its sibling genus *Leptoceratops* had evolved into an evolutionary sideline where the tail
had become very slender from side to side but quite tall from top
to bottom. This tall-tailed condition was probably part of *Protoceratops*'s tactics of intimidation—broadside huff and bluff. The big
North American horned dinosaurs had neither the narrow paws of
parrot dinosaurs not the billboard-type tails of *Protoceratops*. To sum
up, parrot dinosaurs were the granduncles of *Protoceratops*, which
was the granduncle of *Triceratops*.

Altogether the history of dinosaur arms and armor must rank
as one of the most dramatic aspects in the pageant of evolution,
but it poses a question. Was this parade of tanks and dreadnoughts one natural unit of evolution, a single major branch of

the dinosaur family tree, or not? In the early days of American dinosaurology, museum scientists believed that stegosaurs, nodosaurs, and ankylosaurs were closely related and that horned dinosaurs were perhaps cousins. But this idea lost favor among Americans in the 1920s for no good reason. And for the last half century most books place the armored groups—stegosaurs, ankylosaurs, and horned dinosaurs—into separate suborders.

On the other side of the Atlantic, a very different conception was adumbrated by a flamboyant but perceptive Hungarian paleontologist, the Baron von Nopsca. Nopsca was a bright spot during the dreary days of the 1920s when English and American dinosaurology was settling into a muddle-headed and lackluster orthodoxy. Nopsca was a genuine Transylvanian aristocrat with a fondness for dressing in Balkan native dress and for reconstructing the sex lives of dinosaurs. He was also gay, a life style quite at odds with the macho frontier style of American bone diggers. Nopsca was universally recognized as a creative thinker—at his worst no less thoughtful and no less accurate than muddling orthodoxy, at his best head and shoulders above his contemporaries in discerning evolutionary patterns.

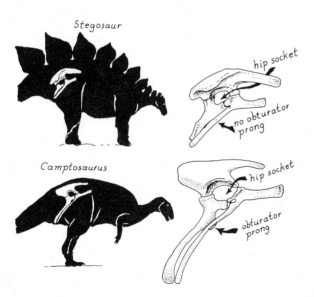

The hallmark of the Thyreophoroids. All the armored dinosaurs were specialized in losing the obturator prong, a short flange of bone that connected pubis and ischium. The nonarmored dinosaurs retained the prong.

Nopsca argued the case for an evolutionary unity among all armor-clad Dinosauria. Testing his evidence in *Geologica Hungarica* in 1928, he coined the term "Thyreophoroidea"—shield carriers—for all armored and spiked dinosaurs. Americans more or less ignored his hypothesis, but since 1975 there has been a Thyreophoroid revival. Walter Coombs, an expert on ankylosaurs, pointed out that boneheads, stegosaurs, and ankylosaurs all shared a most unusual feature, armored eyelids. All three groups had stiff plates of bone embedded in their upper eyelid to protect the eyeball from attacks delivered from above (only the accessory eyelid was armored; the inner eyelid was soft skin and could close over the eye). Such armored eyelids support the baron's theory. The most primitive beaked dinosaurs lacked any such wide, bony eyelids, so all dinosaurs with them could have inherited their armored blinkers from one common ancestor.

If all the shield carriers, domeheads, ankylosaurs, horned dinosaurs, and stegosaurs were related, then evidence of this pedigree should be found in body architecture. And we do find clues. All the primitive beaked dinosaurs' lower hip bones (pubis and ischium) were joined together by a short shelf of bone, called the obturator prong. But all the later shield carriers lacked this telltale shelf—evidence that perhaps one common ancestor had done away with the obturator prong when it diverged from the primitive beaked condition. Other clues of common pedigree can be found in the skull. In the most primitive beaked dinosaurs, the bones of the roof of the mouth (palate) were loosely connected. But in all the shield carriers, the skulls were far more rigid and the palate bones firmly connected to one another.

Altogether, the baron's hypothesis now seems a happy suggestion indeed. The armor-clad "suborders" probably were evolutionary cousins, descendants of one branch of dinosaurs that embarked on the adaptive path leading toward armored resistance, passive and active, against the threats of the meat-eaters. The Early Jurassic *Scelidosaurus* may be close to the Thyreophoroid stem. And so the baron's term, "Thyreophoroidea," should be resurrected as the appropriate label for this grand tribe of armored dinosaurs.

12

DEFENSE WITHOUT ARMOR

Throughout their entire history, dinosaurs and their prey co-evolved in a mutually stimulating arms race. A new defense plan among the plant-eaters would give rise to a new mode of attack among the meat-eaters. In the previous chapter we met the armor-clad tribes. Here we shall review the parade of unarmored plant-eaters and the evolution of their defensive equipment. These dinosaurs with naked hides defended their vulnerable bodies with slashing claws and lashing tails against wave after wave of meat-eating species.

The earliest wave of big herbivores evolved during the late epochs of the Triassic and Early Jurassic periods. These were the long-necked anchisaurs, distant uncles of the brontosaurs. Anchisaurs displayed no body armor, but they wielded huge curved claws on their powerfully muscled thumbs and long pointed claws on their stout hind feet. These plant-eaters therefore had defensive claws both front and rear, a combination unusual today. So wrestling with an anchisaur was a dangerous business. Modern anteaters have hooklike claws on their forefeet, while the most dangerous modern ground bird, the cassowary (a two-hundred-pound flightless creature from New Guinea), has a big hind claw. Together, anteaters and cassowaries demonstrate how anchisaurs fought. Living species of anteaters grow only up to 150 pounds maximum weight, but their hooked foreclaws are potent weapons of defense against

The first dinosaur panzer—*Scelidosaurus*. Early in the Jurassic Period, the scelidosaur clan evolved top and side armor composed of stout bony cones. The twenty-foot-long predator *Dilophosaurus* would have found it hard to deliver an effective bite against this defense (dilophosaurs had two thin bony crests running down their snout—probably a sexual advertising device).

jaguars. When angry and cornered, the anteater stands erect on its hind feet and tail and lashes out with left and right swings of its foreclaws. Knowledgeable zoologists take great care in the face of this attack, for if the anteater strikes full force in a vulnerable area, such as the stomach, its great claws can effect a full disembowelment.

Anchisaurs' tails were stoutly muscled and they could easily have reared up, foreclaws at the ready, to face their enemies. Anchisaur hind claws, especially the one located on the large inner toe, could lash out with even more powerful blows than the foreclaws. Cassowaries jump to strike with the full force of their massive thighs behind their long inner toe. Zoo keepers always treat cassowaries with the utmost respect—these birds are much more dangerous than their bigger cousin, the ostrich. And they are just plain mean, often attacking humans without provocation. Yet anchisaurs grew to much larger sizes than do cassowaries: a half-ton anchisaur could have unleashed a kick five times more powerful than can any cassowary.

The predators that threatened to attack throughout the Jurassic and Cretaceous Periods were the long-legged theropod dino-

Long-snouted *Coelophysis* attacks an anchisaur.

saurs. This clan is best known for *Allosaurus* of the late Jurassic Period and for *Tyrannosaurus* belonging to the last days of the Cretaceous. The very earliest theropod meat-eaters had appeared side by side with the anchisaurs and the other early species of dinosaurs during the Late Triassic.

From the earliest days of dinosaur hunting in the mid-1800s, these predatory dinosaurs, especially those from the Triassic, have constituted the most cherished discoveries of any field expedition. The reason is simple—they are quite rare. Over six seasons in the field digging for dinosaurs, I have personally seen only one predator skull, one battered predator backbone, and one predator claw. On average, in the Jurassic beds, one can't expect more than one *Allosaurus* skeleton at most per ten brontosaurs. This scarcity of predator remains is especially acute for the dawn of the Age of Dinosaurs, the end of the Triassic and beginning of the Jurassic. For over a hundred years paleontologists sought predator skeletons from this earliest epoch with disappointing results.

But two great discoveries during the last thirty years have provided us with a wonderful glimpse of the first predatory dinosaurs. The first was the grandest of all: not just one perfect skull, nor one complete skeleton, but a whole quarry filled with the complete and partial skeletons of one Late Triassic species, all preserved in the red mudstone of Ghost Ranch, New Mexico. Ned Colbert of the American Museum made this discovery, and under his direction the museum technicians have erected quite beautiful displays of these predators and have sent excellent casts of them to dozens of institutions throughout the international community of scholars. Colbert had stumbled upon a most unusual prize: a predator trap, a pocket of mudstone that formed in a peculiar locale where predators had huddled together in death. Predator traps constitute one of the most puzzling enigmas in paleontology. What would have attracted meat-eaters to one small spot a few hundred yards wide, and what had killed and buried them there?

The Tar Pits at La Brea, California, dating from a time late in the Age of Mammals, are the best-known predator traps and the best-studied. And they shed some light on Colbert's Triassic predator trap. La Brea is filled with saber-tooth cats and huge wolves, all jumbled together in tar-soaked sand which dates from about twenty thousand years ago. A few plant-eaters—mammoths, camels, horses, and others—have been quarried out of La Brea, but the overwhelming majority of bones are those of the big meat-eaters. La Brea seems to have been a deathtrap for wolves and big cats, acting much like a giant sticky flypaper surface whose tar-soaked sand entrapped the meat-eating mammals' paws in viscous asphalt, miring them until the exhausted beasts sank down and died. Dead and dying animals would attract more predators to the tar sweeps, unwary meat-eaters who thought they could get a meal with little effort. And each new victim would add to the lure.

Could Ghost Ranch have been such a flypaper trap? The site hasn't yet been analyzed sufficiently to yield any conclusions. The mudstone at Ghost Ranch did not yield the slightest trace of asphalt. But it's not impossible that sticky mud might have served to have the same effect, trapping dinosaur feet in a viscous, inescapable mire.

Colbert's splendid skeletons seem to belong to the same genus Cope had named from fragments in 1880: *Coelophysis,* roughly translated as "hollow-boned beast." Hollow it indeed was—all of the major limb bones and vertebrae were constructed like those of birds, with an outer shell of dense bone rind surrounding an empty core. So perfect are Colbert's skeletons that no guesswork is required to reconstruct these bodies. *Coelophysis* was small compared to its Jurassic nephews *Allosaurus* and *Ceratosaurus*; the fully adult length was only six feet, half of which was tail. Compared to those Jurassic predators, *Coelophysis* was long and slender in the torso and very long in the neck—the neck, body, and tail all seem to flow into one another to create an unusually smooth profile. Although it appeared early in dinosaur history, *Coelophysis* was already a birdlike biped with wide upper hip bones and deep lower hip bones, the whole design providing for ample thigh muscles and quick thrusts of the hind leg. The vertebrae in the neck were angled, producing a natural S-shaped curve, so the head was carried high above the shoulders as a bird's would be.

How did *Coelophysis* hunt? Its graceful yet strong neck could lunge forward for a quick snap at a small prey or for hit-and-run attacks against large prey. Teeth always provide the best biomechanical clues to the killing tactics. *Coelophysis*'s dental pattern was totally different from the killing apparatus we find in mammalian predators—wolves, leopards, and lions. When a wolf or cat bites, the four fanglike front teeth (canines) penetrate deeply into the prey. This bite is precise. The lower canine pair bites just in front of the upper, and the two pairs together lock the prey in a killing grip. Cats and foxes kill rabbits with one bite through the nape of the neck. Lions kill big prey—zebras and buffalo—by clamping their canines down on the throat and holding on until the prey suffocates. *Coelophysis*'s killing teeth were organized for a very different technique. Instead of two pairs of canine fangs, *Coelophysis* had a long row of small, curved, daggerlike teeth, each with the sharp, serrated edge both fore and aft characteristic of nearly all hunting dinosaurs. A bite from such an assemblage of teeth would have left a long, shallow wound across the prey's flesh.

Coelophysis's teeth were designed to slash through flesh, not to hold it. A cross section of this dinosaur's tooth shows a teardrop outline, with a blunt, rounded front edge and a tapered, sharply chiseled rear. When *Coelophysis* bit through its prey's hide, the blunt front edge prevented the prey from slipping away while

Elegantly designed meat slicer—the skull of *Coelophysis*. This lively Early Jurassic hunter epitomizes the light, flexible construction of paper-thin bony sheets and slender struts.

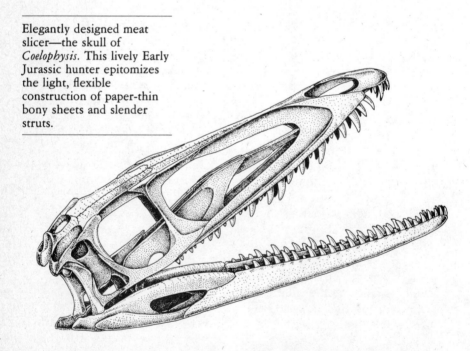

the sharp rear edge slashed through the flesh. The backwardly curved tips of the teeth assisted in driving the whole tooth row backward through the wound. Saw-toothed from top to bottom, the serrated rear edges were designed to assist this backwardly directed slash, since they would allow the entire tooth to saw backward through hide and muscle. But on the front edge the serration was only at the tip. The blunt base along most of the front of the tooth was smooth, and so would hold the prey as it struggled to free itself. All of the structural details were cunningly calculated to permit the tooth to act as both knife and fork, cutting and holding.

Some living species of monitor lizards have teeth like those of *Coelophysis,* and these lizards inflict long, jagged wounds when they bite. Komodo dragons, the biggest monitor lizards alive today, can even kill cows and people with the wounds they inflict. Since both monitor and dinosaur teeth curve backward, the jaw muscles must be arranged to pull the teeth rearward as the jaws

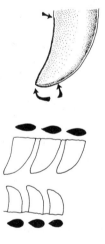

Coelophysis teeth worked like a combination fork and steak knife. All the teeth were backwardly curved blades with saw edges running along the entire trailing edge and the tip of the leading edge. In cross section (shown in black) the leading edge was blunt but the trailing edge was very sharp. Upper teeth were much larger and much sharper along their trailing edges than lower teeth. But lower teeth had stronger, blunter leading edges. So when the biting muscles contracted, the lower teeth held the prey and prevented it from slipping out of the mouth while the sharp upper teeth slashed backward, making a long, nasty wound.

Dinosaurian answer to the electric carving knife—*Coelophysis* biting mechanism. The huge hole in the snout bones was for the big muscle that powered the bite. A much smaller biting muscle was in the hole just behind the eye socket, and the jaw-opening muscle was strung from prongs sticking backward from the head and jaws. The big snout muscle was arranged to pull the skull down and backward (line of pull shown by the black arrow). So when the muscle contracted, the big upper teeth slashed back and downward toward the lower teeth.

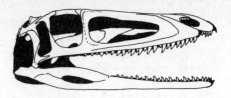

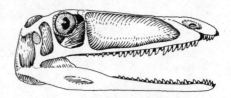

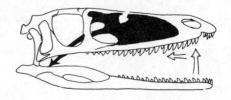

close. The skull and jaws of the lizards feature extra joints rather like a snake's to permit this. Yet *Coelophysis*'s killing bite must have been different from the lizards' in one fundamental way. The upper and lower teeth of Komodo dragons are the same size, but in *Coelophysis* the upper teeth were much larger than the lower. Consequently, more muscle power was required to pull the upper teeth back through prey than was necessary for the lower. The upper teeth were also more sharply edged, so they must have produced more of the cutting action, while the lowers did more of the holding. This sort of dominance of the upper teeth was a characteristic of the dinosaurs; *Allosaurus, Ceratosaurus,* and *Tyrannosaurus* all had much larger uppers than lowers.

The biological engineering behind *Coelophysis*'s bite can be worked out from well-preserved jaw joints and the muscle-attachment sites they reveal. The shape of *Coelophysis*'s teeth indicates the upper row of teeth had to move rearward relative to the lower so the bigger crowns of the upper teeth could be effectively exploited. Since it consisted of a pair of grooves that allowed two knobs on the skull to slide, *Coelophysis*'s jaw joint was indeed arranged so the skull could shift fore and aft in relation to the lower jaw, and the biggest jaw muscle pulled the snout backward. The strong neck would also assist in delivering a killing blow because *Coelophysis* could rake its teeth backward through its prey by retracting its neck muscles. Such contractions of the neck were amplified by the back of the skull, which was strengthened and enlarged to support larger and more powerful neck muscles.

From such evidence, the lethal interplay of predator and prey, anchisaur versus theropod can be fairly clearly imagined. *Coelophysis* stalks the Triassic floodplain, head held high, its large, birdlike eyes scanning the landscape for the slightest movement. A rustle in the conifer bushes betrays an anchisaur, a small, half-grown specimen some four feet long. *Coelophysis* strides to the attack. The anchisaur, its back to a dense stand of undergrowth, rises on its hind legs, brandishing its foreclaws. The combat is joined. *Coelophysis* dances, darting in and out in feinted strikes, weaving to avoid the dangerous counterstrokes from the anchisaur claws. An opening appears—perhaps a momentary error on the anchisaur's part. And *Coelophysis* lunges, its tooth-studded jaws raking some exposed part of its victim.

In a split second the entire series of the predator's jaw and neck muscles fire off in a spasmic, contractile sequence originating in instinctive action unguided by conscious thought. The anchisaur struggles to free itself but its efforts serve only to make the wound longer and more ragged.

The predator's dance continues, punctuated by more feints and quick raking strikes. No one bite is fatal. There is no quick coup de grâce like a lion's. But *Coelophysis*'s prey succumbs after a short time, weakened by trauma and loss of blood. Finally, the anchisaur sinks to the ground, unable to right itself, and *Coelophysis* finishes with a series of slashing bites to the neck just behind the head.

Coelophysis was not the only practitioner of this style of hunting at the end of the Triassic and beginning of the Jurassic. Samuel Welles of the University of California hunted in the red beds on a Navaho Indian Reservation and found several nearly complete skeletons of big predators, between fifteen and twenty-five feet long. Today, these are the earliest complete skeletons of large predatory dinosaurs known. Welles's animal, *Dilophosaurus*, "two-crested-lizard," exhibited a striking similarity to *Coelophysis* in its very long tail and elegant hind limbs. But the two-crested dinosaurs were proportioned for killing much larger prey, with their shorter, more massive necks and skulls and very much larger upper teeth relative to the skull's length. These dinosaurs were strong enough to attack any of the Early Jurassic herbivores, even the largest anchisaurs.

As the long Jurassic Period passed through its middle and late epochs, the dinosaur arms race produced more heavily armored herbivores—the stegosaurs—and the immense brontosaurs with enough strength in their legs and feet to simply crush most predators. Predator strength increased too; the Late Jurassic *Ceratosaurus* was thirty feet long, and *Allosaurus* forty-five. *Ceratosaurus* and *Allosaurus* were both discovered by Professor Marsh in the late 1870s. And for a long time only one ceratosaur's skull and only two or three complete allosaur skulls were known. Then, in the 1940s, a spectacular predator trap, containing ceratosaurs and allosaurs, was found at the Cleveland–Lloyd site in Utah. Sixty or seventy *Allosaurus* specimens at all stages of growth—young, adult, aged—have been quarried from this small area of mudstone. Jim Madsen, state paleontologist of Utah, directs the work at the quarry, and his practical experience with hundreds of predator bones endows him with unequaled expertise on the subject of predator anatomy.

I have spent several unforgettable weeks in Salt Lake City studying Jim Madsen's laboratory full of allosaur and ceratosaur bones. In this astounding treasure house every detail of their biomechanics stands revealed. A most unexpected characteristic of the skulls is how easily they fall apart. A fully adult *Ceratosaurus*'s skull, nearly three feet long in life, was not one tight mass of bones and teeth; it consisted of a loose kit of thin bony struts, flexible bony sheets regularly perforated by holes, ball-in-socket joints, and sliding articulations, the whole bound together with ligaments. After death, the ligaments of course soon rotted and the skull fell apart, scattering its pieces across the mud. Today's largest predatory mammals—polar bears and lions—possess a strong, unified cranial structure that remains solid long after death. Bioengineers who study skulls must consequently refashion their thinking when they seek to reconstruct the mechanics of the loose rod-and-sheet construction found in *Ceratosaurus* and *Allosaurus*.

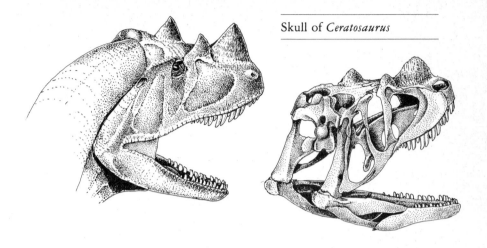

Skull of *Ceratosaurus*

There was a strong central core in the heads of the predatory dinosaurs: their thick-walled braincase. The term "braincase" is a misnomer in dinosaur anatomy, because in fact the brain of larger species was minute compared to the surrounding mass of bone. The primary function of the dinosaur's "braincase" was to provide attachment sites for the neck muscles and to serve as the foundation point for all the thinner, more flexible components of the snout, palate, and roof of the skull.

The biggest surprise found in Madsen's ceratosaur skull was the tooth-bearing bones of the snout. Instead of being firmly attached to the braincase, the tooth-bearing bones were only loosely bound to the top of the snout and the roof of the mouth. Such looseness is repeated all through this skull. The tall strut of bone (called the quadrate) which connected the lower jaw to the braincase shared a hinge joint with the top rear corner of the skull. When this strut swung outward, it splayed out the jaw to the sides. Even the lower jaw was loosely constructed of two sections. The front section carried the teeth, the rear housed the muscles and joint of the jaw. The front and rear complexes met along a quite loose ligamentous junction. At the dinosaur's chin, the right and left lower jaws met at yet another very weak joint held together by ligaments.

So much looseness was baffling to biologists who knew only the mechanics of our own Class Mammalia. If we humans had as

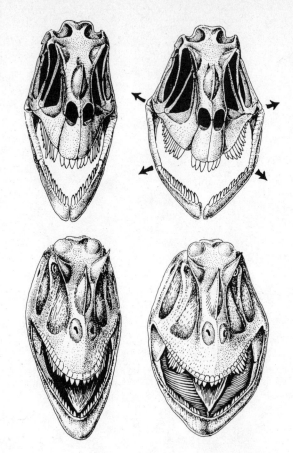

How to swallow something larger than your head—dinosaur-style. Face-front view of *Ceratosaurus*. All the bones of the skull's side were loosely hinged to the skull top, so the head expanded sideways when the beast swallowed an extra-large meat chunk. And a hinge in each lower jaw opened outward, just like a boa constrictor.

loose a skull as the ceratosaur, every time we bit down, our cheekbones would flex inward, the roof of our mouth would contract, and we would feel the rear of our skull swing toward the base of our neck. Anyone who has kept snakes as pets wouldn't be puzzled by ceratosaur heads. The heads of snakes are generally similar in design to those of the dinosaurs—snakes have a central, tightly knit braincase, which acts as the core for the loosely attached jaws, snout, cheek bones, and palate. Snakes also possess backwardly curved teeth, another similarity. When a snake starts to swallow large prey, the jaw muscles pull these teeth into the prey's body and all the loose joints swing apart so that the snake's gullet can accommodate a very large body. The *Ceratosaurus* must have functioned in very much the same way. When a ceratosaur swallowed a large chunk of meat, its capacity would have increased as each loose joint flexed and bowed outward.

The largest prey commonly available to *Allosaurus* and *Ceratosaurus* were the stegosaurs and the brontosaurs. Stegosaurs of course wielded their spike-and-plate armor, but at first sight brontosaurs appeared poorly armed. Most brontosaurs did have short, inwardly curved claws on their fore and hind feet, so the paws were potential weapons. But a more potent defensive weapon was located at the rear end of the whip-tailed genera like *Brontosaurus* and *Diplodocus*. The final ten feet of the tails of these dinosaurs featured slender bony rods in the core of the tail. When these huge dinosaurs swung their hugely muscled tails, the whiplash effect could inflict crippling wounds on an unwary predator.

The ultimate phases of the arms race between predator armament and antipredator adaptations were played out during the Cretaceous. *Allosaurus,* itself a Late Jurassic type, displayed the beginning characteristics of Cretaceous-style hunters, while *Ceratosaurus* represented the older predatory design, little changed from Early Jurassic days. The *Allosaurus*'s skull was more thickly boned than that of *Ceratosaurus,* and its jaws were deeper, providing for larger jaw muscles and a larger, stronger area for neck muscles. Not only was *Allosaurus*'s bite stronger, it was also faster on its feet. The allosaur's hind legs were longer and more compact than those of *Ceratosaurus*. And from an *Allosaurus*-type ancestor developed the last major group of big predators: the most strongly jawed, and fastest runners of all, the Tyrannosauridae of the Cretaceous.

In a glass case on the fourth floor of the American Museum of Natural History in New York resides the single most famous dinosaur head in the world—the *Tyrannosaurus rex* from Hell Creek, Montana. All the biomechanical trends started in *Allosaurus* culminated here. Primitive theropods like *Ceratosaurus* had teeth that were big but delicate and thin in section. *Tyrannosaurus*'s teeth were gigantic and very thick, capable of resisting exceptional forces when biting. Whereas the ceratosaur's head was a loose strut-and-ligament construct, *Tyrannosaurus*'s skull was one unified whole, very solidly constructed, with no moving parts except at the joint of the jaw. The compartments in the tyrannosaur's skull and in the lower jaw that housed the muscles were enlarged more than in any other predator. Its neck too represented an apogee of power. *Tyrannosaurus* had surrendered nearly all the primitive expansion points in the skull. But it compensated in the lower jaw, where the hinge between the front and back sections was much better developed than in the older predators. When *Tyrannosaurus* bolted down huge pieces of meat, the deep lower jaw flexed easily from side to side to widen its gullet.

Tyrannosaurus and its close kin *Albertosaurus* (named for the

Diplodocus defends itself with tail swipes at two allosaurs.

Canadian province) confronted the most heavily armored and armed adversaries—the tanklike nodosaurs and ankylosaurs and the dangerous horned dinosaurs already described. Part of any predator's advantage is the opportunity to make feints and lunging attacks. But such tactics require a good judgment of space and distance,

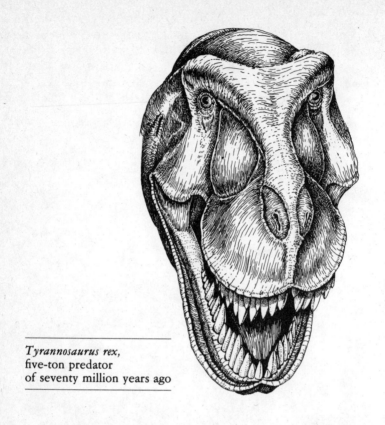

Tyrannosaurus rex,
five-ton predator
of seventy million years ago

and early predatory dinosaurs possessed very little depth perception, because their eyes faced directly sideways. The tyrannosaur's snout was sharply pinched to clear its field of vision. And its eyes faced forward to provide some overlap between visual fields from the right and left eyes. That would have permitted stereoscopic vision. Moreover, evolution had made additional improvements for attacking dangerous prey in the tyrannosaur's limbs. Its hind leg was much longer and more compact even than *Allosaurus*'s. And its torso was shortened to benefit balance and speed. Despite its great size—up to five tons—*Tyrannosaurus* was surprisingly slender-limbed, graceful, and fast.

All these evolutionary increases in the bulk of its jaw muscles and the strength of its limbs seem to demand that something be

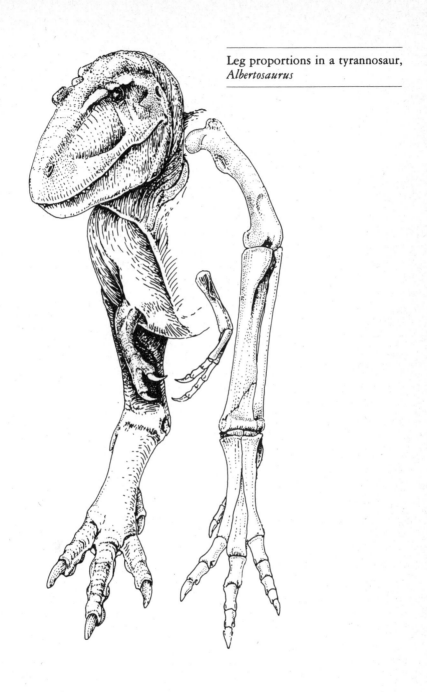

eliminated from the tyrannosaur's design. Forelimbs had to go. *Ceratosaurus* had had short but well-muscled forelimbs. *Allosaurus* had had shoulders of reduced bulk but still had had a strong hand and a fearsome claw on its thumb. Tyrannosaurs reduced their forelimb to such an extreme that it appeared useless, or nearly so. A thirty-foot *Albertosaurus*'s arm was shorter than a man's, and most of the muscle-attachment processes were subdued. So the hand was not only short, it was weak. Strange as it may sound, any average adult human could have won an arm-wrestling contest with a five-ton *Tyrannosaurus*.

Additional weight was saved in the tyrannosaur's hind foot. Very early predatory dinosaurs had had strong claws on the three main toes. But the tyrannosaurs reduced both the size of the claws and the bulk of the tendons and muscles supporting them. Their feet were thus adapted for running and dodging, avoiding counterattacks from the spikes, tail clubs, and horns of their prey. Strong hind claws might have been useful weapons but their weight would have detracted from speed and nimbleness. *Tyrannosaurus* surrendered the attack function of both the hind and forefoot in favor of a concentrated mass of muscles and power in the neck and head.

A final mystery looms large in the story of predator and prey. At present I can offer no solution for this and neither can anyone else. In most places, the most common, large plant-eaters of Late Cretaceous days weren't the heavily armed horned dinosaurs or the armor-clad ankylosaurs. Most common were the naked-skin duckbills, which lacked any sort of obvious defensive weapons. Duckbills had no whiplike tails, long claws, or any type of spike or plate. And their limbs were shorter and designed for lower top speeds than were those of their gracefully long-legged hunters. How ever did duckbills escape their enemies? To date, no one knows. But I am convinced some young paleontologist, perhaps someone reading this book, will one day solve this enormous riddle.

13

DINOSAURS TAKE TO THE AIR

Seventy million years ago a dragon of the air stretched its membranous wings over the Texas delta. Forty feet from wingtip to wingtip, this aerial leviathan possessed a wingspan greater than some twin-engine airliners and was three times wider than the greatest living bird, the Andean condor. The fossil annals in the Texas rocks yield an image as marvelous as any fabrication of the human imagination. Petrified wing bones, vertebrae, and jaws make it possible for us to envision the largest flying creature produced by evolution.

Flying dragons entered the sphere of human knowledge not as giants, but as tiny winged skeletons from the fine-grained limestone of Bavaria. There, quarrymen hewed out slabs carefully, because this limestone was ideal for making lithographic slabs, plates of stone used to print drawings of delicate and subtle tones. The salty Jurassic sea that laid down these limestone beds in shallow bottoms behind reefs preserved tiny skeletons with exquisite fidelity, because the hypersalinity discouraged the activity of scavengers. As the layers of lime accumulated, bodies of horseshoe crabs, shrimp, and insects settled to the bottom and were entered into the ever-growing record. Occasionally, a tiny leather-winged form would fall into these briny waters, and 150 million years later, during the nineteenth century, quarrymen could send a slab containing the raven-size bones to a nearby German scholar. Before long this skeleton reached Baron Cuvier in Paris.

Quetzalcoatlus buzzes two tyrannosaurs.

Cuvier's felicitous name for this creature was *Pterodactylus*—literally translated, "wing-finger."

With the stroke of his pen, the baron evoked a creature without precedent in human knowledge: a large-headed aerialist that supported its batlike wing membranes on a single, elongated finger in each wing. This single finger alone was longer than the creature's head and body. Its long jaws set with small teeth and the general cast of the skull proved that Cuvier's *Pterodactylus* was neither bat nor bird but *sui generis,* a unique and totally extinct order of organic creation.

Cuvier decided *Pterodactylus* was closer to the crocodiles than to any other living family and thus was born the term "flying reptile." No other creature resurrected from the rock by the baron's scholarship gripped the public imagination more than *Pterodactylus* and its leather-winged kin. Nineteenth-century engineers and inventors regarded flight as the highest form of locomotion in the *Scala Naturae,* and they wistfully scrutinized bats and birds as attempt after attempt to build a flying machine failed. But there in the Jurassic strata was a "reptile," a member of the lower vertebrate class, possessing a breastbone keeled for flight muscles and arms designed for powerful flapping. "Flying reptile" seemed a contradiction in terms—by definition reptiles were crawling things, condemned to slinking across the surface. But Cuvier's *Pterodactylus* tore a veil from the present, revealing the unexpected achievements of the Reptilia past.

Early reconstructions of pterodactyls depicted them as dark-hued animals of nightmarish aspect. Even in this century pterodactyls have been cast as villains in prehistoric drama—the oversized wing finger that tried to make off with Fay Wray in *King Kong* gave Kong his chance to show his chivalry in the rescue. But dark colors and darker character were entirely inappropriate for flying reptiles. With few exceptions—the Texas giant is one—the aerial dragons habitually flew over shallow regions of the sea. As hunters of fish and squid, they were therefore the equivalent of shorebirds in today's ecosystem. And shorebirds are rarely somber in plumage. As a group, pterodactyls probably sported the camouflaging color scheme common among shorebirds, a dark topside to hide it from bigger pterosaurs attacking from above, a white bottomside to hide it from prey in the water below. A prob-

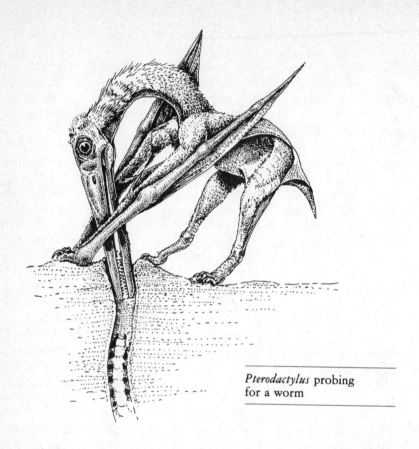

Pterodactylus probing
for a worm

able color pattern for Cuvier's *Pterodactylus* would be puffinlike. When the flying reptiles are portrayed in seabird tones, these Mesozoic fliers lose their malevolent aspect and become positively handsome.

Pterodactyls should evoke awe. But in the most commonly used twentieth-century paleontology textbook, these noble creatures were described as failures in everything they did. They couldn't have flown because, it was asserted, their wing membranes weren't stiff enough and were too crudely controlled by the reptile's muscles. A single finger was deemed far less efficient as a support for the wings than the four fingers bats employ to stretch out their flying surface. It was postulated that pterodactyls couldn't have flapped at all because their wing surface had been too flaccid. Furthermore, the experts decided that pterodactyls were accident-prone. Since there existed no stiff anatomical structure within the wing to prevent a tear from running right across the entire surface, pterodactyls were supposed to have been vulnerable to snagging on branches or rocky outcroppings. Even on the ground pterodactyls were portrayed as clumsy locomotor machines, incapable of walking normally either on two legs or four. All told, the mid-twentieth-century portrait of the pterosaur was wretched: a flying creature that managed to get into the air only when the wind was precisely right, permitting its underpowered, floppy wings to glide passively on updrafts. According to the orthodox theory, these flying reptiles survived only because there was no aerial competition. And when flying birds finally did evolve in the Cretaceous, their elegantly designed feathered wings were so manifestly superior to those of pterodactyls that the avian tribes quickly replaced the obsolete harpies of the Mesozoic.

Nineteenth-century scholars had more confidence in pterodactyl's design. Baron Cuvier certainly believed his little wing-fingered beast could fly. Mid- and late nineteenth-century students of flying reptiles had faith in pterodactyl's landing gear as well. During the last century, many restorations were conceived, showing *Pterodactylus* running successfully about the land on all fours, its long fingers folded back from the wrist raising the wingtips back over the hips. Some modern tropical bats—the South American vampires especially—move in this way and can be quite mobile on the ground. Which view is closer to the truth, that of the nineteenth century or that of the mid-twentieth?

In the 1970s new fossils and fresh studies of the old specimens began to rehabilitate the pterodactyl's image. From the very same Bavarian quarries that yielded the first *Pterodactylus* came specimens with the wing membranes preserved in perfect detail—discoveries that suggested that the theory of the flaccid membrane was wrong. The newly cleaned specimens confirmed that the wing tissue had not been weak, unsupported skin at all. In life, long, stiff fibers of connective tissue had stretched across the wing, and were probably attached to muscles controlling the tension of the wing's surface. Pterodactyls thus were equipped to fine-tune the

Dimorphodon feeding on squid

shape and camber of their wings. A Ph.D. thesis completed at Yale in 1979 argued forcefully that pterodactyls possessed the equipment for flight under their own power; the deep keel of the breastbone showed that the "white meat" muscles were as large relative to the body's size as are those of many flying birds today. It is true that the joints at the pterodactyl's shoulder, elbow, and wrist were not identical to the birds', but the flying reptile certainly could have executed powerful up-and-down strokes with them, and the muscle processes along the arm bones were huge.

In fact, as nineteenth-century anatomists had pointed out, pterodactyls were more fully committed to an active aerial way of life than any modern bird or bat, with the possible exception of swifts or hummingbirds. Every section of its anatomy evinced the drastic remodeling performed by evolution in order to transform the pterodactyl's terrestrial ancestor into a consummate aerialist. Shoulder bones had been reshaped so that the shoulder socket, which usually faced rearward, faced forward and outward like a bird's. Why should the joint have been so totally reorganized unless pterodactyls were actively flapping? In addition, if a powerful rhythm of muscular contractions did propel the pterodactyl's wings up and down in the figure-eight pattern required for active flight, then we would expect the shoulder to be very firmly braced against the torso. And so it is. The pterodactyl's entire torso was highly compact from front to rear and the whole was reinforced by two rigid bony girders. Where the shoulder blade touched the backbone, the shoulder abutted the anterior bony girder, composed of a half-dozen vertebrae firmly stuck together. This girder was a naturally evolved back brace that could support the enormous stresses of the beating wing. In front of the hip joint, the right and left hip bones were fused to another long set of vertebrae, constituting a second back brace. Together, these shoulder and hip braces made the pterodactyl's torso a light but incredibly strong boxwork of bony struts, exceeding in strength the body of the most modern birds. All this evolutionary modeling in the pterodactyl would have made no sense if these creatures had been merely passive gliders. Great strength in the bony frame evolves to resist great forces—in this case, the forces of active and strenuous flapping.

Bird skeletons delighted medieval anatomists because of their lightness and economy. The bones of most flying birds are of a tubular-strut design. All the major limbs are cast in a thin-walled, hollow construction. Just so were the pterodactyl's bones designed. Even the apparently massive upper arm bone (humerus) of the gigantic Texas pterodactyl had only an outer shell of very hard bone a few millimeters thick. And just as avian bones achieve their greatest lightness by being filled not by marrow but by a core of air sacs connected to the lungs, likewise the pterodactyl's bones are constructed to contain air-sac liners. Though lung tissue itself is never preserved in fossils, the presence of air sacs can be detected from the characteristic pores in the bony walls which provided entrance for the air canals. Running one's finger over the smooth-edged pore in the arm bone of a great aerial dragon is like feeling the fossil breath of the giant, now long gone. Through these pores surged the oxygen-rich air each time the stout basketwork of the ribcage drew the Mesozoic atmosphere through the dragon's nostrils.

Arm power was wing power in pterodactyls, as it is in birds and bats. The pterodactyl's upper arm bones were short compared to birds', but its forearm and wrist were longer—a difference that must surely reflect an as yet undiscovered contrast in the mechanics of upstroke and downstroke. Strong fliers among today's birds put greater stresses on the upper arm than on any other bony component. In these birds the humerus is the largest bone in the skeleton. And the distribution of stress in a pterodactyl's skeleton can be discerned by scanning the patterns of girth in the bones— the humerus is always the thickest, usually twice the girth of the thigh. There's no ambiguity here; pterodactyl's evolutionary transformation had concentrated nearly all the body's strength in the flight organ.

Pterodactyl's wrists are evolutionary chimeras, combining mechanical features found today in two or three different animal families. The long bones of the wrist were tightly bound into a single bundle, much like the wrist of the rabbit. This strong wrist structure certainly could have functioned for running and hopping on the ground. Anatomists number fingers from the thumb outward. Counting in this fashion, the pterodactyl's enormous wing finger was number four. But fingers one through three were specialized for grasping, not flying. Each of these three inner fingers was short and flexible, ending in a sharp claw, which was deep from

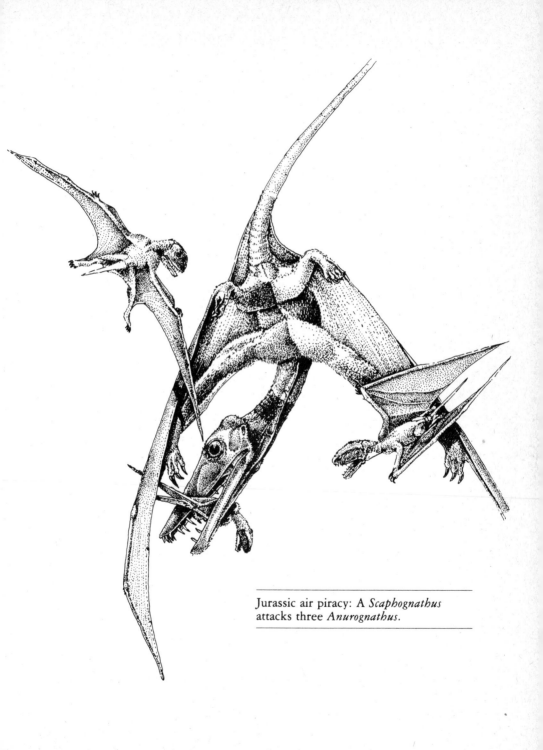

Jurassic air piracy: A *Scaphognathus*
attacks three *Anurognathus*.

top to bottom, thin from side to side. This type of claw proves that pterodactyls could cling to trees. Among living mammals, such a deep, narrow claw shape is the preferred device for clinging in species that roost in trees—the most specialized arboreal squirrels are so equipped. And even closer to pterodactyls in claw pattern is the strange Indonesian "flying lemur," a glider with a wide skin membrane between its front and rear limbs. Flying lemurs use their deep, sharp claws to cling firmly to the back of tropical tree trunks. Pterodactyls must similarly have gripped the trees of Mesozoic forests when they rested from the day's hunt.

The construction of pterodactyl's hind limbs also proves that all its body mechanics—not the front limbs alone—were subservient to its flying organs. Pterodactyl's thighs and shins were long and slender. When hopping, the knee joint's geometry imparted a strong natural bow outward to the legs. If pterodactyls had to cling to trees and cliff faces to roost and breed, then they would also have required unusual mobility in their thighs for maneuvering around tree trunks, branches, and rocky crags. And, indeed, pterodactyl's hips and thighs were the most mobile of any Mesozoic vertebrate's. They mimic the extraordinary flexibility developed independently by modern bats. The hip socket of these animals was shaped as a nearly perfect ball-in-socket, circular in outline— a most unusual configuration. The surfaces of the thigh bone, which fit into that socket, permitted the leg to swing in all directions. Finally, the hind foot's claw matched the forefoot's in shape and gripping strength. So pterodactyls were both agile and strong as they scampered over their roosting sites above the Mesozoic ground surface.

After reading Professor Seeley's *Dragons of the Air,* first published in 1901, I am at a loss to explain why the popular textbooks of the 1940s and the 1950s portrayed pterodactyls as so faulty in design. Seeley summed up his thirty years of firsthand study by describing pterodactyls as precision-crafted flying machines. What could account for the later change of view? The best guess is probably that after 1920, few careful scholars interested themselves in pterodactyls. Graduate students were steered away from the Mesozoic monsters of any sort and into more respectable fields—such as the horse's evolution, fossil beavers, and the tiny mammals of the Mesozoic. The textbook writers drifted into the view of pterodactyls as inefficient, influenced by the generally anti-dinosaur atmosphere among English-speaking scientists. Nowadays the flying dragons are enjoying a renaissance. Innovative and vigorous young scientists in China, Europe, and the Americas are discovering winged dragons of unexpected sizes and shapes. The reappraisal of the pterodactyl's flying prowess is returning to the nineteenth-century view.

Several years ago I began my own investigations of pterodactyls' prey-catching devices. Most studies, naturally enough, have concentrated on their flight mechanics. But the pterodactyls' heads and necks spin an intriguing tale of flying dragons altering their hunting tactics through evolutionary time.

Sharp fangs and large jaw muscles marked the head design of the most primitive pterodactyls. The muscle arrangement resembled those found in many of the primitive small birdlike dinosaurs: the jaw muscles of the skull compartment behind the eye were small, but those in front of the eye in the snout were large. The strongest bite belonging to any pterodactyls must have been owned by the big-snouted *Dimorphodon,* described with loving care in 1840 by the superb Victorian anatomist Sir Richard Owen. Until the discovery of this fierce-looking species from the black shales of Lyme Regis on the Dorset coast of England, all the pterodactyls had been found in France and Germany. But Owen's research located *Dimorphodon* (the name means "beast with two sizes of teeth") in the key position near the base of pterodactyl's family tree, a position it occupies in modern opinion. *Dimorphodon*'s sharp teeth jutted directly upward from the lower jaw and directly downward from the upper. Therefore its strong bite must have been delivered by a quick, simple snap of uppers and lowers together.

But *Dimorphodon*'s prey-catching devices weren't limited to that simple snap of the fangs. *Dimorphodon*'s neck was also constructed to deliver rapid lunges. All pterodactyl necks were long and graceful, and all had joints between the vertebrae of the neck which allowed their owners to hold their head and neck in an S-shaped curve. Since they all possessed large skulls compared to their body size, this S-shaped curve allowed them to fly with their heavy heads held far back over their shoulders. And this posture permitted better distribution of weight both for flight and for walking, exactly as modern pelicans tuck their big heads over their shoulders

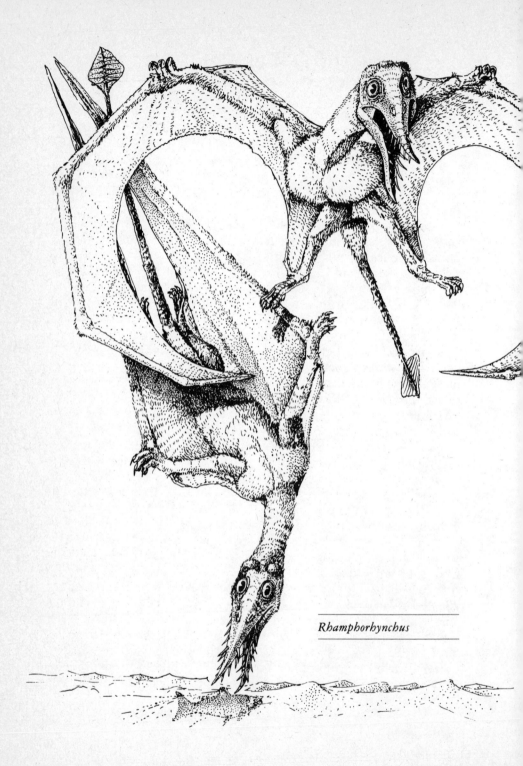

Rhamphorhynchus

to gain better balance. But the S-shaped neck also provided *Dimorphodon* with the ability to lunge; it could rapidly flip its head and neck forward to make a quick grab at prey.

Other early pterodactyls shared the *Dimorphodon*'s basic design but with variations. One Italian species had three-cusped molar teeth interspersed with tall fangs—a very strange arrangement—crowded together along its jaws. The entire row of teeth could slice fish into strips in a matter of seconds. Swallowing big prey is a challenging biomechanical problem for today's shorebirds, because most can't easily tear up fish carcasses with their beaks, and speed is of the essence because there are always thieves around trying to run off with the prey. Most birds solve this problem the same way dinosaurs did—the joints of the skull and jaws expand sideways to enlarge the gullet's capacity. Pterodactyls faced the problem but couldn't expand the rear of their jaws because the skull's bones were too tightly knit. However, at the halfway point of their lower jaws there was a zone of weakness that might have allowed the jaws to bow outward so a large fish could slip down the throat.

Fish and squid are tricky prey for an aerial hunter. These quick-moving, slippery creatures can detect a pelican or puffin's dive just before it strikes the water, and the whole school of them may scatter in all directions. The pterodactyl's hunting tactics evolved to maximize the quickness of its strike. In the lustrous, lithographic limestone of Bavaria are preserved the Late Jurassic squadrons of flying dragons arrayed with a wide variety of head and body shapes. Still in evidence are the straight-toothed biters similar to *Dimorphodon* in design. But other tribes exhibit features for a tactic newly evolved among flying reptiles—spearing with the head. Most numerous of the tern-sized Bavarian pterodactyls is spear-headed *Rhamphorhynchus* ("beaked-jaws"). The S-shaped neck of primitive pterodactyls was accentuated in this animal so that the head could be carried coiled tightly against the shoulders. *Rhamphorhynchus* had jaws and teeth shaped exactly like the fishing spears used by some Amazonian Indian tribes today. Long, sharply tapered teeth in both its upper and lower jaws were bent forward, so all the points would converge to form a thrusting fish trap. Even the tip of the snout and chin tapered to deadly points to form the apex of the spear. Amazonian Indians hurl their spears at the heads of the fish: the intermeshing set of points snags the fish's body, and its struggles to escape only serve to drive the barbs in more deeply. Just so *Rhamphorhynchus* could dive toward a fish, suddenly uncoil its S-shaped neck flexure, and hurl its spear-shaped head at its prey, impaling a hapless fish in the intermeshing barbs. This aerial fishing spear must be ranked as one of the most effective fish traps ever to evolve.

Baron Cuvier's *Pterodactylus* hunted for its daily ration in the same reef-fringed waters as *Rhamphorhynchus*. But it had a totally different apparatus for snaring prey. Extremely long, gently tapering jaws terminated in a cluster of short, straight teeth. *Pterodactylus*'s jaws looked just like the barbed tweezers used to manipulate squirmy invertebrates in today's zoology labs. And quite possibly *Pterodactylus* was an airborne worm tweezer. It may well have probed the sand flats like a Jurassic sandpiper, poking its long snout into the burrows of polychaete worms, shrimplike crustaceans, and sand fleas.

There's excellent evidence that one rather rare Argentine species, the bristle-toothed pterodactyl, pursued a flamingolike style of life. Modern flamingos derive their pink coloration from the pigments stored in the tiny shrimplike creatures they filter from the shallow salty waters. The shrimp, in turn, get this pigment from tiny algae that they filter through their leg bristles. Captive flamingos fade to off-white when given prepared zoo food, much to the disappointment of curators and public alike. Fortunately, the natural pigment can be replaced by simple food coloring (the same kind used to dye Easter eggs) added to the flamingo's diet, so most zoos can keep their birds in the pink. Shrimplike crustaceans are a very ancient group, as are the red algae that are the ultimate suppliers of the pigment. Salty pools must have hosted red algae blooms in Jurassic days exactly as they do today. Then, as now, both algae and shrimp were an excellent source of food for any larger animal equipped to sieve them out of the water through an anatomical strainer. Was there, then, a pink strainer pterodactyl? Probably. The Argentine pterodactyl in question possessed a flamingo-shaped mouth with a dense row of thin, bristlelike teeth. Without question this bristle-toothed pterodactyl pumped water through its mouth with its tongue, straining out tiny food particles in the process. And since blooms of red algae were common in

Rhamphorhynchus fishing technique

briny water, it's reasonable to suppose it would often filter both algae and shrimp, and behold, a pink pterodactyl!

Flamingos appear especially awkward when they are hard at work feeding because their filtering bristles are in the upper jaw and their head must be upside down to perform its function in the water. Bristle-toothed pterodactyls didn't have to perform handstands, because their filtering apparatus was located in the lower jaw, so the head could be lowered right side up into the water. Curiously enough, flamingos and bristle-toothed pterodactyls aren't the only examples of algae-straining aerialists to evolve in the history of life. Forty-five million years ago a long-legged duck with flamingolike bristles in its lower jaw waded through the salty lakes of Wyoming, Colorado, and Utah. Fossil beds laid down in these briny lakes preserve the skulls, skeletons, footprints, and even some mound-type nests and the eggshells of the bristle-beaked duck. These three filter feeders are an extraordinary example of how evolutionary processes can shape unrelated clans into one and the same specialized ecological mode.

While on the subject of color, *Dimorphodon*'s snout deserves comment. Seeley reconstructed this animal with a dark snout. But more likely *Dimorphodon*'s face was positively gaudy. Its high-snouted profile was more like a puffin's than that of any other bird, and puffins employ their tall beaks to advertise their social status. Juvenile puffins start out quite drab in the snout, but adults are marked by faces run riot with white, red, and orange stripes and splotches. *Dimorphodon*'s deep snout cannot be explained by any hypothesis involving its jaw muscles or teeth. It's quite likely that *Dimorphodon*'s snout evolved its unique high contours to advertise its owner's rank in pterodactyl society. Since pterodactyls were highly visual creatures, with large eyes and bulbous optic hemispheres in their brain, it's very probable in fact that colorful devices evolved many times among the various branches of the family

The flamingo pterosaur
—*Pterodaustro* from
Argentina

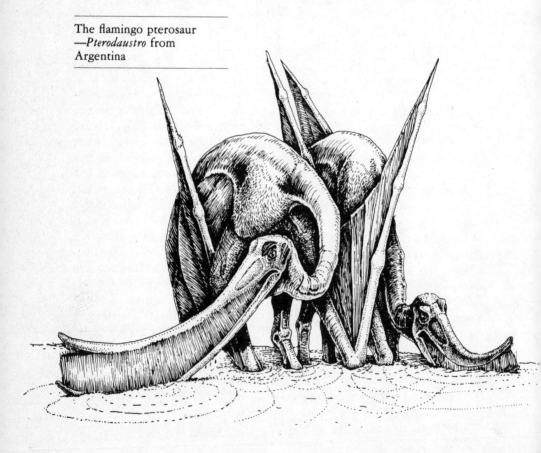

tree. Of course there is no direct evidence allowing reconstruction of the color pattern for any particular species of pterodactyl, but their color must have evolved to brighten Jurassic and Cretaceous skies in many ways.

Everywhere in the world's ecosystems the transition from the Jurassic to the Cretaceous was marked as a time of disaster, disturbance, and extinction. Flying dragons did not escape this vast ecological shake-up. The experts divide all pterodactyls into two great tribes: the long-tails and the short-tails. After the Late Jurassic extinctions wiped out most of the previously dominant long-tails, the short-tailed species moved in to fill the Cretaceous skies. *Rhamphorhynchus,* with its fishing-spear head, was a long-tail; Baron Cuvier's *Pterodactylus* was a representative short-tail. By and large the long-tailed species did have fairly long tails—and at least some of them possessed kitelike tail rudders. Professor Othniel Marsh of Yale bought a superb *Rhamphorhynchus* from German fossil dealers in the 1880s and subsequently announced to the envious Europeans that his skeleton possessed a tail rudder, previously unknown. *Rhamphorhynchus* carried a vertical diamond-shaped fin at the end of its very long tail. The fin consisted of tough skin reinforced by rods of connective tissue. The entire tail could be energetically swished by muscles at the base of the tail. The precise aerodynamic effects of this intriguing equipment aren't yet understood, but the kite-tailed pterodactyl must have exercised precise control over its maneuvers, at least at slow speeds.

Short-tailed pterodactyls generally had more specialized skulls, longer necks, and longer forearms than their long-tailed forebears—implying a fundamental change (still not well understood) in flight mechanics. This short-tailed Cretaceous dynasty certainly won an undisputed place in the book of aerial records, for it included the largest flying creatures ever to evolve. Marsh made headlines in the 1880s when he announced short-tailed pterodactyls from Kansas with wingspans of twenty feet or more. But the mind-boggling pterodactyl was yet to come. In the 1970s, Professor Wann Langston led teams from the University of Texas into the scorching badlands of Big Bend National Park where the Rio Grande makes its huge loop on its way to the Gulf of Mexico. When Langston discovered a Cretaceous pterodactyl at Big Bend, its upper arm bone measured twice the size of the next-largest known, and its jaws indicated a head eight feet long. Preliminary reconstruction, based on the wing plans of smaller species, produced an estimated wingspan of up to twenty meters—sixty-three feet, greater than the wingspan of the old twin-engine DC-3 airliners. *Quetzalcoatlus* was featured on the cover of *Science,* the most widely read scholarly journal in the United States. Immediately after their paper came out in *Science,* Wann Langston and his students were attacked by aeronautical engineers who simply would not believe that the Big Bend dragon had a wingspan of forty feet or more. Such dimensions broke all the rules of flight engineering: a creature that large would have broken its arm bones if it tried to fly. Quite a flap erupted over whether the Big Bend pterodactyl could even have powered its wings in the up-and-down strokes necessary for active flight. Under this hail of disbelief, Langston and his crew backed off somewhat. Since the complete wing bones hadn't been discovered, it was possible to reconstruct the Big Bend pterodactyl with wings much shorter than fifty feet.

I believe Langston and his Texans were right—the Big Bend aerial leviathan was stupefyingly large. Mechanical engineers go often astray when analyzing the strength of skeletons. The most common difficulty with their method is that they calculate the strength of an arm bone as though the bone by itself had to withstand all the stresses of flapping the wings. If the pterodactyl were a man-made machine, the wing skeleton would indeed bear all the stress. But naturally evolved arms are far superior to mechanical ones. The bundles of muscles sheathing the arm bones of birds or humans contract to reorient stresses when the body is exercising vigorously. Such contractions are automatic, since the muscles are sent a constant flow of orders from the posture-control centers of the nervous system. Therefore a live *Quetzalcoatlus* was stronger than an engineering analysis of its bones might indicate. Moreover, calculating the stresses in a sixty-foot pterodactyl's wing is also subject to extreme variation—the Big Bend animal may well have flapped its wing from the wrist and not from the shoulder, for example. And in fact until the joints of the wing are clearly understood, any attempt to calculate stresses remains dubious at best. In general, I believe it dangerous to argue *a priori* that *Quetzalcoatlus* couldn't have been as big as seems indicated. The theories of bioengineering relating to flight in live mammals are

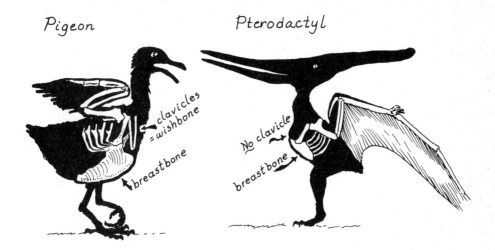

Pigeon

Pterodactyl

clavicles = wishbone

breastbone

No clavicle

breastbone

still too crude to yield anything more than imprecise boundary conditions that set limits only on the most extreme possibilities. Based on the proportions from the wings of other Cretaceous pterodactyls, the best estimate of the wingspan for the Big Bend dragon remains, in my opinion, the original fifty feet plus.

Cretaceous pterodactyls from North America are notable not only for their size but for their flamboyant head crests as well. *Pteranodon* ("wing without teeth"—a reference to the toothless beak), Professor Marsh's big Kansas specimen, had a long, narrow, bony prong sticking out rearward from the top of its skull. What was the function of this extraordinary cranial ornament? Some have suggested this prong was a sort of rudder; others that it was a bony banner for display and intimidation. Closely related species that are very similar in body outline have little or no crest, so this problem is complicated. And no bird, living or extinct, possesses anything even remotely similar to *Pteranodon*'s headgear.

On the subject of pterodactyls, two questions are enjoying considerable debate: their warm-bloodedness and their relation to the dinosaurs. Professor Seeley summed up the nineteenth-century view: If pterodactyls flapped actively during flight, the heat generated by their muscles would have warmed their bodies to temperatures higher than that of the air. Seeley was almost certainly correct (he usually was).

Recently developed measuring devices such as those sensitive in the infrared and ultra-violet allow zoologists to measure heat production and body temperature in all sorts of creatures engaged in various exercise. The various heat detectors and oxygen analyzers placed on creatures of small size have yielded startling results. Nineteenth-century zoology reckoned "warm-bloodedness" as the highest level of adaptation, reserved for the top rungs of *Scala Naturae*. Birds and mammals were clearly warm-blooded, snakes and insects clearly weren't. But this view was wrong, as the delicate apparatus of the late twentieth century reveals. Hawkmoths are powerful, nocturnal flying insects whose torsos are covered with dense, hairlike scales. Elegant experiments show that hawkmoths heat themselves with their own flight muscles. Before they begin their mighty flights, hawkmoths send shivers of contraction through their powerful flight muscles, generating waves of body heat. After the moth has raised its temperature above that of the air, it takes off—a warm-blooded, fur-covered flier. As long as it keeps flying, the hawkmoth keeps its body temperature high through the heat of its own movements. Powered flight requires a high, continuous output of energy. Since pterodactyls definitely were powered fliers, it's reasonable to suppose that Baron Cuvier's *Pterodactylus* warmed its own flesh with the heat of its own exertions as it flapped through the Jurassic skies.

There's one serious stricture against Seeley's view that pterodactyls were self-heating. Small animals—such as *Pterodactylus* and most of its Jurassic kin—lose heat rapidly through their skin. Hawkmoths succeed as self-heating fliers because their flight-muscle compartments are insulated by the outer covering of hairlike scales. The small pterodactyls would have required a coat of hair, or some good substitute. However, most paleontologists have assumed, since pterodactyls were classified as "reptiles," that they had a naked, scaly skin. Such images certainly dominated the restorations of pterodactyls until 1970, when a startling report arrived from a Russian paleontologist: Hairy pterodactyls had been found in Russia. Professor Sharov had been engaged in separating the slabs of Jurassic lake beds which preserved delicate leaves and insects. In one split slab lay a pterodactyl—not unusual in itself. But highly unusual was the near-perfect preservation of the pterodactyl's body covering—a dense coat of long, hairlike scales.

Sharov's pterodactyl had been as warmly clothed as any hawk-moth.

Other specimens have confirmed Sharov's discovery. Flying dragons as a group seem to have been insulated. Were they alive today, it is very much to be doubted whether biologists would place them in Class Reptilia. The dragons of the air lived their lives with adaptations beyond the traditional limits of the term "reptile."

How, then, were the pterodactyls related to dinosaurs? Pterodactyls show some clear signs of sharing a common heritage with crocodiles, dinosaurs, and birds; for example, their snouts included a large hole in the bones of the face, a hole filled in life by the anterior jaw muscles. Most authorities on the history of vertebrate evolution place pterodactyls as uncles of the dinosaurs—not ancestors of dinosaurs, but relatives of their ancestors. I suspect, however, that the true relationship between pterodactyls and dinosaurs was far more intimate. A few years ago, José Bonaparte, a very sagacious Argentine paleontologist, published a keynote paper about a small beast from the Triassic beds of Argentina. This graceful creature was named *Lagosuchus* ("rabbit-crocodile") on account of its long, rabbitlike legs. (The Greek root *suchus* shouldn't be taken too literally as crocodile; in scientific jargon *suchus* is used for any sort of reptilelike creature and has even been applied to some froglike amphibians.) Bonaparte pointed to one feature of the rabbitcroc especially reminiscent of all flying dragons: the *Lagosuchus*'s neck was long and had joints arranged to produce a natural S-shaped curve. True crocodilians never have such a sigmoid flexure in their necks, and this adaptation, considered with *Lagosuchus*'s light overall build, makes it a good candidate for proto-pterodactyl. It's possible to imagine rabbitcrocs bouncing over the landscape, scurrying up trees, leaping from branch to branch, and—just maybe—evolving a wing membrane.

Rabbitcrocs offer other evidence of their potential ancestral status, not just for pterodactyls but for dinosaurs too. Early dinosaurs of all sorts had the S-shaped neck posture. *Lagosuchus* also had a head shaped like that found in both early dinosaurs and early pterodactyls, particularly in the way the supporting strut for the lower joint of the jaw is arranged.

José Bonaparte was kind enough to send me a copy of his paper before it was published. I was surprised at how similar our ideas

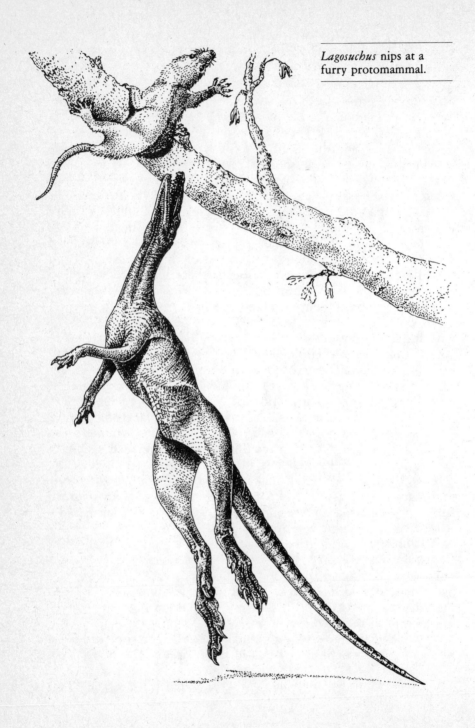

Lagosuchus nips at a
furry protomammal.

about the pedigree of dinosaurs were—surprised because both of us had independently concluded that the orthodox view of the dinosaur ancestry was incorrect. According to orthodoxy, the two great dinosaur clans, the beaked dinosaurs and the meat-eaters, had evolved from quite different ancestors (brontosaurs supposedly evolved from early meat-eaters). By 1920 this view had gained wide acceptance, although the issues were never debated thoroughly. José Bonaparte proposed that the truth was in fact closer to the older nineteenth-century idea that all dinosaurs were one natural group derived from the same ancestor. And, according to Bonaparte, *Lagosuchus* was that ancestor. I had already come to the same conclusion. In the dinosaurs' family tree, *Lagosuchus* was the ultimate evolutionary grandparent, and, therefore, deserved the label of "first dinosaur."

Could it be, then, that *Lagosuchus* was the ancestor of both dinosaurs *and* pterodactyls? Taxonomically speaking, an exhilarating thought, because it would mean that the warm-blooded pterodactyls evolved from a very primitive ancestor of the dinosaurs! If warm-blooded pterodactyls had evolved from early dinosaurs, perhaps the early dinosaurs themselves had already become warm-blooded. The case for the evolution of pterodactyls from *Lagosuchus* or from some very similar early dinosaur is fairly good. The shoulders and ankles of pterodactyls display the same unusual evolutionary modifications found in rabbitcrocs. Very primitive reptiles of all kinds had collarbones (clavicles) that braced the shoulder blades (humans retain this primitive bony strut, as do most lizards). All dinosaurs, including *Lagosuchus,* either lost the collarbone entirely through evolution or had a drastically reduced one. And pterodactyls likewise were without this collarbone. In all primitive reptiles and true crocodiles a long bony strip, the inter-collarbone (interclavicle), lies on the chest between the shoulders. Lizards generally retain this inter-collarbone, but all dinosaurs lost it or reduced it to very narrow splint. And pterodactyls too lack the inter-collarbone.

Head, neck, and shoulder therefore all suggest that pterodactyls evolved from a primitive dinosaur, and so does the ankle joint. All dinosaurs had a hingelike ankle that allowed the foot to flex forward and backward relative to the shank. But the dinosaur-bird type of ankle didn't allow the foot to twist much, so dino-

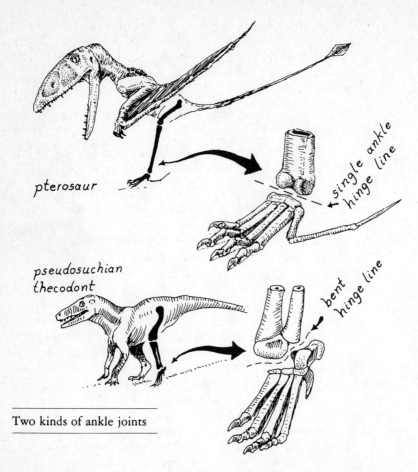

pterosaur

pseudosuchian
thecodont

single ankle
hinge line

bent
hinge line

Two kinds of ankle joints

saurs couldn't turn the soles of their feet inward as can monkeys, most infant humans, and very agile human adults. Pterodactyls had an identical basic plan in the structure of their ankles. Crocodilians and their kin have a bent ankle-hinge line and so did most Triassic uncles of the dinosaurs (the thecodonts).

Reconstructing the ancestry of a clan like the pterodactyls remains an especially difficult challenge. Flying dragons seem to burst into the world like Athena from the mind of Zeus, fully formed. Even the earliest skeletons of pterodactyls already display fully developed wings and the specialized torso and hips so characteristic of the entire order. Cases like this in paleontology—and there are many more—persuade many scholars that evolution doesn't work slowly and continuously at one even pace. Instead, there appear to be times when evolution speeds up and suddenly produces totally new adaptive configurations. Pterodactyls must have emerged in one of these creative spurts of the evolutionary process. As of today, no fossils have been discovered to show how the pterodactyl's forelimbs became transformed into wings. But the S-shaped neck, the simplified shoulder structure, and the bird-type ankle are excellent clues to the ultimate ancestry of the dragons—the quick and agile early dinosaurs of the Triassic Period.

14

ARCHAEOPTERYX PATERNITY SUIT: THE DINOSAUR-BIRD CONNECTION

In 1892, Congressman Herbert directed an energetic diatribe against the fledgling United States Geological Survey, a new agency given the task of mapping the landscape and exploring the rock formations all over the continent. One-armed Colonel John Wesley Powell (left arm lost to a Confederate minié ball at Shiloh) was the vigorous driving force in the Survey office. Powell moved so fast and far in Washington circles that his success bred all manner of envy. His enemies sought any excuse for attack. The Survey office had just published a monograph describing its discovery of birds with teeth. Herbert railed against wasting tax money on godless nonsense about "birds with teeth." These birds, so infuriating to the congressman and all the Survey's enemies, were the fossil seabirds from the Kansas chalk deposits, birds from the Late Cretaceous age, discovered and described in loving detail by Othniel Charles Marsh, professor at Yale. Marsh was supported in part by government money and his monograph was issued as part of an official government series. But Congressman Herbert's accusations of fraud and boondoggle were totally wrong. Marsh had in fact personally paid most of the costs of research and all the extra expenses of running gilt-edged copies of the monograph, special gift editions he sent to scientists all over the world. An attack against "birds with teeth" was, in any event, guaranteed to bring chuckles from the assembled House of Representatives; pure science was a

Archaeopteryx

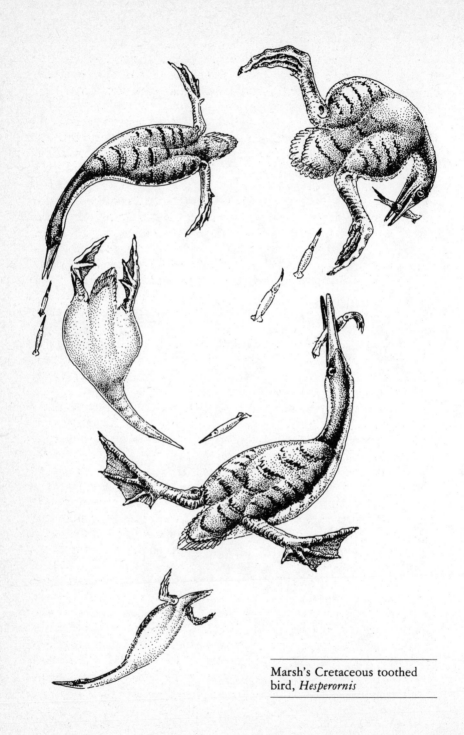

Marsh's Cretaceous toothed
bird, *Hesperornis*

safe target for election-year calumny. Surveys of flooding and coal-mining areas were proper red-blooded topics for the Geological Survey, but of what earthly use was a thick book about long-dead seagulls that supposedly had teeth?

Across the Atlantic, in the great university towns and museums, Marsh's monograph enjoyed a totally different response. "One of the strongest proofs of my theory," wrote Charles Darwin about the fossil Kansas birds. Thomas Henry Huxley could barely contain his delight: Marsh's birds were a devastatingly effective weapon for beating down the prejudices and half-truths published daily by the anti-evolutionists. Meticulous German anatomists nodded their agreement about the conceptual blow struck by Marsh's Odontornithes ("toothed birds"—both the title of the monograph and Marsh's name for his new order of ancient tooth-bearing birds). Even European anti-evolutionary scholars (and only a few good ones were left in 1885) regarded Marsh's toothed birds as the long-sought anatomical intermediate between advanced reptiles and modern birds.

Marsh's Odontornithes deserved every bit of the scientific attention they stirred up. The battered skeletons of fish-eating birds from the ancient tropical seas of Kansas became key arguments for the capabilities of the evolutionary process. Darwin had insisted evolution could alter organic forms to such a degree that it could bridge the great gulf between the Class Reptilia and the Class Aves. Anti-Darwinian critics retorted that Natural Processes might be able to change one species into another closely related form—create a wolf from a coyote, for example. But God's law supposedly forbade a lizard or a crocodile from transmutating into an ostrich or a whippoorwill. Even in our own day, the Creation Science group in California grinds out pamphlets bearing the same message: One warbler "species" might transform into another, but all birds have always been true birds and the first sprang full-blown from the creative hand of God.

Since all birds in the modern world are toothless, the tooth-less condition was regarded as an immutable part of the definition of Class Aves. Marsh's fossils had undeniably been birds—the smaller ones, *Ichthyornis* ("fish-bird"), possessed powerful wings constructed nearly exactly according to the plan found in living avian species. Marsh's bigger bird, *Hesperornis* ("western bird"), had clearly been flightless—only the remnants of wing bones remained—but its vertebral column and hind legs were of typically avian architecture. *Hesperornis* had clearly swum like a loon, with powerful strokes of its hind feet, in a thoroughly avian manner. So far so good as far as the creationists were concerned. Marsh's Mesozoic birds had had certifiable avian characteristics. But both *Ichthyornis* and *Hesperornis* had had teeth set in their jaws—sharply pointed, curved teeth with big roots, just like those of crocodiles. And Marsh detected other more subtle remnants of reptilian ancestry: the upper wing bone (humerus) of *Ichthyornis* featured a wide crest for supporting the flight muscles, and this bony crest more resembled the one along a dinosaur's arm bone than the structure found in any modern birds. *Ichthyornis* also had simple dinosaur-style joints between its neck vertebrae—the vertebral bones met at flat bony surfaces, unlike the strongly involuted, saddle-shaped joints of all modern birds. Clearly, Marsh's Cretaceous birds bridged the gap between bird and dinosaur.

These birds were part of the one-two-punch avian paleontology delivered against creationism in the 1860s and 1870s. In 1861, the lithographic limestones in Bavarian quarries yielded a fossilized bird from the Jurassic Period, *Archaeopteryx* (ancient wing). The Bavarian discovery consisted of a nearly complete skeleton of a dinosaurlike animal, strongly resembling *Ornitholestes,* with long hind legs and a very long tail. But there, on the carefully chipped-out limestone slabs, impressed into the fine limy mud before it had hardened, were also the unmistakable impressions of long flight feathers attached to the forearm and wrist and big tail feathers trailing behind.

Inveterate creationists, then or now, never allow their faith to fall victim to facts. But to any careful, unbiased observer, it was clear that the fossil bird from the Age of Reptiles consisted of a genuine missing link between classes. The fossil bird of 1861 displayed one undoubtedly obvious reptilian feature: a bony tail that was very long and not the abbreviated stub found on all modern birds (any long tails of modern species consist of feathers only; the tail bones are always stumpy like a chicken's). At first, the details of *Archaeopteryx*'s skull and jaws remained obscure, because the head was the worst preserved part of the skeleton. Certainly no one expected the Jurassic bird to have teeth. Though Marsh's birds from Kansas were much younger geologically than *Archaeop-*

teryx, he discovered the teeth in their jaws in 1872 and so received credit for the first discovery of toothed birds. After that, of course, evolutionists projected that *Archaeopteryx* too would prove to have been equipped with teeth. Then, in 1877, the Bavarian quarries yielded a second *Archaeopteryx* skeleton, and detailed analysis uncovered the startling pro-Darwinian evidence: *Archaeopteryx* too had had teeth, and everywhere the structure of its joints and muscle processes had been much less birdlike and far more primitive than those of the Odontornithes of the Cretaceous.

For paleontologists to accept an animal as a real "missing link" between classes, the fossil is not only required to display an anatomical structure intermediate between two distinct classes, but it also has to appear in the "correct" sequence of time, intermediate between the two classes. If the Darwinists were right, and birds had evolved from a long-acting process, then the fossil record had to read correctly, from bottom to top: The strata had to show (1) the first primitive reptiles, then (2) advanced "reptiles" (dinosaurs), then (3) primitive birds with teeth, then (4) more modern birds with teeth, and finally (5) totally modern, toothless birds. Now, primitive reptiles had been found low in the strata, in rocks from the Coal Age and the Permian Period. Dinosaurs with birdlike bodies made their entrance in the next-higher strata of the Triassic Period. *Archaeopteryx,* a very primitive bird with teeth, showed up in the next strata, the Jurassic. Marsh's toothed birds had been more advanced than *Archaeopteryx* and appeared in the next period, the Cretaceous. And, finally, truly modern birds without teeth made their debut at the very end of the Cretaceous and the beginning of the next epoch, the Paleocene. So it all fell into place exactly as evolutionists would have predicted.

The stratigraphic proof for a Darwinian origin of birds appeared incontrovertible—the rocks preserved the stages of development in the exactly proper sequence through time. Any impartial observer might conclude that if God had really created birds, he must have been going out of his way to fool humanity into believing in evolution.

Dinosaurophiles had reason to celebrate *Archaeopteryx* and Marsh's birds. Evolutionists and nonevolutionists agreed on one

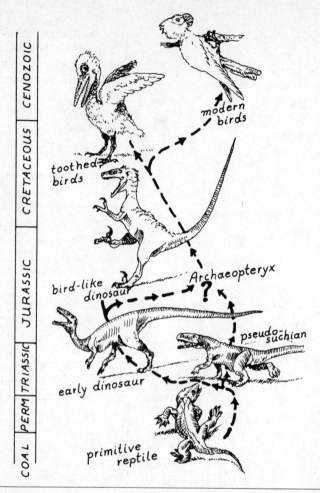

Birds evolved direct from pseudosuchians or direct from dinosaurs.

point: the group of "reptiles" closest to the birds had been the extinct Dinosauria of the Mesozoic. Sir Richard Owen discerned avian patterns in the feet of *Iguanodon*. Edward Drinker Cope believed that the ankle of a New Jersey duckbill dinosaur was so birdlike that he named the beast *Ornithotarsus*—literally, "bird-ankle." Thomas Henry Huxley summarized arguments in favor of a dinosaurian origin for birds in one of his most famous and lyrical pieces of scientific prose, published in the quarterly review of the Geological Society of London. Huxley knew birds—he had worked out a detailed classification of avian families based on the bony structure of the palate, a classification still highly regarded today. And Huxley knew dinosaurs and their kin. He had personally examined and discussed dozens of specimens. Huxley scrutinized the anatomy of dinosaurs from nostrils to tail, claws to hips, revealing that advanced meat-eating dinosaurs approached the true birds in nearly all the details of their anatomy.

Huxley's case was impressive in documentation, persuasive in argument: (1) Only in dinosaurs did he find the distinctive bird type of ankle joint, where movement had been concentrated into a single hinge running between the two rows of ankle bones. (2) Only in dinosaurs had he found the great expansion of the upper hip bone (the ilium) so characteristic of all birds. (3) Only in some dinosaurs had the hind foot been arranged in a birdlike fashion where the inner toe turned backward and the three main toes pointed forward to produce the unmistakable footprint of birds. In fact, some dinosaur tracks were so birdlike that they had been mistaken for bird tracks when discovered in 1830. (4) Only advanced dinosaurs displayed the compact bipedal body fundamental to avian anatomy—the very short torso, massively braced hips, long and highly mobile neck, and long hind legs. (5) Only in dinosaurs and pterodactyls had Huxley noted holes in the vertebral bones for the air sacs which connected to avian-style lungs, and the pterodactyls had been far less birdlike than advanced dinosaurs in most other regards. (6) Only in some dinosaurs had the pubic bone been turned backward exactly as in birds.

Over in America, Marsh for his part accepted the concept of a dinosaurian ancestry for birds. And he pointed to the very birdlike pattern of the small Bavarian dinosaur *Compsognathus,* a chicken-sized predator preserved in the same lithographic limestone that yielded *Archaeopteryx.* Huxley favored a beaked dinosaur as the most likely ancestor for the birds because the beaked clan had had a backwardly reoriented pubis. In general, then, there existed a firm consensus among the best paleontologists—Old and New World scholars alike agreed that dinosaurs had been a birdlike clan and birds were direct descendants of those dinosaurs.

Since the birdlike nature of advanced dinosaurs won wide acceptance, late nineteenth-century naturalists couldn't dismiss the Dinosauria as merely cold-blooded dead ends. If the proto-bird had been a dinosaur, it wasn't unreasonable to suppose that some dinosaurs had had a proto-birdlike physiology. The resulting speculations about the dinosaur's metabolism therefore continued to appear occasionally in respected European publications right up until the 1920s. But all this changed in the late twenties. The claims dinosaurs had to the paternity of birds were largely forgotten. And the interrelated case for the warm-bloodedness of dinosaurs was concomitantly dismissed for alleged lack of evidence. How had such a revolution come about? Strangely enough, dinosaurs seem to have lost their claim to having been the ancestors of birds because of the best, most thorough book written about avian ancestry, a book that got nearly everything right except its final conclusion.

Gerhard Heilmann's *The Origin of Birds* was first issued in 1925. It won nearly instant acclaim as the finest work on the subject, far wider in scope than any of Marsh's papers on toothed birds and far more detailed than Huxley's essays. Heilmann remains a popular and widely accepted author among paleontologists of all stripes even today. Trained both as an ornithologist and anatomist, Heilmann's eye for evolutionary architecture was further sharpened by his draftsman's skill—he penned his own drawings of bones and muscles, depicting some of the liveliest and most accurate restorations of dinosaurs in the flesh. Heilmann performed Huxley's task over again, and did it better, demonstrating how dinosaurs did indeed exhibit extraordinarily birdlike adaptations of their torso, neck, hip, ankle, and forelimb. But Heilmann was also deeply disturbed by the possibility that the evolutionary argument had been carried too far. As far as he could make out, all the birdlike dinosaurs had evolved too far in one direction; they had lost or greatly reduced the long collarbone of their proto-dinosaurian ancestors. Yet all flying birds possess just such a very large collarbone resting against either side of their shoulders, which fuse together at their lower end to form that avian hallmark—the wishbone.

Birds simply cannot fly without wishbones. The flexible, U-shaped strut formed by the fused collarbones braces the entire shoulder against the stresses of flapping. Even *Archaeopteryx* had already evolved a modern grade of wishbone, with long left and right collarbones fused together beneath the chest. Heilmann was convinced that the direct ancestor of birds must have had large collarbones, so that evolution could transform these structures into the avian pattern. It was inconceivable to him that the bird's ancestor could have undergone evolutionary reduction of the entire collarbone set. Were that the case, Heilmann would have been forced to believe in a monumental reversal of the evolutionary process. Like most paleontologists, Heilmann was a confirmed skeptic when it came to evolutionary reversals. Minor reversals were conceiv-

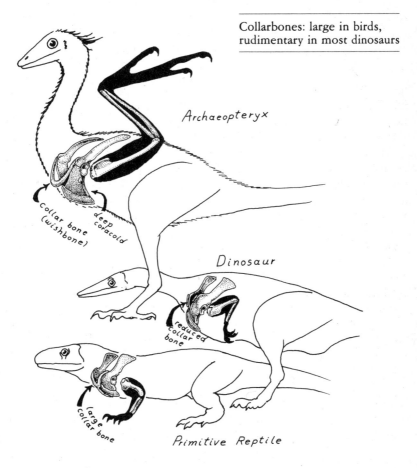

Collarbones: large in birds, rudimentary in most dinosaurs

Archaeopteryx

collar bone (wishbone)

deep coracoid

Dinosaur

reduced collar bone

large collar bone

Primitive Reptile

able. But complete loss of the collarbones followed by a re-evolution of a very large set of them seemed totally incredible.

The very dinosaurs Marsh and Huxley had identified as most birdlike were precisely the ones possessing little or no collarbone. The small, long-legged carnivores such as *Compsognathus* were certainly birdlike in overall design and in the details of their joints, but most skeletons revealed no trace of a collarbone. At most, there might be a tiny splint of bone running alongside the shoulder blade. And such a splint *might* at most represent only the remnant of a highly reduced collarbone. What was Heilmann to do with this apparent paradox? The most birdlike dinosaurs had been totally unbirdlike in this one indispensable characteristic. Heilmann had a most astute sense of the general lessons taught by fossil history. He knew that many times in geological history two distantly related tribes had progressed along very similar evolutionary pathways, so that after twenty or thirty million years the final products were quite similar. An incautious observer might even be fooled into believing in a close relationship between the two lines. Tasmanian wolves are the classic example of this. They look very much like true wolves. But the Tasmanian was a pouch-bearing marsupial that evolved its wolflike configuration from a 'possumlike ancestor in Australia. This was an evolutionary line totally separate from the evolution of true wolves in Europe and North America. Tasmanian wolves and true wolves aren't closely related at all—each of their similarities evolved independently.

Faced with what he regarded as an impasse, Heilmann made a very reasonable suggestion: He argued that both birds and dinosaurs had evolved from a common ancestor. This ancestor already had some of the advanced characteristics of both dinosaurs and birds, such as long hind legs, but hadn't undergone reduction of the collarbone. From this source two great evolutionary columns had advanced through time, marching along parallel adaptive tracks. One track led to the birds, the other to the dinosaurs. Therefore dinosaurs weren't the ancestors of birds, but only distant cousins.

There existed a fossil clan from the Triassic rocks that seemed to fulfill nearly perfectly the role Heilmann ascribed to the common ancestor of dinosaurs and birds—the predatory reptiles known as "pseudosuchians" (literally, "false crocodiles"). The false croc-

The Argentine ornithosuchid *Riojasuchus*, a group
supposedly close to bird ancestry

Deinonychus was a two-hundred-pound Cretaceous predator with a wickedly enlarged hind claw for disemboweling prey. The *Deinonychus* family ranged from 140 to 67 million years ago and left superb fossil skeletons in Wyoming and Mongolia. These powerful hunters had an almost incredibly birdlike anatomical structure from nose to tail—in fact, based on bone anatomy, we would be justified in classifying them as either birdlike dinosaurs or dinosaurlike birds. So avian were these dinosaurs, it is quite probable that they had already evolved feathers for insulating their bodies.

odiles' hind limbs were longer than their forelimbs, their general build was rather light, fast, and lively—precisely the type of locomotion to be expected in the forerunner of both the dinosaurs and the birds. False crocodiles also displayed sharp teeth set in sockets just like the dental arrangement found in toothed birds and dinosaurs (an important point because many primitive reptiles had teeth that were fused to the jaw bone). Most important of all, the best-preserved pseudosuchian skeletons contained big collarbones. Therefore false crocs had had precisely the right mix of primitive and advanced characteristics to serve as the ancestors of birds. Heilmann indicated one false croc in particular as a suitable ancestor for the birds, *Ornithosuchus* from the Triassic red sandstones of Elgin, Scotland. *Ornithosuchus* means "bird-croc," and its name was coined in the 1880s by a British geologist who had been impressed by the beast's narrow, birdlike snout.

Heilmann's hypothesis won the day. Almost overnight, textbooks converted to the theory of the false crocs as the evolutionary ancestors of the birds. Dinosaurs were demoted from patriarchs to distant cousins. And since dinosaurs were no longer considered in the evolutionary line of the birds, they lost any claim to a genuinely avian physiology. Heilmann certainly deserved credit for a theory carefully worked out with attention to the nuances of evolution. But, unfortunately, most scholars missed the main point— Heilmann had substituted one implausibility for another. To be sure, it was implausible that a collarbone would atrophy and then re-evolve. But it's equally implausible that *Archaeopteryx* and dinosaurs like *Compsognathus* would evolve separately yet end up incontrovertibly similar in nearly all characteristics. Heilmann's theory required parallel evolution in unusually large doses, far larger doses than would be required even for the Tasmanian wolves.

Heilmann's theory reigned as fact down into the 1960s. In 1964, John Ostrom at Yale discovered the very advanced predatory dinosaur *Deinonychus,* a long-armed Early Cretaceous carnivore with a cruel-looking killing claw on its hind foot. Ostrom spent two years carefully analyzing *Deinonychus*'s place among the meat-eating dinosaurs. Unknowingly he was being led directly to the final and correct solution of the mystery of the origin of birds. Biomechanical analysis applied to *Deinonychus*'s bodily configuration yielded evidence for exceptionally high levels of locomotive activity: both running speed and maneuverability. Quite clearly, *Deinonychus* had had a great deal of *birdness* built into its limbs, a *birdness* that would have expressed itself in life by a daily metabolic regime more fitting for a ground bird such as a cassowary than for the orthodox view of any cold-blooded dinosaur.

After completing his monograph on *Deinonychus,* Ostrom planned for his next project—a study of pterodactyls. His two projects, the first on *Deinonychus,* the second on pterodactyls, seemed totally separate endeavors, and neither appeared to have anything at all to do with the origin of birds. Yet his investigation of pterodactyls was destined to lead John Ostrom to the discovery of a new fossil. And that would lead him directly to the dinosaur that was the true ancestor of all birds.

His serendipitous master stroke befell Ostrom in a Dutch museum where he found a set of bony fingers on a limestone slab out of those famous Bavarian quarries. The slab supposedly contained yet another fragmentary pterodactyl skeleton, not an important find because dozens of complete specimens were available from the same localities. But those long bony fingers, tipped by needle-sharp claws, had been misidentified. They were not pterodactyl at all but the rarest of the rare, the most sought after of all Bavarian fossils, an *Archaeopteryx.* After fully a century of quarrying, only those two early skeletons of 1861 and 1867 were known. John Ostrom's was the third.

Alternating images flashed before his mind's eye as he scrutinized the Dutch specimen. He recognized the bony hands with their three long, clawed fingers as belonging to *Archaeopteryx.* But he also recognized in that hand a miniature version of *Deinonychus*'s. *Archaeopteryx* had been pigeon-sized, its hand four inches long; *Deinonychus* had been as heavy as an average man and could stretch its hand a full nine inches. Yet the small bird hand and the dinosaur hand were virtually identical in shape. Each finger and wrist bone had been molded to the same peculiar biomechanical pattern, an adaptive plan totally unknown anywhere in the animal kingdom outside the Dinosauria. There was an important message on this Dutch slab, and Ostrom read it correctly. *Archaeopteryx* and *Deinonychus* had been very closely related. And birds were indeed the direct descendants of dinosaurs.

Back at Yale, Ostrom constructed an overwhelming argu-

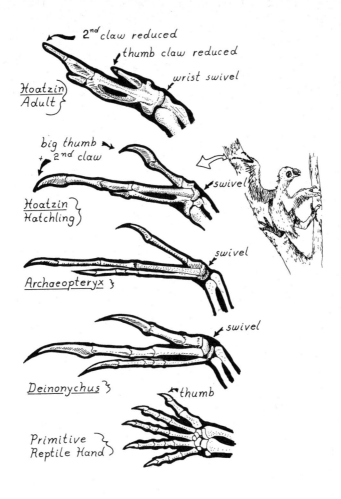

2nd claw reduced

thumb claw reduced

wrist swivel

Hoatzin
Adult

big thumb
+ 2nd claw

swivel

Hoatzin
Hatchling

swivel

Archaeopteryx

swivel

Deinonychus

thumb

Primitive
Reptile Hand

ment to prove his case. Between *Archaeopteryx* and *Deinonychus* the long bony fingers were not the only things identical. So was nearly every detail of their shoulder, hip, thigh, and ankle. As a consequence, questions about dinosaurs which had long baffled the scholars could be solved when *Archaeopteryx* was used as a standard of comparison. *Deinonychus,* for example, had had a strange lower shoulder-blade bone (coracoid), unique among dinosaurs because of its great depth from shoulder socket to breastbone. But *Archaeopteryx* had had a coracoid of the same deep pattern as *Deinonychus.* Clearly both the bird and the dinosaur had evolved the unusual shape to increase the size of their breast muscles. *Deinonychus* had had a peculiar upper hip bone (ilium) compared to other predatory dinosaurs, but *Archaeopteryx* exhibited the same peculiarities. And in its wrist *Deinonychus's* similarities to birds were nothing short of astounding.

The wrist bones and hand—the structures that had first caught John Ostrom's attention—were, in fact, the single most persuasive part of his argument. The wrists of primitive reptiles were simple devices, merely a flexible mosaic of squarish bones held together by ligaments. When a primitive reptile pressed its wrist against the ground, the mosaic of bones could bend or twist but couldn't produce any precisely controlled movement. *Archaeopteryx* and all modern birds are different. The joint surfaces of wrist bones are complexly curved, and the large central bone has an elegantly designed joint surface of semicircular shape that guides the animal's hand in a precisely controlled flexing movement. Prior to Ostrom's discoveries, most scientists tended to believe that the bird type of wrist had evolved rather suddenly, in the first true birds. Neither primitive dinosaurs nor the false-croc reptiles Heilmann favored as the ancestors of birds possessed anything like the bird type of wrist. But Ostrom found that *Deinonychus's* wrist bones were identical to those of *Archaeopteryx,* and that *Deinonychus's* wrist would therefore deliver the very same sort of precise flexing movement in the entire set of fingers. Moreover, the long fingers so distinctively characteristic of *Deinonychus* had been designed identically in *Archaeopteryx.* Both *Archaeopteryx* and the dinosaur had had three fingers only—not the five found in primitive dinosaurs. And the proportions of the fingers had been the same: A short, stout thumb and two longer outer fingers, with the outermost of the three very slender, bowed outward, and closely bound by ligaments to the middle finger. This unique pattern can still be recognized in a modern bird's wing; the three fingers are all firmly fused together in an adult bird, but in an unhatched chick, the bones are not yet fused. In a chick the separate wrist and hand bones are clearly discerned, exactly as they had been in *Deinonychus* and *Archaeopteryx.*

There exists today one species of bird that retains its finger bones unfused and flexible into the first weeks of life in the nest. This bird, the hoatzin of South America, allows us to surmise how the *Archaeopteryx* worked. As birds go, an adult hoatzin exhibits nothing special in the anatomy of its wing. But the young nestling is a genuine evolutionary throwback, an ugly little chick that climbs through the vegetation by grasping with its three-fingered, claw-tipped hands designed to the *Archaeopteryx* blueprint. The hatchling can thus escape predators—snakes and hawks—by using its wing-claws to climb out of its nest and work into the labyrinth of vines surrounding it. Heilmann had drawn diagrams of wrists of these hoatzin chicks next to those of *Archaeopteryx* in his book—the anatomical identity was so stunning. But Heilmann didn't have *Deinonychus* for comparison. If he had lived to see Ostrom's discoveries, I'm certain that Heilmann would have converted to the dinosaur-bird theory.

Hoatzin chicks also force a rethinking of the idea that there could be no big reversals in the evolution of birds. Evolutionary reversals unquestionably were necessary to make a hoatzin. Hoat-

A hoatzin chick climbing with its wing claws

zin's relatives all have much weaker wing claws in the chick stages of life than hoatzins themselves have. Most ornithologists therefore conclude that hoatzins evolved from some ancestor with the "normal" pattern of growth in which the chick never possesses strong, flexible, unfused fingers for climbing. According to this view, the hoatzin chick evolved by means of a Darwinian U-turn—the strong, *Archaeopteryx*-like flexible fingers were recalled from genetic storage.

Genetic storage is a nuance of evolution too often ignored. Many paleontologists believe that when a bone disappears in evolution, the genetic blueprint for that bone is also erased. Hence, when dinosaurs lost their clavicle, their genetic code also supposedly lost the instructions for making collarbones. If evolution really occurred in this fashion, Heilmann would unquestionably have been right when he maintained that re-evolution of a lost clavicle was most implausible. Re-evolution of a lost set of genes for making clavicles would entail a highly unlikely swarm of mutations and natural selections.

But in fact evolution does not occur in this fashion. Hoatzin's ancestors never lost the genetic blueprint for producing *Archaeopteryx*-style clawed fingers. In essence, they merely turned off the physiological switch that ordered genes to produce organs according to the encoded information. Recent advances in genetic research reveal that most species carry such blueprints that are "switched off" and can't express their code as fully formed tissue. In other words, when an organ has been "lost," most of the time its blueprint is still there, in genetic storage. Hoatzin's ancestors were "normal" modern birds that employed a modern blueprint to produce a wing in their nestlings that was like a chicken's, with stiff, fused fingers. Hoatzins evolved their distinctive *Archaeopteryx*-like clawed fingers by the process of turning off that blueprint for its nestling and turning back to the older one to reexpress itself.

A wealth of evidence supports this theory of reexpression by genes that have been turned off for millions of years. Most of it occurs in throwbacks (what nineteenth-century scientists called atavisms), the rare appearance of ancient organs in species that, as a whole, had lost the anatomical features millions of generations earlier. A good example is multi-toed horses. Modern horses belong to the same general group as tapirs, and tapirs have four toes on each forefoot. The single-toed modern horse evolved from a four-toed ancestor. Every so often a healthy, normal, single-toed mare gives birth to a colt that has little extra toes sticking out beside the big main toe. Zoologists point to this multi-toed foal as a case where natural processes allow a bit of the ancestral blueprint to show through, letting ancient ancestral traits reexpress themselves.

Whales offer a more spectacular case. Modern whales have no hind legs at all, and even when all the blubber and muscle are flensed from the hip region, there is no remnant of the hip bones except a small splint representing the ilium. Even the oldest-known fossil whales display only slightly enlarged hip bones and some remnants of thigh and knee. But way back in their ancestry whales did have big hind legs, at a stage when they were land-living predators. And every once in a while a modern whale is hauled in with a hind leg, complete with thigh and knee muscles, sticking out of its side. These atavistic hind limbs are nothing less than throwbacks to a totally pre-whale stage of their existence, some fifty million years old.

Such throwbacks even occur in human infants. Hospitals occasionally register an entirely modern-looking baby characterized by all the expected organs, plus an unexpected tail, a long, caudal appendage protruding beyond the buttocks for two or three inches. Some of these tails are even bigger than the average caudal remnant displayed by our close kin, the chimps, gorillas, and orangutans.

Genetic experiments have revealed that these throwbacks are controlled by suppressor genes. We now know that most complex pieces of anatomy—such as the clavicle and its muscles—are controlled directly and indirectly by scores of genes that interact and can suppress each other. We also know that the full genetic blueprint in any single species is rarely, if ever, fully expressed. Instead, much of the genetic information is stored in the "inactive file," genes that don't produce their potential impact because some other gene prevents them from turning on. When an anatomical feature disappears during evolution, its genetic blueprint is not erased. Some new combination of genes has evolved to suppress the still-present blueprint.

Birds with teeth may have appeared ridiculous to creationists, but in point of fact modern birds do carry the ancestral genetic code for making teeth tucked away in their inactive file. No living species of bird manufactures teeth. But recent surgical manipulations of bird embryos demonstrate clearly that the potential is still there. In 1983, experimenters transplanted tissue from the inner jaw (dental lamina) of an unhatched chick to an area of the body tissue, where the graft could grow. In the transplanted position, the chick's dental lamina started to produce tooth buds! Birds with teeth could grow right in the twentieth century.

Suppressor genes solve Heilmann's paradox, the problem of evolving birds with big collarbones from dinosaurs with atrophied collarbones. Evolution at some point must have been able to remove the genes that suppressed collarbones from the dinosaur that was ancestor of the birds. This is not farfetched, nor even mildly implausible. The scenario might have run like this: A long-armed dinosaur, such as *Deinonychus*, might evolve extra-long arms with extra-long scales (feathers are modified scales) and begin to jump from branch to branch, using its arm scales to gain a few extra feet of glide, much like a flying squirrel. This proto-bird has no collarbone, but its ancestors long before did. The genetic code for a collarbone remains in the dinosaur, stored in its inactive genefile. Once the proto-bird uses its forelimbs for gliding, a bony strut in front of the shoulder blade becomes an advantage. Any mutation that removes genes suppressing the clavicle now becomes favored by natural selection. In a few hundred generations, the proto-bird could therefore re-evolve a collarbone, rearranging it a bit to produce the distinctive V-shaped wishbone characteristic of birds.

Why would dinosaurs begin to fly and thus cross the threshold into the avian class? What was *Archaeopteryx*'s niche? Most paleontologists have leaned toward an analogy with flying squirrels. Proto-birds are supposed to have been tree climbers who evolved wings first for gliding, then for powered flight. But there's an alternative possibility—the speedy jogger. Birds might have evolved flight first by running at high speeds over the Mesozoic landscape, employing their arms, outfitted with protofeathers, as airfoils for increasing ground speed. According to this theory, hypothetical proto-birds finally evolved a speed fast enough to become airborne. John Ostrom champions this speedy-jogger theory. He was unhappy with the traditional restoration of *Archaeopteryx* as a tree-climbing glider and flier. Among other details, he had observed that *Archaeopteryx*'s foot couldn't get the same grip on a branch as can modern birds. Climbing birds have an inner toe that faces backward and flexes forward to grasp a branch against the other three toes. For the most efficient performance, all four of these toes must be long and their base joint must be at the same level, located at the very bottom of the long ankle bones (metatarsals). *Archaeopteryx*'s foot was not so built. The toe facing rearward was too short and too high up on the ankle, so that its grip on a branch wouldn't be anywhere near as effective as a modern bird's.

As an alternative, Ostrom suggested that perhaps wings first evolved as catching devices. Today, small birds and bats use strokes of their wings to sweep prey into their mouth. *Archaeopteryx* wasn't a strong flier—its major feathers weren't fused to its arm and wrist the way they are in modern flying birds. So maybe *Archaeopteryx* had been a land-running predator that used its feathered, *Deino-nychus*-type arms to coerce prey.

I accepted this hypothesis in an article about the renaissance of dinosaurs I published in *Scientific American* in 1975. But accumulating evidence has forced me back to the orthodox view of *Archaeopteryx* as a climbing and gliding flier. The aerodynamic shape of its flight feathers is the first consideration. Flying birds today have asymmetrical feathers—the leading edge is narrower and stronger than the trailing edge. This is a necessity for powered flight because air pressure is greater along each feather's front edge. Recently, an ornothologist from North Carolina took the obvious step of carefully examining *Archaeopteryx*'s feathers—the first time anyone had done so since the initial discovery in 1861. There was no doubt, the wing's main feathers were asymmetrical. Therefore *Archaeopteryx* very probably did indulge in powered flight, even though it must have been a noisy, slow, and inelegant performer in the air. Furthermore, even though *Archaeopteryx*'s foot didn't have as precise a grip as the most specialized modern perching birds do, it did have as much grasping power as many modern birds that climb adequately. And *Archaeopteryx* wouldn't have had to rely on its hind feet alone for effective climbing because its wings also had hooklike claws. *Archaeopteryx* certainly could have clambered through the ancient Bavarian vegetation as efficiently as any hoatzin chick. Finally, if *Archaeopteryx* were a ground jogger, its hind claws would have been blunt like those of a modern ground bird. In fact, the *Archaeopteryx*'s feet ended in needle-sharp claws. And if it had run about on such pointed hind claws, it would have worn down their horny outer sheath. Yet the fossils display hardly any wear even on the delicate points of the claws.

More clues as to how *Archaeopteryx* developed flight come from considering its teeth and claws combined. *Archaeopteryx* and the Cretaceous birds from Kansas had teeth that terminated in thick, barrel-shaped roots, like crocodiles'. Teeth so shaped are special

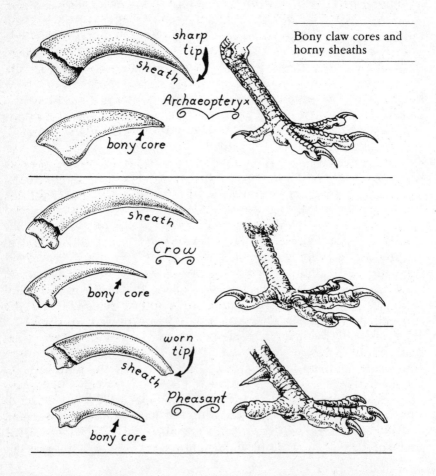

Bony claw cores and horny sheaths

adaptations and are evidence for a seafood diet. Over a century ago, Sir Richard Owen demonstrated that such teeth were a trademark of the fish-eaters of the Mesozoic seas—the ocean lizards (mosasaurs), and the fast-swimming fish lizards (icthyosaurs). *Archaeopteryx* and the Kansas birds were preserved in saltwater deposits full of fish, squid, shrimplike crustaceans, and other seafare. The strong-winged *Ichthyornis* probably dove at fish from the air, while the flightless loon-footed *Hesperornis* must have chased fish underwater. *Archaeopteryx* is usually portrayed as a land feeder, swooping down on oceanside prey along the beach from its roosts in the seashore trees. But the shape of its teeth requires a different hypothesis, a fishier one. Many modern fish-eating birds—puffins, penguins, snakebirds—swim with their wings. Hoatzin fledglings also swim underwater with strong strokes of their wings. *Archaeopteryx*'s hoatzinlike wings would have been fine for submarine propulsion. And its needle-sharp claws would have been perfect for snaring slippery aquatic prey. Maybe *Archaeopteryx* sometimes hunted like present-day fishbats, occasionally snagging fish with its hind claws as it swoops and glides over the surface of the sea.

It must be said, restoring *Archaeopteryx* to its proper place in the dinosaur's family tree has been a great boost to the morale of dinosaurophiles. No open-minded observer of the fossil sequence, from Coal Age reptiles with stubby legs to the birdlike dinosaurs of the Jurassic, can be other than convinced that our present glorious array of feathered creatures is truly the direct descendant of those primitive land creatures via the intermediary agency of the dinosaurs. There are over eight thousand species of birds alive in today's ecosystems, and each one, from the hummingbird to the ostrich, is incontrovertible evidence that the bloodlines of the dinosaurs are still full of evolutionary vigor.

The story of *Archaeopteryx* is a boon to dinosaur-lovers in another way as well. According to the orthodox theory, remember, dinosaurs didn't have enough metabolic energy to walk fast, let alone fly. But both pterodactyls and birds had to evolve high-pressure hearts and lungs before flight could be achieved. Pterodactyls most probably were the descendants of very primitive dinosaurs, of the bunnycroc variety, while birds were surely products of the advanced dinosaurs. If both branches possessed a high-pressure, hot-blooded metabolism, then it's not impossible to suppose that the entire stock of primitive dinosaurs was already equipped for high metabolism before either aerialist tribe evolved. In other words, it's quite possible that flying dragons and birds inherited their high-capacity hearts and lungs from their dinosaur forebears and that powered flight was simply one application by evolution of the fundamental bioenergy of the dinosaurs.

PART 4

THE WARM-BLOODED METRONOME OF EVOLUTION

15

SEX AND INTIMIDATION: THE BODY LANGUAGE OF DINOSAURS

Ever since the first Mesozoic fossils came to light, there have been features of them that appear to defy explanation, at least in terms of the usually considered aspects of the Natural Economy—eating, drinking, preying, avoiding predators. As each new species was excavated, the list of prehistoric anomalies grew: sail-backed reptiles from the Coal Age, horned amphibians from the ancient red beds, battering-ram skulls on the protomammals, baroque crests on the heads of duckbill dinosaurs. American paleontologists traditionally favored a strictly utilitarian interpretation of these things; bones should be shaped to perform a useful function for procurement of food or defense. Bones of nonutilitarian shape were therefore puzzling in the extreme. Faced with a bewildering variety of crests and cranial ornamentations, the older generation of American paleontologists sometimes advocated a moralistically motivated theory of racial decadence: As an evolutionary family approached its time of extinction, its species would indulge in nonadaptive decoration. Like biological ancient Romans, they had supposedly lost control of their adaptive sense and hastened to their doom amid orgies of useless ornamentation.

Until the 1970s, few American scientists referred to sex when they analyzed dinosaur skeletons. But evolution is full of sex. And natural selection favors structures that produce results in winning or enticing mates and discouraging rivals. The beauty of nature is

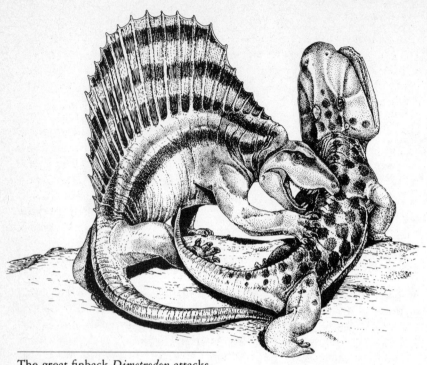

The great finback *Dimetrodon* attacks
the amphibian *Eryops*.

not spoiled by the great influence wielded by sex and intimidation. Nor does it lessen the fascination of fossils to suspect that much of the most extraordinary bony paraphernalia may have served as enticements to prospective mates.

The early chapter of the sexual epic can be read back in the steamy days of the Coal Age, long before there were any dinosaurs. The primitive vertebrates with legs of that period would be classed as amphibians in the reproductive sense of that term, for they laid eggs, frog-fashion, in water. It is certain they reproduced in water, because aquatic hatchlings are common fossils in the dark, carbon-rich shales laid down in the lakes of the Coal Age. Often the skin of these ancient larvae is outlined in the stone where the slow decay distilled the living body tissue into an oily stain surrounding the skeleton. These larval amphibians are fossils of unexcelled loveliness. Dark organic outlines mark out each limb and, behind the head, the long branched filaments of their gills.

Gills just like these, pulsing with oxygen-rich blood, can be found in the throats of modern salamander larvae. Holding a live salamander in a handful of pond water is like looking back into a past 300 million years old, back to a time when the evolutionary tree of land vertebrates had just taken root.

For the first fifty million years of life on land, all of the vertebrates with legs were amphibians of one tribe or another. How did they court each other? Since they reproduced in water, their pre-nuptial displays must have taken place in ponds and quiet backwaters or along the banks of ancient waterways. Living amphibians feature some of the richest sonic symphonies in today's ecosystem—the chorus of mating frogs. But another amphibian family, the salamanders, far more primitive than frogs, is nearly mute. Some salamanders (the newts) substitute dance for song. The male newt waves his tall, bright red tail in a kind of underwater flutter-dance as he minces before his prospective lady love. The fossil records from the earliest Amphibia do turn up some eel-like tails that could have been used in this fashion. But what about sound? When did amphibians evolve that marvelous capacity for serenade so characteristic of modern frogdom?

The early fishes did not hear airborne sounds, and their ears were used mostly to maintain body balance. Ears for hearing on land require a taut membrane in the skull to pick up airborne vibrations. Living species of frog have such a membrane shaped like a tiny drumhead, constructed of special skin. A deep notch in the frog's skull holds the eardrum (known technically as a tympanum), and between it and the brain stands an air-filled chamber: the middle ear. To transmit sound to the brain, a slender ear bone runs from the eardrum to the canals of the ear in the side of the braincase. If it could be discovered when this type of ear first evolved, it would constitute an important clue about when the sexual chorus first began.

The eardrum doesn't preserve in fossils, but the notch for it in the skull does. Earliest of all amphibian fossils is the famous *Ichthyostega* from the lake beds of Greenland (its name means "fish with a roof," a reference to its primitive fishlike structure and the thick roof of its skull). This Ur-amphibian has no definite notch for an ear, and couldn't have possessed any special auditory adaptations. Therefore, when *Ichthyostega* and its kind waddled over the primeval land, they must have marched into a silent world where the humid stillness was broken only by the rustling of ancient rushes in the wind and the near-silent footsteps of Ur-spiders hunting in the leaf mold.

But it did not take long for the fledgling land vertebrates to evolve greater sensory complexity. Early in the Coal Age, quite large notches for eardrums appeared prominently in the skulls of the keyhole amphibians (loxommatids), a tribe of aggressive, sharply fanged predators with alligatorlike skulls. ("Keyhole" refers to the peculiar shape of their eye socket; an enlargement at the front of it may have housed a gland.) Keyhole amphibians clearly could hear airborne sounds, and therefore quite possibly used their voices to bluff and challenge and court. Since their heads reached a length of two and a half feet, they would surely have uttered a croak that would command respect. Other ear-equipped amphibians evolved

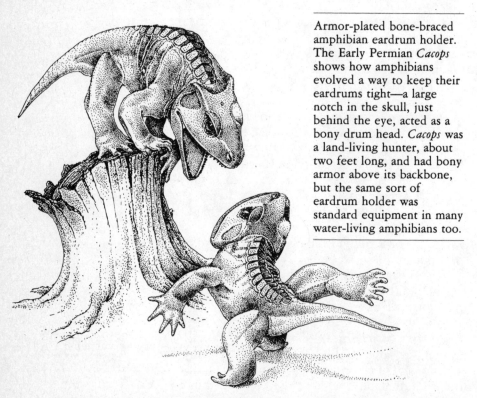

Armor-plated bone-braced amphibian eardrum holder. The Early Permian *Cacops* shows how amphibians evolved a way to keep their eardrums tight—a large notch in the skull, just behind the eye, acted as a bony drum head. *Cacops* was a land-living hunter, about two feet long, and had bony armor above its backbone, but the same sort of eardrum holder was standard equipment in many water-living amphibians too.

Quick history of butting, bluff, and intimidation. It all started with the evolution of eardrum skull notches in the Coal Age, proceded through the Age of Finbacks in the Early Permian, and then up to the head-butters of the Late Permian-Triassic, the tooting and butting Cretaceous-Jurassic dinosaurs, and the modern mammals of the Cenozoic.

after them as the Coal Age continued, so the spring mating season probably witnessed a diversified range of timbre and tone.

Reptile ears are built to the same general pattern as are amphibian ears, but the details of how the nerves pass through the auditory apparatus are different. Most paleontologists presently believe that reptiles evolved their ear independently and did not simply inherit their auditory machinery from amphibian ancestors. Today, the ears of lizards work much like the ears of frogs, but the ears of Coal Age reptiles are biosonic puzzles. A good notch for the eardrum evolved in some reptile tribes very early, yet the bone of the middle ear was thick and ponderous, not the delicate, thin bone absolutely required for hearing mid and high frequencies. Massive ear bones wouldn't transmit most vibrations from the eardrum to the brain, and some of these early reptile ear bones are as big as a man's thumb. What could these ancient reptiles have heard, if anything? It remains a mystery. Some anatomists have suggested that the heavy ear bone was suspended by delicate ligaments and acted as a kind of seismograph for detecting very low-frequency sound. This suggestion evokes visions of a mating dance in which the courting couple stomps about producing minor earth tremors to communicate their lust. Reptiles did not evolve ears of high sensitivity until late in the Permian Period, long after the Coal Age, and the Reptilia certainly weren't equipped to transmit and receive airborne melodies before then.

Sex is not all melody. Pushing and shoving, intimidation, have their place too. Frog suitors often try to kick their rivals off the back of a female in the mating pond. The rhinoceros iguana lizard of Cuba indulges in male-male wrestling contests. Males push each other with their snouts, grab loose skin in their teeth, and may clamp on each other's mouths in what is technically known as jaw wrestling. Not many living amphibians have specialized organs for sexual wrestling, but among the Coal Age fossils there is one spectacular case.

Most amphibian skulls are designed quite straightforwardly. With few exceptions, they can be explained in terms of jaw muscles, bracing for teeth, sucking in prey underwater, or other purely dietary needs. But diet can't explain the most grotesque amphibians skull ever evolved—the boomerang-shaped head of *Diplocaulus* ("two-tailed," a reference to its double-spined tail bones).

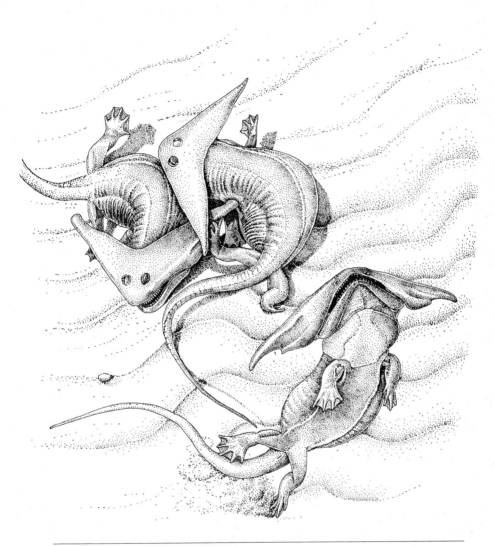

Battle of the boomerang-heads. Swoosh, clunk, and thud on a Texas stream
bottom during the Early Permian Period. *Diplocaulus,* a three-feet-long flat-
bodied amphibian, probably used its grotesque head horns for sideways
slugging matches. Eyes faced directly upward, so the underwater head-
bashing had to be done by touch.

The jaws, teeth, and face of this animal were quite "normal"; its snout was flat top to bottom and its eye sockets faced directly upward. The animal was probably a pond and stream predator, lying in ambush on the murky bottom, awaiting unwary prey. This was an ecological role that evolved many times in separate amphibian tribes. Probably all of *Diplocaulus*'s life—including courtship and mating—was spent underwater. Its young exhibited standard cranial geometry for the role of bottom-predator: a generally wide, rounded skull without significant protuberances. But as it grew into adolescence, a transformation carried it into an exceedingly unusual development of the skull. The rear corners grew outward at great speed. Well before it was fully adult, its skull had become twice as wide as it was long. And even faster grew the hornlike devices at the extreme ends, until at maturity the head was finally four to six times wider than long. Viewed from the top, these heads resembled nothing so much as organically grown boomerangs.

Paleontologists tried to explain the boomerang shape as an adaptation for swimming; supposedly it worked like an underwater wing, imparting hydrodynamic life as the beast swam at high speeds through the Coal Age streams. But *Diplocaulus* was not a strong swimmer. Its body and tail were too flat to have borne the muscles needed for fast underwater propulsion. So hypotheses based on locomotion just don't seem plausible for the boomerang shape. To my knowledge, no one has suggested an hypothesis based upon sex and intimidation as the biological function that might make sense of the grotesque cranial shape.

Diplocaulus was at a disadvantage in evolving organs for intimidating other members of its species. Its life was spent looking up, so it couldn't easily see its neighbors lying alongside. It was moreover an animal that lived in flat areas—hence its very low and wide skull and body, no doubt to help it hide in ambush. And the most vigorous movement available to it was rather awkward undulation along the pond bottom. So how was evolution to work to create a sexually impressive *Diplocaulus*? The evolution of intimidation devices usually operates by modifying preexisting patterns. *Diplocaulus* moved by wriggling across pond bottoms. During its evolution, males and females must often have bumped into each other in so doing. Genes that favored wider corners of the skull could therefore yield an advantage. The longer the hornlike extension, the greater the range of the bumping action. Boomerang-heads couldn't see one another, but they could reach out and bump someone.

While *Diplocaulus* was evolving in the waters of the Late Coal Age and the following epochs of the Early Permian, evolution on land was producing a spectacular show of its own. In water and on land, walking tall has very frequently served as an effective sexual advertisement. Several kinds of modern lizards grow exaggerated spines from the backbone to provide themselves with a dominant profile. The dorsal crests displayed by the Jesus lizards of Mexico, for example, are impressive indeed. But none of this display can compare with that of the long-spined clans that began in the Coal Age. And most dramatic of the Permian finbacks was the predator *Dimetrodon,* a genus that included species up to seven feet long. *Dimetrodon* means "two sizes of tooth"—its razor-sharp dentures varied from large fangs in front to short molars behind. Its jaws were designed like *Tyrannosaurus*'s (long before that creature saw the light of day) and its lethal combination of jaws and teeth made it the king of the Permian deltas.

However, *Dimetrodon*'s formidable head was not its chief characteristic. Far more impressive was the unusually long row of spines rising from its neck, torso, and hips. Complete *Dimetrodon* skeletons are rare, but there is some suggestion that one gender's (presumed to be female) spines were shorter than the other's. Even at its shortest, however, *Dimetrodon*'s spiny back-sail made a splendid spectacle as it strutted slowly, puffed itself up, and displayed itself broadside to potential mates and sexual rivals. *Dimetrodon* couldn't waggle its crest, because each supporting spine was rigidly anchored to a vertebra. But the skin stretching between those spines might well have contained some message indicated by the pattern of scales. Maybe, like many amphibians and reptiles today, *Dimetrodon*'s scales changed color during the mating season.

Dimetrodon's spines generated a hundred years of debate within the paleontological community—a great deal of it unnecessary, in my opinion. The learned Edward D. Cope of Philadelphia made a tongue-in-cheek suggestion: the fin on *Dimetrodon*'s back was a sail to allow it to scud across Permian ponds like a scale-covered racing yacht. Al Romer, who spent a lifetime studying *Dimetrodon* and related clans, believed the sail worked well as a radiator. In the morning, when the creature was cold from the night air, its sail would be turned toward the sun to soak up the warming rays. When the noonday Permian sun became too hot, and *Dimetrodon* was in danger of overheating, the sail could be turned into the breeze for a cooling effect. Two quantitative paleontologists developed elegant mathematical models to show how blood could flow to and from the skin of *Dimetrodon*'s sail to provide both solar heating and wind cooling.

This heating-cooling hypothesis is widely accepted, but it has weaknesses. The chief problem is that *Dimetrodon* had a close relative, *Sphenacodon*, that didn't have a dorsal sail. *Sphenacodon* was identical to *Dimetrodon* in all the details of its anatomy. Only the spines of its back differed. *Sphenacodon*'s spines were only very slightly elongated. If we accept the heating-cooling theory, it would have to be concluded that *Sphenacodon* was very different from *Dimetrodon* in its thermoregulatory adaptation. That implies a most unusual evolutionary development. In today's ecosystem, closely related species usually exhibit far greater differences in their courtship behavior than in the way they use heat. In other words, evolution usually works faster in changing display behavior than in changing thermoregulation. Several genera of lizards alive today include some species with backbone crests and others with no elongation of the spines. Except for the spines, these clusters of closely related species are adaptively very similar. On balance, therefore, it's far more reasonable to interpret *Dimetrodon*'s sail as a display for sex and intimidation. It might indeed have been a radiator—anything sticking out from the body might be. But the overall pattern of evolution implies that display organs evolve more rapidly into grotesque shapes than do such utilitarian devices as radiators.

At least four other early Permian creatures carried equally extraordinary display sails on their backs. A distant relative of *Dimetrodon* was *Edaphosaurus* ("earth lizard"), a small-headed, barrel-bellied reptile, up to eight feet long, that preferred swampy habitats. *Edaphosaurus* evolved its fin totally independently of *Dimetrodon* and even featured extra ornamental devices—knobby crosspieces sticking out sideways from the long spines. *Edaphosaurus* itself had a close relative with simple spines.

Permian amphibians were hardly upstaged by the skeletal

Body billboards in the armadillo toad. *Platyhystrix,* a close relative of *Cacops,* strutted around the Early Permian landscape with a sexual billboard constructed from armor-plated vertebral spines. The three-feet-long *Platyhystrix* and its smaller kin *Cacops* both belonged to a very successful family of strong-legged amphibians that sported armor plate over their spinal columns, a trend that gave them the nickname "armadillo toads." (*Platyhystrix* went even further by evolving armor over some of the ribs.)

theatrics evolved among the finback reptiles. The amphibians evolved an outstanding finback of their own. A strong-legged, three-foot-long amphibian, *Platyhystrix* ("flat-spine"), evolved a dorsal display piece every bit as baroque as the edaphosaur's. Just as with *Dimetrodon, Platyhystrix* had close relatives that hadn't evolved such a crest. A single quarry at Rattlesnake Canyon, Texas, has produced specimens of *Platyhystrix, Edaphosaurus,* and *Dimetrodon.* What was so special about Early Permian times that they should have produced so many giant dorsal displays in such profusion? No one knows.

The Golden Age of the finbacks is, however, a sobering discovery for scientists who believe in the principle of extreme uniformitarianism. This theory insists that evolutionary processes work the same way at all times. But there has never been another age of finbacks to compete with the Early Permian. And why this episode in the evolution of body organization should have remained unique is one of the great unsolved mysteries in the history of life.

The Age of Finbacks ended suddenly about halfway through the Permian Period, and the Age of Head-Butting began. Protomammals (generally called mammal-like reptiles) took over the leading roles in the land ecosystem, and they generally did not indulge in extravagant visual displays—a curious state of affairs because protomammals descended from some close relative of *Dimetrodon.* With them, courtship and intimidation evolved along very different lines. From their very beginning, mammals evolved hornlike growths, thickened skull roofs, or knobs and bumps to cover their faces. Earlier paleontologists were at a loss to explain these cranial excesses (of course the theory of racial decadence was invoked). But in recent years Herb Barghusen, a Chicago anatomist, has developed a strong case for head-to-head butting matches during the mating season as the most likely explanation.

Some of the head-butters had wide, flat snouts, enabling two males to indulge in a pre-mating shoving match. The most extraordinary protomammals were the dinocephalians, the "terrible heads." These animals evolved skulls that, at a distance, looked a lot like a bowling ball with a snout attached. All the skull bones around the forehead, cheeks, and eyes were enormously thickened, endowing the beast with a bony puffiness all over its face. There can be no doubt that such heads were designed for butting. Their necks entered their skulls from a right angle, so a charging male could lower its head and bash its opponent with the mass of its forehead facing forward. When two charging males collided, at full speed, the Late Permian air must have resounded with loud, clunking crashes.

Dome-headed protomammals raise an interesting question

concerning the evolutionary connection between sex and warm-bloodedness. Recently, popular health and diet magazines have discovered what physiologists have known for a century—sex can be a strenuous exercise and often an important way to burn up calories. Sexual practices embrace not only the physical act of copulation, but all the pre-mating ritual, strutting, dancing, brawling, and the rest of it. They can consume enormous amounts of energy. Successful bull elephant seals are tattered, scarred, and exhausted after the long mating season of wrestling matches against rivals. A male moose may lose weight during the rut, because of the exertions expended when running into other males. Mammals can afford to squander vast amounts of energy during courtship and mating since their warm-blooded system produces a huge amount of energy. In comparison, a totally "cold-blooded" animal produces a tiny amount of energy and therefore cannot expend as much effort in sexual athletics. When a five-hundred-pound moose spends 50 percent of its total energy in mating contests, the total calories burned are ten times greater than those burned when a five-hundred-pound tortoise uses 50 percent of its energy for sex, because the tortoise starts out with much less.

Protomammal head-butters of the Late Permian: the one-thousand-pound *Tapinocephalus*

Knobby-snouted protomammal of the Late Permian. *Aulacephalodon* was a plant-eating two-tusker of the Tartarian Epoch and may have had a warm-blooded pre-mating style, complete with vigorous competitive butting and pushing. It was about four feet long and two hundred pounds adult size.

Judging from a consideration of the evolution of available energy, the contrast between the Age of Finbacks and the Age of Head-Butting is intriguing. The amphibians and primitive reptiles of the Early Permian adopted a low-energy approach to sex and display; the tall sails on their backs were visual signals that really didn't require violent body language to be effective. Head-butting was something else again. The robust construction of its head and neck testifies to the vigor of the protomammal's physical effort. There may be an important clue here to a major increase in available energy that had occurred during the evolution of the first protomammals. Quite possibly they expended much larger totals of calories than do typical reptiles today. As we'll see in a subsequent chapter, there is excellent evidence from other sources that suggests the Late Permian head-butters were indeed the first warm-blooded animals ever to evolve. Whatever their metabolic rate, in any case, it must be recognized that the Late Permian head-butters were certainly the first land vertebrates to escalate sexual gymnastics into a high energy level.

Protomammals continued to butt their heads until the end of the Triassic Period, when dinosaurs took over the roles of large herbivore and carnivore on land. The dinosaurs' approaches to sex and intimidation ran the entire gamut from elaborate dorsal displays to head-butting and perisexual symphonies. Largest of the dinosaurs resorting to display was the appropriately named *Spinosaurus,* the "spine lizard," a forty-foot predator probably related distantly to *Allosaurus*. All specimens of *Spinosaurus* are frustratingly fragmentary, but it's clear a tall sail decorated its back, rising six to eight feet above the backbone. A strutting *Spinosaurus* must have been a singular sight—striding on its long hind legs, its head twenty feet above the ground, turning broadside to dare its rival to test its potency. Sex also probably explains the tails of duckbill dinosaurs. Those tails were very deep from top to bottom and well suited for conveying messages. Some duckbills even evolved true sails constructed from vertebral spines over the base of their tail.

Torso and tail were not the only sites of sexual adornment. The Early Jurassic carnivore *Dilophosaurus* evolved a striking cranial profile: two tall crests, very thin from side to side, rose from the edges of its skull from snout to forehead. Lower, thicker crests in the same location decorated the heads of the Late Jurassic *Allosaurus* and *Ceratosaurus* and the tyrannosaurs of the Cretaceous. Dilophosaur crests were so thin that they could have been only for visual effect. But the bony crests of the later meat-eaters were heavy and covered by stout layers of horny skin. *Allosaurus* and its relatives probably butted heads during confrontations on the field of sexual valor. To be sure, a pair of male allosaurs, driven by their hormones, could have bitten each other to death. But I suspect such terminal contests were relatively rare. Evolution tends to favor the sexual soldier who can win multiple contests and who therefore, by implication, does not run the risk of being dismembered in his first bout. Less than lethal horns would confirm this

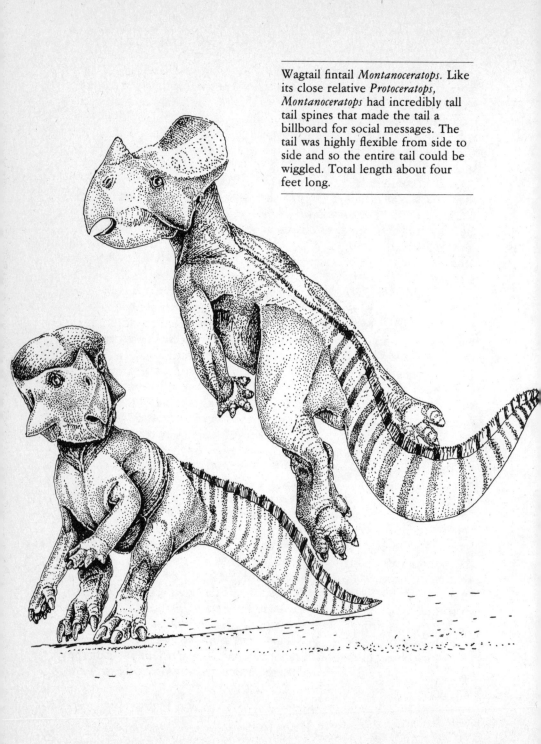

Wagtail fintail *Montanoceratops*. Like its close relative *Protoceratops*, *Montanoceratops* had incredibly tall tail spines that made the tail a billboard for social messages. The tail was highly flexible from side to side and so the entire tail could be wiggled. Total length about four feet long.

general theory. The allosaurs' ancestors possessed little in the way of crests or horns but they did have dangerously sharp teeth and quick-biting jaws. These earlier carnivores could have resorted to biting to settle mating contests, but they were probably restrained by genes that programmed for less dangerous encounters. And the success of genetic changes that increased the disposition toward head-butting among the later, larger carnivores indicates that butting, not biting, was the best strategy for maximizing success in mating. Big mammals show the same pattern—clans with dangerous teeth often evolve nonlethal horns.

Dinosaurs as a class must have owned large quantities of energy to pour into the rigors of courtship and mating, because head-butting evolved several times in different families. *Tyrannosaurus rex*'s massively thick skull edges, covered with horn, represent the acme of the evolutionary trend toward head-ramming among carnivores. Pachycephalosaurs, thick-headed dinosaurs described in a previous chapter, evolved huge, bowling ball–shaped skulls very much like the ones carried by the dome-headed protomammals of the Late Permian. Some Polish scientists have suggested that head-ramming was too powerful to be used against sexual rivals and that the head-down charge at full tilt must have been employed against predators. However, the domes both on dome-headed protomammals and on dome-headed dinosaurs really resembled bony boxing gloves built atop the skull. The boxing glove, even one of bone, delivers a blunt, stunning blow. If evolution had really been working to produce a deadly antipredator weapon, a spearlike point on the head would have been much more effective than a blunt dome. It seems far more feasible therefore to envisage the domes as ideal weapons for sexual contests without incurring the danger of escalating the match to the point where even the winner goes away mortally wounded. A mortally wounded winner doesn't win in the game of evolution because dead heroes can't mate.

Quite the reverse, however, must hold true for the headgear of *Triceratops* and the other long-horned dinosaurs. Recently, a pair of American paleontologists, apparently caught up in the rush to reinterpret dinosaur features as organs for sexual display, suggested the horns and frill on *Triceratops* were not for defending against *Tyrannosaurus* but for display to other *Triceratops*. But the head ornaments of *Triceratops* were simply much too deadly to serve a sexual purpose. Those long, sharply pointed horns were for kill-

Nonlethal butting crests of the meat-eating dinosaurs. *Tyrannosaurus* had thick, low butting ridges, *Allosaurus* had shorter, sharper crests, and *Dilophosaurus* had tall, thin snout crests.

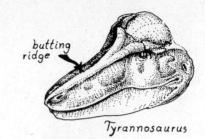

Tyrannosaurus

Allosaurus

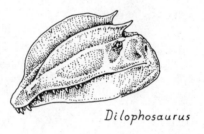

Dilophosaurus

ing. The wide frill, sometimes edged with horn-covered spikes, served for protection against dangerous bites. The little horned dinosaurs, *Protoceratops* and its relatives, probably had started out designed for nonlethal jousting. Their snouts were strong, but only the slight suggestion of a blunt horn grew above their nostrils. Large suites of *Protoceratops* skeletons from the red Mongolian sand dunes indicate that males had larger heads and stronger nose horns than females. So springtime probably did bring thoughts of sex and snout-butting into the minds of the little horned dinosaurs. But that could no longer have been true of their larger, far more lethally endowed descendants.

By far the most spectacular devices for sex intimidation were evolved by the duckbill dinosaurs. Some duckbills evolved large,

Mammal head-butters—the uintathere *Loxolophodon* from the Eocene epoch (above). Early uintatheres—like *Bathyopsis*—had small horns and big, dangerous canines. Later species—like *Uintatherium*—evolved big, blunt nonlethal horns (below).

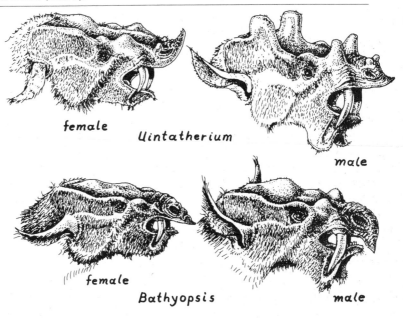

female   *Uintatherium*   male

female   *Bathyopsis*   male

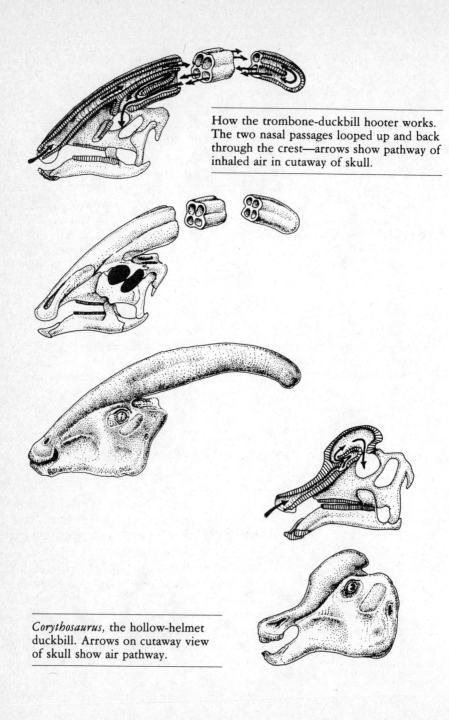

How the trombone-duckbill hooter works. The two nasal passages looped up and back through the crest—arrows show pathway of inhaled air in cutaway of skull.

Corythosaurus, the hollow-helmet duckbill. Arrows on cutaway view of skull show air pathway.

display tails, but in general, they reserved the most vigorous expression of their evolutionary changes for their heads. The duckbills actually divided into four different subclans evolving ever greater cranial specialization. Perhaps the most primitive display was *Kritosaurus*'s Roman nose. This animal had enlarged compartments around its nostrils and probably amplified its bellows and snorts through resonating nasal chambers. A bit more complex was *Saurolophus,* which combined sight with sound: A solid spike of bone jutted backward from its head and probably supported a wide flap of skin; meanwhile its nasal compartments were huge, implying great resonance when it snorted. A strictly audio approach was favored by *Edmontosaurus*. Its head was large and its nasal compartments comparatively huge. The most complex headgear of all among the duckbills belonged to *Parasaurolophus*. Each nostril started with a separate trombone-shaped tube leading from the nose up to the top of the skull, then out and behind the very long crest, a sharp U-turn and back down the crest, then down along the head, and through to the windpipe. Since each nostril had a complete tube of its own, a crest in section reveals four separate chambers—two ingoing and two outgoing.

Hollow-crested duckbills are widely regarded—certainly with good reason—as head-hooters, amplifying and modulating their cries through their crests. All of the varied, hollow cranial ornaments were specialized outgrowths of the normal air tract. In a primitive duckbill, like *Kritosaurus,* as the animal inhaled, the air would pass through the nostrils, then through a short passage in the snout, to the rear of the throat into the windpipe at the base of the tongue. A hollow-crested duckbill complicated the course the air had to follow: in through its nostrils, up and back through special bony tubes growing backward from the nose, up and above the eyes into a huge bony compartment, then down and forward into the throat and windpipe. With all their loops and extra chambers, the hollow-crested duckbills could reproduce in bone some of the qualities instrument makers seek to design into brass and wood today. Duckbill springtime choruses may well have been the loudest and richest cacophony evolution has ever produced. Being large conferred great lung power. A male *Parasaurolophus* would have weighed three or four tons. In the fossils of the Judith Delta in Alberta, six different duckbills were found within a small area, each with its unique nasal amplifier. If all started playing their sexual overtures together, the din must have been thunderous.

The great emphasis dinosaurs placed on auditory messages correspondingly demanded an efficient, sensitive hearing system. And, as a group, the Dinosauria were indeed equipped with good to excellent hearing machinery. All dinosaurs had the skull notches to hold a taut eardrum. And all dinosaur middle-ear bones were thin and delicate, like a bird's, for picking up higher frequencies. In the fluid-filled canals of the brain, the dinosaur's ear was rather like a crocodile's. And since crocodiles today have the most sensitive hearing of any "reptiles," the dinosaurs were certainly tuned in to a wide range of airborne sound.

Taken as a whole, the dinosaurs' adaptations for sex and intimidation simply don't seem to fit the orthodox definition of their cold-bloodedness and lethargy. The abundance of head-butting devices, the extraordinary exuberance of the cranial hooting and snorting apparatuses, are powerful arguments for the idea that the Dinosauria as a whole put a lot of evolutionary energy into mating. Modern lizards, crocodiles, snakes, and turtles simply do not show so many strongly modified organs for high-energy aggression. Male alligators, for example, have deep loud voices but have never invested in cranial remodeling to achieve a wider range of tones. Male rhinoceros iguanas butt one another, but their skulls display nothing to compare with the highly specialized ramming devices of the dome-headed dinosaurs.

How warm-blooded were the habits of dinosaurs when they mated and defended their territories? The cranial evidence strongly points to high energy, to a Mesozoic world where grunting and crashing alternated with hooting and bellowing to rend the tropical silence as multi-ton monsters vigorously hurled their muscular bulk at one another in pain, victory, and frustration.

16

THE WARM-BLOODED TEMPO OF THE DINOSAURS' GROWTH

An animal's metabolism is inextricably connected to many of its characteristics—one of the most important being its rate of growth. Some of the very best evidence for warm-bloodedness in dinosaurs is supplied by the study of their patterns of growth, research begun fifty years ago but nearly totally overlooked by professional paleontologists until very recently. In 1972 I stumbled upon some fairly old monographs about the warm-blooded texture found in the bones of dinosaurs. I was in the process of working on my hypothesis about warm-bloodedness in dinosaurs, relying on my own data about predator-prey ratios and some speculative ideas about the posture of limbs. But I had been ignorant up to that point of clues that came from evidence about bone texture discovered in the 1930s. From 1930 till 1970 the warm-blooded style of the dinosaurs' growth had stood as a potent support for the nonreptileness of the Dinosauria. But since orthodoxy suffocated dissent, no one paid much attention to the data derived from growth rates. In the early 1970s, however, Armand de Ricqlès attacked the problem of growth with such vigor that it became impossible to ignore.

In today's ecosystems, warm-bloodedness leaves its unmistakable mark on the patterns of birth, adolescence, and adulthood. Warm-blooded mammals grow quickly. A German shepherd pup weighing five pounds will become a nearly full-sized adult of 120

Six years from egg to five-ton adult? The giant eight-spiked stegosaur, *Stegosaurus ungulatus,* grew very fast, judging by the bone texture preserved in juvenile specimens, as fast or faster than a warm-blooded rhino or water buffalo.

pounds one year later. And birds grow even faster. Ostriches grow at astonishing rates, from egg to 150-pound bird in as little as nine months. But a young, reticulated python of five pounds in the zoo requires ten to twenty years to reach 120 pounds. And a reptile in the wild grows even more slowly. Box turtles reach sexual maturity at a weight of about four pounds, the size of a small adult cat. A cat reaches breeding weight within half a year after birth, but a wild turtle usually needs five to ten years. Alligators too are slow growers. In its native Florida habitat, the Mississippi alligator requires ten to twenty years to reach two hundred pounds, a weight a lioness can reach in two years.

Our own human species is not a good example of warm-blooded patterns of growth—we are exceptionally slow-growing compared to nearly all members of the Class Mammalia. We linger in drawn-out adolescence, using up fifteen or twenty years to reach our final adult size. A four-year-old hyena, white-tailed deer, or porpoise is already adult and weighs 120 pounds. A four-year-old human weighs about thirty pounds and has just begun to pass through the many stages on the path to full physical and social maturity. The explanation for slow growth in humans probably relates to the bewildering complexity of our adult society. We have to grow slowly in order to absorb the myriad dos and don'ts of our parents' culture. It's much simpler for a hyena to be socially mature because hyena society contains but a few rules and regulations.

Even the most socially complex reptiles—probably the alligators and crocodiles—are still far less subtle psychosocially than the average bird or mammal. An alligator therefore can't blame its overly long prepubescence on its need to accumulate the wisdom and social nuance of 'gator culture. In fact, from an evolutionary point of view, their slow growth is a mistake. Alligators would be much better competitors if they could match the rate of growth of mammals or birds. The primary Darwinian goal for each and every species is to breed—breed early, breed often. In the swamp, there is only a limited supply of food to eat or burrows to hide in or logs to bask on. And the species that fills the swamp with offspring monopolizes the natural economy. Moreover, fast rates of reproduction are powerful evolutionary weapons; they provide an enormous advantage in coping with predators or surviving climatic catastrophes.

The surest method of speeding up rates of breeding is to become warm-blooded. Why do alligators and tortoises continue to grow slowly if this is an inferior evolutionary tactic? There is no defect in their biomechanical system. Turtles and alligators rely on the same basic systems of enzymes employed by mammals. If those systems were exploited at full capacity, an alligator would be able to grow as fast as an ostrich. But reptiles cannot exploit their full potential for growth, because their cold-blooded physiology makes them less effective in gathering food in the wild than a warm-blooded creature. Their fluctuating body temperature forces them to operate their food procurement and growing processes at levels far below maximum for much of their lives. Warm-blooded birds and mammals, on the other hand, may be absorbing nourishment into their digestive systems at rates very close to the biochemical maximum.

A lot of direct evidence proves that present-day Reptilia in the wild usually operate their growing apparatus far below capacity. Wildlife biologists generally study the stomach contents of their specimens in order to study the animals' diet. What is found in alligators is surprising—on average, big crocodilians are empty, or nearly so. Compared to the average lion or hyena, a Nile crocodile spends most of its life fasting. Lizards tell the same story—on average, lizard stomachs are less full of food than are mammals'. The ultimate proof that reptilian growth usually works far below maximum capacity comes from what happens when reptiles are kept in cages warmed to their favorite temperature and are continuously provided with food. This turns out to be the only way to accelerate an alligator's rate of growth to the maximum: Keep it warmed all day long, seven days a week, and keep forcing protein-rich food into it. Most research scientists couldn't afford to perform such an experiment, but the private sector has come to the rescue. Alligator and crocodile skins sell to a lucrative market for shoes and handbags, and since conservationist measures restricted hunting of wild specimens, enterprising businessmen started to farm them. Others have even tried turtle farming, because giant sea turtles produce highly esteemed meat. On all these farms, crocodiles, alligators, and turtles grow almost as fast as warm-blooded mammals. The only side effect the reptiles suffer is an occasional attack of gout from the combination of rich diet and lack of exercise.

Did the dinosaurs have a fast-growth weapon in their adaptive arsenal? Did *Tyrannosaurus rex* grow to breeding weight in five years? Was part of the reason dinosaurs enjoyed such unchallenged dominance throughout the Mesozoic that they bred earlier and bred faster? A most intriguing question. Genuine mammals were present during that time and were potential ecological threats as their later development demonstrates. But mammals never did evolve to large size until after the dinosaurs had died out. Maybe the dinosaurs were just too good at growing quickly?

How can the dinosaurs' growth be measured? An accurate estimate can be derived from the texture of fossil bone. A thin slice can be cut from a fossil-bone chip and glued to a glass plate. It can then be ground so thin that light shines through. The slice under the microscope will allow an observer to see precisely how the bone crystals were arranged as the bone grew. This transparent thin section, as it is called, is standard today for analyzing the structure of the widest variety of hard natural substances—rocks, metals, sin-

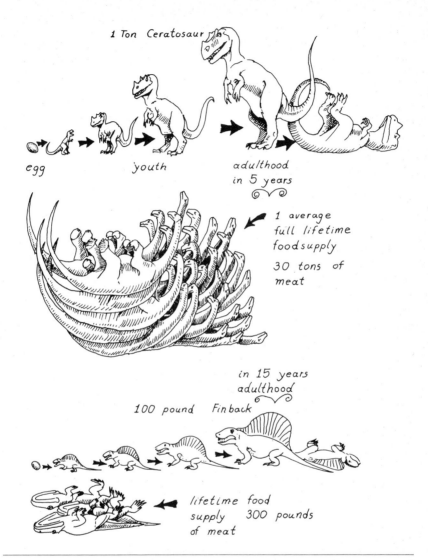

1 Ton Ceratosaur

egg   youth   adulthood
in 5 years

1 average
full lifetime
food supply

30 tons of
meat

in 15 years
adulthood

100 pound Finback

lifetime food
supply 300 pounds
of meat

Dinosaurian inefficiency. If dinosaurs were truly warm-blooded, then it would take thirty tons of meat to raise a one-ton ceratosaur from egg to adult. But a cold-blooded finback from the Permian Period would be much more efficient—three hundred pounds of meat would raise a one-hundred-pound finback. Still, the much higher metabolism would let the dinosaur grow much faster.

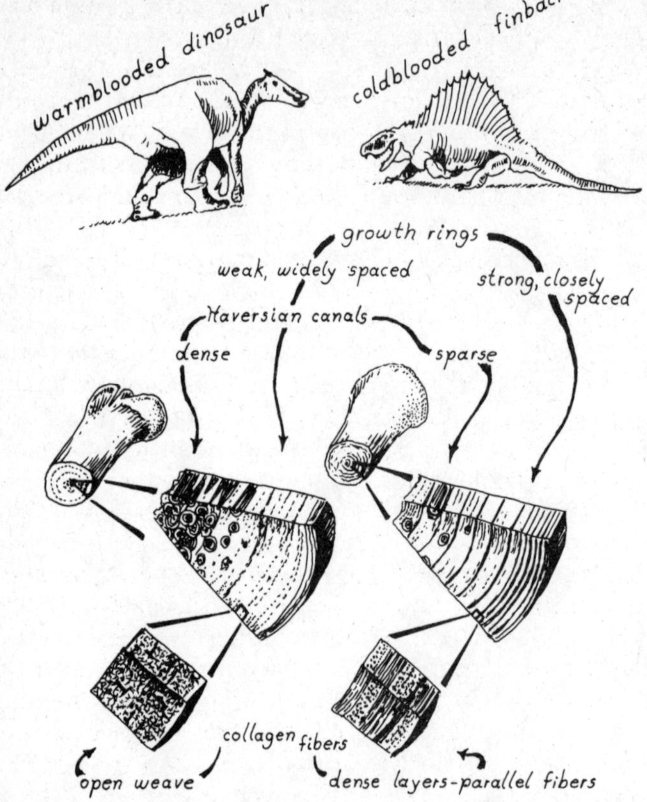
warmblooded dinosaur
coldblooded finback
growth rings
weak, widely spaced
strong, closely spaced
Haversian canals
dense
sparse
collagen fibers
open weave
dense layers-parallel fibers

How dinosaur-bone microtexture differs from the texture in primitive cold-bloods.

gle crystals, and bone from living species. Geologists originated the thin-section technique in the 1830s, and it wasn't long before paleontologists took it over for fossils. Since bones grow by adding crystals of mineral, the microtexture of bone indicates how fast the body grew.

When nineteenth-century scientists examined slices from fossil bones and teeth, they found that dinosaur bone looked very like mammal bone. Both dinosaurs and mammals possess many tiny channels for blood vessels running through their bone, and both have the curious structures known as Haversian canals—long cylinders, pointed at both ends, where bone mineral had been dissolved and then redeposited in concentric layers. When cut in cross section, Haversian systems look like tiny onions sliced across the middle. Cut lengthwise, they resemble tiny, multi-layered electrical cables.

Using this technique, early twentieth-century scientists assembled an impressive body of histological data about the entire 400-million-year history of vertebrates from the earliest fish to Neanderthal Man. And all the dinosaur bone slices looked more like mammal bone than reptile. These studies were masterfully summarized in a series of papers published in the early 1950s by two histologists from Texas, Enlow and Brown. But their labors had astoundingly little impact. The standard textbooks on dinosaurs had hardened into "cold-blooded" orthodoxy. And so the work done by these histologists remained in relative obscurity.

The material concerning the texture of the dinosaurs' bones and their rates of growth burst upon the world in the 1970s thanks to two independent rediscoveries of the old published work. By purest chance I ran across some articles dealing with the texture of dinosaur bones and subsequently was led to the wealth of information published by Enlow and Brown. They had cut samples from dozens of dinosaurs and concluded that the animals may have been warm-blooded. In 1972, I published a paper in the journal *Nature,* calling attention to all this forgotten material. Meanwhile, in Paris, Armand de Ricqlès had also rediscovered the question of bone texture and had inaugurated a massive research project involving hundreds of new thin sections. I've cut a few fossil thin sections myself, but de Ricqlès is the unchallenged bone-slicing champion of all time. He has cut and polished samples from nearly every type of prehistoric vertebrate. And the evidence he provides for warm-blooded growth patterns in dinosaurs is overwhelming and incontrovertible.

After de Ricqlès and I had published our first papers, several biologists and paleontologists published critical reactions that were, to be polite, difficult to take seriously. I had cited Enlow and

Brown's argument that Haversian canals were evidence for warm-blooded dinosaurs. A student from Duke University disagreed, arguing that some primitive cold-blooded reptiles had Haversian canals and therefore that the presence of them in some later dinosaurs proved nothing about warm-bloodedness one way or the other. But this critic had missed the point. Enlow and Brown hadn't been convinced by a few isolated Haversian canals, they were impressed by the enormous *abundance* of them found in some dinosaurs, an abundance far exceeding that typical of reptiles and matched only in big mammals. Dinosaur canal systems are often so tightly intergrown that the thin slice of bone looks like whole clusters of those onions cut in cross section. Now, some old large crocodiles develop a few scattered Haversian canals, and the very primitive fin-backed reptiles from the Permian often showed a few. But only mammals and dinosaurs possess whole swarms of Haversian systems. All through their adult life these animals grow new canals. The dense crowding of the systems forced the newly growing canals to cut into the old ones. No cold-blooded animal, past or present, has ever evolved such densely packed Haversian systems.

Although Haversian canals are somehow connected to high metabolism, no one knows precisely how they work. Adult humans display very densely packed canals, but in young people they aren't as abundant. It can be argued, then, that densely packed systems are needed more by adults than by the young. Yet some mammals and some dinosaurs have no Haversian canals at all, even when they are fully mature. In general, the canal systems are better developed in meat-eaters and omnivores than in strictly vegetarian species, but there are many exceptions. Haversian canals keep dissolving and redepositing bone mineral as they form, so maybe their purpose is to maintain some of the minerals in a fluid state so that the calcium ions can enter the bloodstream quickly if some bodily organ needs calcium in a hurry. Whatever their role, densely packed Haversian systems are clearly marked "for warm-bloods only."

The argument from Haversian systems for warm-bloodedness is only one part of the case that can be made from bone texture. Some dinosaurs lacked Haversian canals, as do some big mammals. But all dinosaurs show direct evidence of fast growth rates.

Bone consists of two materials: (1) the bone mineral, crystals of calcium phosphate; and (2) strands of tough connective tissue called collagen (the same material that also gives strength and elasticity to our skin and muscles). When bone grows slowly, the collagen fibers are wrapped in layers one atop the other, all around the outside surface of the bone. In any one layer, all the strands tend to lie parallel to one another, but the direction of the strands alternates from one layer to the next. Bone mineral forms within the collagen as long, pointy-ended crystals that lie parallel to the strand. The geometric result of slow growth is what de Ricqlès called "lamellar" bone: each subsequent layer of fiber contains densely packed crystals all oriented in one direction. When this type of bone is cut in thin section across its grain and put under the microscope, the alternations in the direction of the fibers catch the rays of polarized light and show up as alternating circles of bright and dark—quite a pretty light show. Crocs and turtles, and most other big reptiles, display this texture. If orthodoxy were correct, dinosaurs should also have this cold-blooded style of texture in their bones. But they don't.

Fast-growing bone has quite a different microtexture. When a young bird or mammal goes through the characteristic warm-blooded spurt of growth, its bones grow so quickly that the collagen fibers aren't given the time to be laid out in neat parallel rows. They are thrown together in an irregular jumble of loosely packed bundles going every which way. De Ricqlès called this "woven bone," because under the microscope the crystal rows resemble a loosely woven fabric. Did dinosaurs have such woven bone? Absolutely. Fossils of young dinosaurs routinely display the texture characteristic of fast growth. And dinosaurs must have kept growing fast until nearly full-sized, because woven bone is the dominant microstructure found in most subadult specimens as well.

Brian McNab is a very good, very thoughtful environmental physiologist at the University of Florida. He has published classic work on how animals of different sizes use their metabolism to meet the challenges of climate. He has written, for example, a superb paper on the world's smallest mammal, the pygmy shrew, a dynamo weighing two grams (one fifteenth of an ounce), ten times smaller than the average white mouse. But when it came to evaluating the texture of fossil bone, McNab was misled by theories of mass homeothermy. He wrote a paper claiming that the only reason dinosaurs displayed a mammal-style bone texture was that they were so big, their bulk alone allowed them to maintain their body temperature more or less constantly high without the need for warm-blooded physiology. His argument completely ignored the fact that giant cold-blooded crocs and turtles never develop a fast-growth bone texture. Really huge living crocodiles can weigh half a ton, as big as the average *Allosaurus*. But in the wild they never possess fast-growth bone texture and never have densely packed Haversian systems as adults. Giant tortoises never develop fast-growth bone either. It is therefore impossible to argue that the texture of the dinosaurs' bones was simply the result of their great size.

McNab's argument also overlooked all the dinosaurs that were not gigantic. Many—both vegetarian and carnivorous—reached adult size between ten and a hundred pounds, no larger than scores of modern croc and turtle species. All these medium-sized dinosaurs also had a mammal-style bone texture, whereas crocs and turtles, and snakes of the same bulk don't. In fact, all dinosaurs of all sizes had a mammal-style bone texture, while all crocs, turtles, and lizards of all sizes have typically reptilian textures.

The final argument from the texture of bones derives from the growth rings. Most people are familiar with growth rings in oak and pine: thin, dark lines are winter wood; wide, pale bands are summer wood. Probably not too many people know that growth rings also form every year in animals. When winter comes, the snapping turtles burrow into the pond bottom to escape freezing. The bones nearly stop growing and lay down a thin, dark layer. The following spring, the turtles start eating and growing and laying down a thick, light layer of growth in their bones. Deer also stop growing in the winter, and the slowdown is marked by a thin, dark line in the bone. Game wardens, in fact, use growth rings in the roots of teeth and in bones to enforce laws against shooting underage bears, moose, coyotes, and beaver. If the warden suspects foul play, he can have the growth rings counted at a lab, and obtain a conviction on the basis of them. Winter isn't the only circumstance that can stimulate growth rings in bones or teeth; anything that cuts off food or water will have the same effect.

Could growth rings tell whether extinct animals were warm-

Brontosaurus
mum – 25 tons

newborn
500 lbs.

Protoceratops
mum
80 lbs.

hatchling
½ lb.

pelvic outlet

LEFT: At birth, a brontosaur was about one fifth adult height and about one one-hundredth adult weight. Bone texture shows that growth was faster than elephants today. (The brontosaur young were too big to be laid in eggs, so the newborn probably passed alive through the mother's large pelvic outlet.)

RIGHT: *Protoceratops* laid relatively large eggs, and the young grew as fast as ostriches do today.

hatchling
bounces
Duckbill
mum   3 tons

Largest growth gap—a hatchling duckbill weighed only one sixteen-thousandth as much as its mother, but bone texture shows that growth was so fast, adult size was reached in a few years.

A few nondinosaurs had fast-growth bone texture in the Mesozoic. The fish-lizards—the lichthyosaurs—were fast growers. (Shown here is the twenty-five-foot *Temnodontosaurus* from the Early Jurassic attacking *Plesiosaurus*.)

blooded or cold-blooded? Maybe, under careful analysis. Both warm-blooded and cold-blooded animals today can develop growth rings in habitats where winter becomes severely cold. But in warm climates where the dry seasons aren't too extreme, cold-blooded species tend to have better-developed rings than warm-blooded species. So, if fossils came from an ancient habitat with a warm climate, it could be expected that warm-blooded animals would have more poorly developed rings—on average. It must be remembered only the average condition is really significant because some warm-bloods will have well-developed rings. Now, in many of the bones Armand de Ricqlès cut from the primitive reptiles and amphibians of the Coal Age in Europe and North America, he found growth rings. The Coal Age environment was warm, tropical. He therefore concluded these growth lines were the products of typically cold-blooded physiology. But growth rings were much less common in dinosaurs, and so he concluded that dinosaurs must have had a more mammal-style rhythm of growth.

Yet some dinosaurs did have yearly rings. A pair of Canadian paleontologists found them in the teeth of duckbill dinosaurs and tyrannosaurs and loudly declared their evidence proved the dinosaurs were cold-blooded. Their conclusion was hardly justifiable since they hadn't taken into account the fact that growth rings are very common in the teeth of some warm-blooded mammals living in tropical habitats (—lions and hyenas in East Africa have such rings—) and that these mammals usually have more sharply defined rings in their teeth than in their bones. Moreover, if we compare the *average* condition of crocs and dinosaurs from any one habitat, the crocs invariably have better-defined growth rings and more of them, just as East African crocs today exhibit better rings than the mammals in the same locale. Finally, the Canadian dinosaurs actually showed the mammal-style pattern: rings in the teeth but not in the bones.

Some other scientists have found growth rings in the limb bones of dinosaurs—in one specimen of *Allosaurus,* in one brontosaur from England, and in another excavated in Madagascar. A great deal was made of each of these specimens with rings, but all the hundreds of dinosaur specimens with no rings whatever were ignored. Were some dinosaurs cold-blooded, then, while others were warm-blooded? A theoretical possibility. But the evidence from growth rings certainly does not *prove,* as orthodoxy would have it, that *any* dinosaur was cold-blooded. Growth rings merely prove that growth stopped during one part of the year. The only useful way to derive evidence from them must come from a broad survey: If, on average, dinosaurs were more warm-blooded than crocs or turtles, then in any one fossil habitat more and better-defined growth rings should be found in the crocs and turtles. And that is exactly what is found. At Como Bluff, all the turtles and crocs display sharply defined growth rings, but the dinosaurs only rarely. The same is true in the Late Cretaceous deltas of Montana and Alberta.

Paleontology's treatment of the evidence from bone texture is an example of what I call the "harrumph-and-amen" syndrome. Enlow and Brown and others pointed to many dinosaurs with a warm-blooded type of bone texture, and the orthodoxy snorted, "Harrumph—all that means nothing." But when a few growth rings were discovered in dinosaurs, then orthodoxy responded with a fervent "Amen, we knew it all the time—dinosaurs were cold-blooded reptiles."

A piece of fossil bone is rich in textural meaning—a labyrinth of canals left by blood vessels, a three-dimensional basketwork of crystals, a diary of the animal's life written in the layers of mineral fabric. Good times and bad are written there, seasons of plenty and seasons of drought. These ancient diaries can be opened and the stories of dinosaur lives read, their youthful exuberance in growth, the pulse of blood flow in maturity. Ever since the 1830s these diaries have been telling the scientific community about dinosaurs' growth and their life style. And the message is clear—not the story of one or two isolated cases, but the chronicle of whole dynasties. Defenders of orthodoxy may quibble over a growth ring here or an isolated Haversian canal there. But the overall point cannot be ignored. Dinosaurs grew mammal-fashion; they grew fast and bred early. And their dynamic approach to quick maturity must have been one of the most powerful weapons in their adaptive arsenal.

17

STRONG HEARTS, STOUT LUNGS, AND BIG BRAINS

Paleontology is a hard science. The objects of our study are crisp, solid bones, free of any soft bits such as blood vessels or muscle tissue that will only rarely indicate their existence in fossil remains. But the soft bits, especially the heart, lungs, stomach, intestines, and brains, do have their place in studying dinosaurs. The gizzard stones and intestinal digestive systems have already been discussed in an earlier chapter. Here we must investigate the other soft organs of the Dinosauria—the heart, the lungs, and the cerebral equipment.

The important thing about hearts and lungs is that evolution designs them to withstand the stress of prolonged intense activity—what physiologists call "exercise metabolism." Metabolic rates during exercise are always many times higher than the rate of average standard metabolism. When we humans sit doing nothing, our metabolism works our heart-lung apparatus at only one twentieth of the maximum capacity of a well-trained athlete. Human hearts and lungs are powerful organs, and to set them going full throttle we have to engage in prolonged, strenuous exercise—cross-country skiing, long-distance swimming, or intense gymnastic exercise (bowling doesn't do it). Then our thickly muscled heart and minutely compartmentalized lungs extract oxygen at maximum rates and send it to all the exercising organs at full speed. A human lung, or a dog's or horse's, is full of tiny cells so that the tissue area is

Dinosaurs with two brains? Stegosaurs—like these *Kentrurosaurus* from the Late Jurassic of Tanzania—had an enlargement of the spinal cord in the sacrum.

maximized for the exchange of gas from air to blood, and vice versa (oxygen must be vented in, carbon dioxide must be vented out of the bloodstream). Such a system is necessary for surviving among the vigorous confrontations of a fully warm-blooded ecosystem. The physical arrangements of heart and lung among birds are often unique, but the avian system has the same high capacity for exercise found in the most advanced mammals.

Modern lizards have no need of a heart-lung system with anywhere near as much capacity. An iguana's lung is a simple sac, a sort of limp-walled balloon that can supply oxygen at rates only one tenth as high as a dog's or chicken's of comparable body size. Iguana hearts too are of low capacity, much smaller than a heart from a mammal with the same size of body. Some other lizards— the monitors, especially—have more thickly walled hearts and more complex lungs than do iguanas, but no present-day reptile can match the heart-lung system typical of active, modern, advanced mammals and birds.

How, then, was the heart and lung system of the dinosaurs arranged? According to orthodoxy, the dinosaurs' physical exertions alternated between episodes of stolid shuffling and long periods of somnolent inertia. But the dinosaurs' limbs, as we have already seen, were not constructed for that kind of movement. *Tyrannosaurus*'s legs were built for speed, vigorous prolonged exercise. To keep such huge muscles functioning, tyrannosaurs would have needed powerful hearts. There may even be some direct evidence for the size of those hearts. Today, whenever a species has a very small heart, the front end of the ribcage can be extremely narrow, because the only internal organ filling it is the heart muscle. Iguanas are a good example: their hearts are tiny and their ribcage seems very constricted in the front end. The ribcages of dinosaurs, by comparison, tended to be very deep, and the anterior ribs were often thick and long. What filled those deep noble chests? Quite probably thickly walled hearts of heroic dimensions.

To maintain bodily activity at high levels, dinosaurs would have needed lungs with a capacity to match the output of their hearts. And those lungs have left some clear traces in their skeletal anatomy. Many dinosaurs had hollow cavities in their vertebrae. A single bone of a *Brontosaurus*'s spine is so full of holes and indentations that the actual bony tissue is reduced to thin partitions, often a few millimeters thick, folded and convoluted many times to produce the major structural contours. *Allosaurus* and other meat-eaters also had such hollowed-out vertebrae, though to a lesser extent than in the brontosaurs. What filled the vertebral holes and hollows is not difficult to see because very similar hollow backbones can be found in today's birds. In them, the hollows are filled by air sacs connected by tubes to the lungs. Avian lungs are exceptionally efficient, better at extracting oxygen than mammal lungs, and this is due to the air sacs and the resultant pattern of air flow. The lungs of mammals and lizards have one fundamental flaw— they're dead-end organs. Air must be sucked into the alveolar sacs and then squeezed directly out again. Such a method is inefficient. The lung would work better if air flowed in one direction only, across the lung's surfaces and then out the throat. Birds solve this problem by means of their complex system of air sacs. They draw the air first into the system of air sacs. Then they pass it from the sacs through the lung tissue proper, so that the exchange of gases between air and blood can happen as the air flows in one direction, on its way out. Physiologists call this a "countercurrent exchange" (the blood in the lung's surfaces flows in the direction opposite to the flow of the air). Countercurrent exchange allows oxygen extraction and carbon dioxide venting at much greater efficiencies than are possible with our mammalian dead-end lungs.

The dinosaurs' vertebral hollows are so similar to birds' that there can be little doubt an avian-style system of air sacs was at work in these Mesozoic animals. Moreover, the holes in the bones represent only the periphery of the total system. Birds locate their largest air sacs between their flight muscles and in their body cav-

Deep dinosaurian lungs. *Brontosaurus* had a very deep chest that must have enclosed large lungs and a large heart. A crocodile of the same weight would have a much shallower chest and far weaker cardiac and pulmonary machinery.

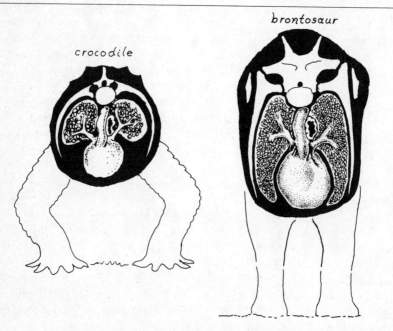

crocodile

brontosaur

ity. *Allosaurus* and *Brontosaurus* must have had huge air sacs distributed throughout their viscera. And it's a good supposition that they had evolved lungs fully capable of intense, prolonged exercise. No typical cold-blooded reptile has ever evolved a dinosaur-bird type system of air sacs.

Some dinosaurs (duckbills, horned dinosaurs) exhibited no vertebral hollowings, but I suspect they had located air sacs fully within their body cavity. Most primitive dinosaurs had some hollowed-out vertebrae, so the ancestral dinosaurs were probably equipped with air sacs. Many modern species of bird that have air sacs have solid vertebrae without any hollows (like the duckbills and horned dinosaurs). In these avian species, the big air sacs remain nonetheless well developed within their body cavity. So duckbill and horned dinosaurs, though they lacked peripheral sacs inside their vertebrae, may well have had the main system of sacs operating between the other internal body organs.

Myths about dinosaurs die hard. As a child, I first heard the tale of the dim-witted, double-brained *Stegosaurus,* a fable created in the 1880s and still popular a century later. This story had two sources: the pea-sized brain of *Stegosaurus,* and its legendary after-brain.

Let us consider the problem of brain size. Large animals need large brains, because the mass of their active cells requires many nerve channels to carry signals to and from the brain, and because the brain needs adequate capacity for processing information about all this physiological activity. An elephant is roughly as "smart" as a dog—they have equivalent powers of learning. But the elephant requires a three-pound brain to achieve this intellectual level while the dog needs only four ounces. To gauge a stegosaur's intellect, therefore, it is only necessary to compare its brain size to that of a mammal of the same body size. It's not hard to measure brain size in dinosaurs. Their braincase was nearly completely enclosed by thick bone. When the first stegosaur skulls were discovered by Professor Marsh in 1878, scientists at Yale simply sawed open the braincase, and measured its volume. Calculating the brain's live weight was straightforward. In big animals, the brain fills about half the braincase; the other half is filled out by connective tissue, a sort of cerebral padding, packed between the outer surface of the brain and the inner walls of the braincase.

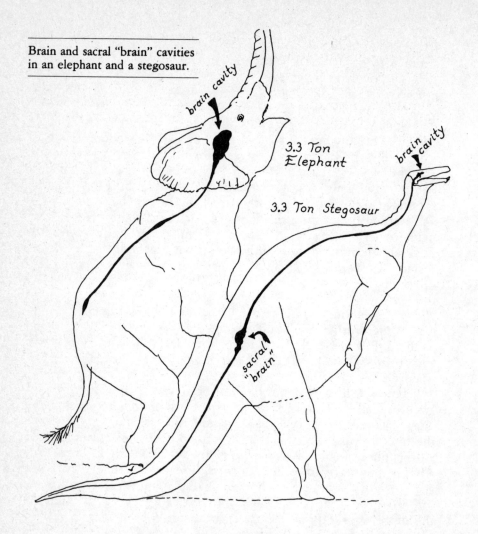

3.3 Ton Elephant

3.3 Ton Stegosaur

brain cavity

brain cavity

sacral "brain"

Stegosaurus's live brain size occupied about half its braincase. Calculations made on the basis of that suggested *Stegosaurus* had been monumentally underbrained compared to modern mammals of the same size. This dinosaur had been endowed with two ounces of brain cells at most. An elephant has a brain at least thirty times larger. So dim-wittedness is a judgment hard to avoid when thinking about *Stegosaurus*. But the cranial end was only half the story of its neurological system. Between its hip sockets, deep within the backbone where the pelvis attaches to the vertebral bodies, was a huge enlargement of the spinal cord, a swelling of nerve tissue thirty times larger than the volume of the brain itself—the stegosaur's "second brain." Professor Marsh found sacral enlargements in other dinosaurs, too, but none as spectacular as in *Stegosaurus*.

What did *Stegosaurus* accomplish with this afterbrain? Enlargements of the spinal cord came as no surprise to experienced anatomists—nerves enter and leave all along its length from head to tail. Each vertebral segment is endowed with its own set of outgoing and incoming signal lines, and wherever organs are especially big and complex, the cord is swollen by additional nervous tissue to help organize preprogrammed reflexes. For example, extra nerve centers are needed to regulate the sequence that makes all the muscles operate in the proper order to carry out the smoothly coordinated movements of a leg or a tail. *Stegosaurus* had huge hind legs and especially huge tail muscles, complexly subdivided and capable of immensely powerful movements. Enlargements of the spinal cord in the hip area therefore made perfect sense. Ostriches are similarly heavy in the area of their locomotor muscles, because their wings are atrophied and most of their contractile tissue is concentrated in the hind legs. And ostriches have an enlargement of the spinal cord inside the hip vertebra. *Stegosaurus* weighed twenty times more than the largest ostrich, so its hind-end demand for coordinated nervous activity was far greater. The other large-tailed, big-rumped dinosaurs also tended to develop very sizable sacral enlargements—*Brontosaurus* was especially well equipped in that regard.

Big-rumped dinosaurs certainly did not "think" with their afterbrain. "Thinking" is usually defined as the highest level of neurological exercise, encompassing analysis of incoming stimuli in the context of experience stored in the memory, and decision making that draws upon the inborn instincts and the immediate perception of circumstance. In all the Vertebrata, only two types of brain tissue carry out these functions, and both types are found only within the braincase, inside the skull. Mammals think with their cerebrum, the part of the midbrain that first evolved to handle information coming in from the sense of hearing. Most present-day mammals have enlarged cerebral compartments, so large that the cerebral lobes, which grow upward from the midbrain, expand forward to cover the underlying forebrain completely. Birds think with expanded midbrain tissue too, but the avian intellectual apparatus develops from a different layer, the corpus striatum. Both bird and mammal thinking organs, however, look alike from the top. And if a bird's brain is dissected, its overall shape strongly resembles a mammal's: a pair of expanded "thinking" lobes covering most of the brain stem.

Never, absolutely never, does "thinking" tissue develop in the sacral enlargements of modern birds or mammals. If the midbrain lobes were removed from an ostrich, it could still run in circles for a while, because the system of muscle-coordinating relays would still be intact. But it certainly couldn't think, learn, remember, and most certainly couldn't make decisions. Therefore the sacral nervous tissue in the stegosaur's rump would have helped it move gracefully and swing its spiked tail with dangerous precision. But this "afterbrain" wouldn't have added even one small storage area to its capacity for remembering and deciding.

As Professor Marsh's laboratory staff examined brain after brain, from dinosaurs, fossil mammals, alligators, he espied a general trend in cerebral history, a common thread running through 400 million years. His observation became known as Marsh's Law. It stated that on average, any evolutionary line of birds' or mammals' brains grew steadily in size over millions of years. And, on average, for any given body size, present-day species were brainier than their ancient ancestors. The modern jaguar has twice the brain size of the jaguar-sized saber-toothed cat that stalked the Nebraska woodlands thirty million years ago. And the modern loon possesses twice the brain of the loonlike birds Marsh excavated from the Cretaceous sediments of Kansas. Most of this evolutionary upgrading was focused in the centers of higher learning—in mammals, the cerebral lobes had enlarged under the guidance of natural selection; in birds, it was the corpus striatum. Primates— the monkey-ape clan—scored especially high in the cerebral sweepstakes through the ages. A chimpanzee has a brain four times larger than a jaguar of the same weight. But the example ne plus ultra of Marsh's Law is ourselves. The average human brain is seven or eight times larger than that of the average modern mammal of the same size. And our bulging cerebral lobes are forty times the size of those found in the average 120-pound mammal of the Paleocene Epoch, sixty million years ago. Since all the dinosaurs' higher intellectual functions had to be carried out inside their tiny braincase, *Stegosaurus* and all the other large dinosaurs must have been single-brained and dim-witted—or so it must be supposed.

But Marsh's law doesn't apply to true Reptilia. The brains of living alligators are no larger than those of their most ancient crocodilian ancestor of Mid Triassic times, over 200 million years ago. Turtles, too, took the low road in the cerebral race. Frogs, snakes, salamanders, and most fish have been content with the low-capacity mental equipment that was in vogue back in the Coal Age, 300 million years ago. Since turtle, lizard, frog, and snake species together outnumber mammal species three to one, it must be concluded that dimwits can be part of an ideal adaptive mode, at least for small creatures.

Dinosaurs—with few exceptions—showed few evolutionary tendencies toward developing greater intellectual prowess. Most dinosaurs had brains no greater in size than a turtle or croc of the same bulk. And the pin-headed dinosaurs with their tiny skulls relative to their body mass had outstandingly small brains.

Big-headed dinosaurs, such as *Triceratops,* had larger brains than stegosaurs but were still far short of the cerebral capacity of any large modern mammal. Were dinosaurs then incredibly stupid? Most popular works today still say so. And it is difficult to deny that most dinosaurs would seem dullards compared to a good Labrador retriever, circus elephant, or jaguar. There is a rough correlation between brain size and intelligence today, and there must have been in the Mesozoic. Humans, chimps, large dogs, and adult alligators are roughly the same size. Humans, if they're careful, can outwit chimps. Chimps routinely outwit large dogs. And large dogs can learn more and make cleverer decisions than can alligators. Most dinosaurs probably would have fallen in the category of the alligators, and no living species with brains as small as the dinosaur's would be called clever.

As soon as the dinosaurs' average brain size became widely known, some paleontologists jumped to the conclusion that tiny brains proved cold-bloodedness. At first sight, this notion seems reasonable. No living cold-blooded reptile has a big brain. Birds and mammals are the only warm-blooded vertebrates in the present ecosystem, and both have high average brain-to-body weight. Some physiologists even went so far as to claim that large brains *caused* warm-bloodedness—presumably the physiological coordination required to balance heat production, sweating, panting, and blood flow implied large areas of cerebral circuitry. But this hypothesis is demonstrably wrong. Mammals and birds don't use their expanded midbrain lobes for thermoregulation; that chore is carried out in the brain stem, the oldest part of the brain in evolutionary terms. There would be no difficulty in guiding a warm-blooded system with an alligator-sized brain stem.

The link between being warm-blooded and having big brains must be, at best, an indirect one. Brain tissue is vulnerable to changes in temperature. Human brains addle when heated to 108°F even for a few minutes, and higher cerebral functions become erratic when the brain is chilled below 90°F. So warm-bloodedness and a constant body temperature are prerequisites for a large brain. It may well be that warm-bloodedness evolved first, and the evolution of large brains followed. Warm-bloodedness would have been an advantage even to an animal with an alligator's brain power, because a constant high body temperature speeds growth, accelerates reproduction, optimizes muscular output, and increases digestive efficiency. And once acquired, it might have launched an evolutionary tendency toward larger brains.

More importantly, a survey of brain sizes in today's creatures leads me to conclude that large brains aren't essential for having a high metabolism. Humans have the biggest brains ever evolved for our weight class. But we don't possess a higher metabolism than a German shepherd, which has a brain one seventh the size of ours. Ostriches have tiny heads for their bodies, and ostrich brains are only one fortieth the size of that of a human of the same weight. Does an ostrich therefore have a metabolism one fortieth that of a human? On the contrary, its metabolism is *higher,* pound for pound, than ours. It is thus very difficult for me to believe that metabolism and brain size evolved in a kind of evolutionary lock step.

I would contend that the only effective way to analyze the connections between the evolutions of brains and of metabolism is to reconstruct the fossil history of intelligence separately from the fossil history of warm-bloodedness. Then the two stories may be compared side by side. The microtexture of bones can be employed to compute metabolic levels. Judging by such criteria, even the primitive dinosaurs of the Triassic Period had a metabolism as high as a modern mammal's. The debut of warm-bloodedness in the Dinosauria certainly occurred when their brains were small. The same pattern emerges in the history of our own Class Mammalia. Bone texture demonstrates that the ancestors of mammals—the protomammals of the Permian Period—had already evolved the essentials of warm-bloodedness quite long before the first large-brained mammal made its appearance. This similar history in both the dinosaur and mammal lines makes a good case for warm-bloodedness coming first, followed, much later, by larger brains.

If dinosaurs were warm-blooded as far back as the Triassic, it could be expected that at least one of their later lines might have evolved some sort of higher intelligence. And indeed some did. Dale Russell from the Canadian National Museum discovered the top of the braincase of a turkey-sized predator named *Stenonychosaurus,* in the Judith River sediments of Alberta. Clearly impressed into the inner roof of this braincase were indications left by a pair of bulging midbrain lobes. Russell concluded that his Alberta dinosaur had possessed a brain at least as large as that of many present-day birds of the same size. The dune sand laid down in Mongolia during the Cretaceous has preserved several skulls of small dinosaurs closely related to *Stenonychosaurus*. The Mongolian species also seem to have carried brains far larger than those of alligators and lizards of comparable weight. These large-brained dinosaurs were evolving quickly in many of their adaptive compartments. And they probably were every bit as endowed as the Late Cretaceous mammals that scampered over those very same sand dunes.

Why didn't these dinosaurs of Alberta-Mongolia continue to evolve ever larger cerebral systems? Why didn't they eventually produce super-intelligent species capable of making stone tools, smelting iron ore, programming computers, or writing master's theses about Dino-Proust? Dale Russell believes they could have if the dinosaurs had been given longer to live. Unfortunately, the larger-brained dinosaurs were among the last, and the merciless hand of extinction fell upon them just as it fell on all the Late Cretaceous groups.

But Russell has indulged in a bit of "what-if" paleontology.

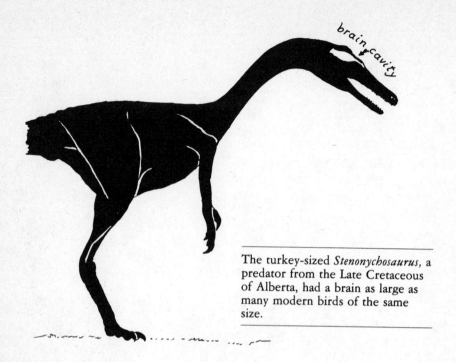

The turkey-sized *Stenonychosaurus*, a predator from the Late Cretaceous of Alberta, had a brain as large as many modern birds of the same size.

What if the Cretaceous extinctions hadn't wiped out the dinosaurs? What if the Alberta bigbrain had continued to evolve? Russell has reconstructed the final evolutionary product which he believes the large-brained dinosaurs would have produced had they survived until the present: a hundred-pound biped with bulging forehead, scaly skin, and clawed hands capable of cleverly manipulating objects. One could quibble about details, but Russell is probably correct in general. Moreover, those large-brained dinosaurs were certainly clever for their time and probably hunted the rat-sized mammals of the period. Russell believes, in fact, that they were the chief predators on Cretaceous mammals, and I tend to agree. As long as they existed, the mammals could not and did not evolve to any size larger than a cat. And if these dinosaurs had continued to evolve past the end of the Cretaceous, it's a fair bet they would have continued to suppress mammal evolution. The

Big-brained ostrich dinosaurs. Close relatives of *Deinonychus* and stenonychosaurs and tyrannosaurs, ostrich dinosaurs like *Struthiomimus* had brains as large as modern ostriches of the same weight.

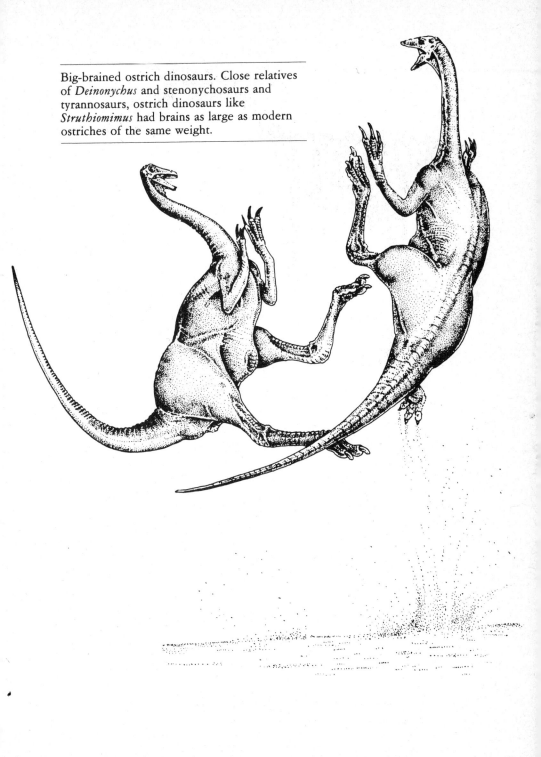

dinosaurs would then have continued their own history of adaptive proliferation, right down to the present era.

Then how would our ecosystem be organized if Russell's scenario had been real rather than hypothetical? You and I, dear readers, would probably be members of some tiny species, eking out a terrified living under the ever-present shadow of a dinosaurian overlord. And this book would have been written by a super-intelligent dinosaurian—a member of the elite species that had evolved four-pound brains, invented language, and built printing presses—on the subject of his own history. If dinosaurs had evolved to write their own history, they certainly wouldn't make the mistake of believing their Mesozoic forebears were cold-blooded.

18

EATERS AND EATEN AS THE TEST OF WARM-BLOODEDNESS

A final test for the theory of warm-blooded dinosaurs is what they ate. A warm-blooded animal consumes ten times as many calories per year as a cold-blooded creature of the same size. If a seven-hundred-pound *Allosaurus* were producing metabolic heat every minute of its life at a rate as high as a modern seven-hundred-pound bear, its meat consumption would have to be enormous. But if that allosaur operated like the traditional cold-blooded dinosaur, then it could bask in the warm Jurassic sunshine, soaking up the solar calories until it reached its preferred body temperature without squandering energy derived from food. Which hypothesis comes closer to the truth?

By 1970, my studies of dinosaur limbs had already persuaded me that dinosaurs were designed for high levels of locomotor activity. I had also suspected that the dinosaurs' metabolism more closely resembled a giant bird's than a giant tortoise's. How else could they have suppressed the evolution of mammals for more than a hundred million years? But how could anyone measure metabolism in a fossil? It seemed a completely forlorn prospect.

Sometime in 1970, Elwyn Simons, professor of primate paleontology at Yale (now a member of the National Academy of Science), provided me with an invaluable insight. He was discussing the fossil mammals he had been excavating for a decade in Wyoming, in Egypt, and in India's Siwalik Hills. He observed that

How much do you feed a sentry-guard lizard? Warm-bloodedness is wasteful—so much body energy is spent on keeping warm. A one-hundred-pound guard dog (plus puppies) demands one thousand pounds of wet dog food per year for an active outdoor existence. But cold-bloodedness is far cheaper. A one-hundred-pound guard lizard (plus hatchlings) is happy with only one hundred pounds of wet lizard chow per year.

numerous large predators were never found in the fossil record; they were always rare. This was because the big meat-eater subsisted at the very top of the ecological pyramid. Its food had to come from the plant-eaters below. And it took roughly a hundred zebra to maintain the supply of meat for one lioness and her cubs. I realized his remarks about the scarcity of predators would apply perfectly to dinosaurs. If predatory dinosaurs required as much meat per week as warm-blooded mammals, then they would have to be rare. The predator-prey relationship might well serve therefore for the calorimeter I was looking for.

The theoretical concept is straightforward: The higher the metabolic needs of a predator, the scarcer in number it will be. To determine the allosaurs' metabolism, all that was required was a count of the number of specimens and a comparison with the number of prey specimens found in the same strata. If allosaurs were always rare compared to all their prey, as rare as lions are relative to zebra and antelope, it would provide direct evidence that the predatory dinosaurs needed a very large weekly ration of meat. But if allosaurs were very common, say ten times more abundant relative to their prey than are lions, tigers, or hyenas, it would provide strong support for the orthodox view that dinosaurs shared the leisurely metabolic style typical of snakes and other cold-blooded animals.

I determined on making a predator-to-prey census through the entirety of geological history, from bottom to top, beginning with the very primitive reptiles of the Coal Age, through successive levels of dinosaurs, and ending with the game parks in Africa and India today. So far my studies have taken ten years, but I believe they have been amply justified by the results. They have revealed a spectacular story of metabolic evolution, a saga of hunters and hunted stretching throughout the 300-million-year record left by evolving ecosystems—one that at last places dinosaurs in their proper place in the grand progression of evolution.

Before starting the count of fossil fauna, I sought some confirmation of the idea that the metabolism of predators does indeed regulate their scarcity and abundance. Interactions in nature are often so complex and unpredictable that perhaps counting predators and prey would yield no reliable information about metabolism. For example, even if a species of allosaur was cold-blooded,

Dinosaur energy budgets—the food-chain restaurant metaphor. Imagine that a family of ceratosaurs lived their whole lives, generation after generation, in a restaurant where all the garbage was fossilized in a nearby river. As they died, the ceratosaurs would be dumped into the river along with the chewed remains of all the prey they had eaten.

and therefore could have existed in relative abundance, diseases might keep its number low, much lower than the maximum hypothetically permitted by its metabolism. And many ecological agents could depress the numbers of top predators: parasites, bad weather, fighting between predators, competition from the scavengers. What was needed was at least one test case from living ecosystems to show that predator-to-prey ratios might work as calorimeters for cold-blooded predators.

Spiders came to the rescue. They can be thought of as eight-legged, hairy lizards, for they are perfectly cold-blooded, operating at a very low metabolic level. Because spiders do not hunt over large territories—a few square yards are the entire dominion of a big wolf spider—spider predation is fairly easy to study in detail. Wolf spiders catch most of their prey on the move as they prowl their turf. Hence the analogy between a wolf spider and a cold-blooded vertebrate predator such as a carnivorous lizard or a possibly cold-blooded *Allosaurus* is a good one.

If metabolism determined the abundance of predators, then spiders should produce huge populations compared to their prey. In Africa's game parks, the ratio of predator to prey among mammals is 1 percent or less—that is, there is roughly one lion or hyena for every one hundred large prey animals (zebra, wildebeest, warthog, bushbuck, etc.). Spiders have such a low metabolism that they could, in theory, reach a ratio ten or twenty times higher. And so they do. Study after study showed spider populations achieving levels of 10, 15, and 20 percent of their prey populations, impossibly high by mammalian standards. There's no doubt, of course, that spider ecology is complicated, and that they suffer from the usual share of parasites, diseases, and disastrous die-offs from bad weather. Nonetheless, their cold-blooded metabolism does, on average, show through this overlay of ecological noise. Predator-to-prey ratios work for spiders; they correctly indicate cold-bloodedness.

Would such ratios test equally well for big, cold-blooded vertebrate carnivores? In today's world there exists no predator-prey system in which both predator and prey are large, cold-blooded vertebrates. Pythons (cold-blooded) feed on deer (warm-blooded), and Komodo dragon lizards (cold-blooded) kill pigs and tourists (warm-blooded), but nowhere does a giant lizard or snake feed on giant lizards or snakes as its principal prey. To test the predator-to-prey method of analysis for this case, I had to go into the fossil record, back to the earliest land vertebrates that evolved into the role of large top predator. A top predator by definition is a carnivore that eats the flesh of the largest available prey. It must therefore develop adaptations for dismembering the carcasses that are too large to swallow whole. The earliest vertebrates that evolved the requisite meat-slicing teeth were the fin-backed reptiles, which first appeared very late in the Coal Age, about 300 million years ago.

Finbacks provided the ideal fossil test case for the predator-to-prey concept. In the first place, they were unquestionably cold-blooded. Their anatomy was still on a very primitive level, more primitive in most aspects of their limb and backbone than today's lizards. Even the structure of their bone appeared emphatically cold-blooded under the microscope. The canals left by blood vessels were few and far between, proving that metabolic activity proceeded at a very modest pace.

In the second place, the finbacks were large—early species grew to the size of wolves and leopards, forty to eighty pounds, and later species were larger still, up to two hundred pounds or more, the size of the average lioness. These big finbacks seemed large enough and well enough armed to deal with any animal in their ecosystem. Their heads were proportionately large, and armed with strong killing teeth in front and razor-sharp rear teeth for cutting up even the largest carcass. Moreover, these creatures are found in nearly every fossil habitat: swamps, lakes, streams, swampy floodplains, dried-out floodplains. With such ecological diversity, it was possible to determine whether the predator-to-prey ratio changed from habitat to habitat. And finally, all the species of prey available to them were incontestably cold-blooded as well.

By the time I had fixed upon the finbacks as a key test for my method, I left Yale for Harvard. There I met Al Romer, the world's leading expert on finbacks. He was even fond of them—especially of one genus, *Dimetrodon*. On his office door he kept a cartoon that featured *Dimetrodon* digging up a human skull. Although he didn't assume the role of quantitative paleontologist in his published articles, he was always careful to pick up the skulls and limbs of every creature he found, and so he built up a great store of unexploited data other scientists could use for quantitative research. (Some excavators will "high-grade" a deposit, collecting only the rarer species, thus ruining the sample for reconstruction of the entire ecosystem.) In his office were the results of his life's work—forty years of expeditions to the richest finback-bearing strata of Texas and New Mexico. Romer was always gracious and generously shared even his unpublished information. He supplied me with precise details of what and how much he had found in the quarries.

Even before I started counting, it was obvious that the finbacks' predator-to-prey ratio was more like the spiders' than the

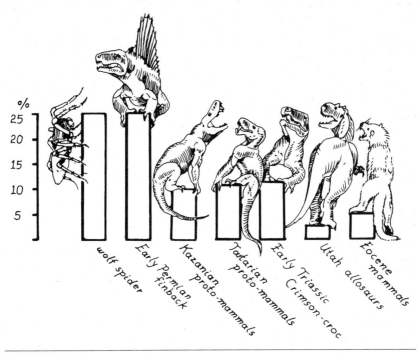

Some predator-prey ratios. Present-day spiders have very high scores—25 percent or more—and so did the finback predators of the Early Permian. Dinosaurs and mammals scored much lower, mostly ranging between 1 and 5 percent. Protomammals and crimson crocs had scores intermediate between the fully warm-blooded and the fully cold-blooded.

lions'. Mammalian top predators are always rare, but finbacks were overwhelmingly abundant. Romer noted that in one of his first papers published about Permian faunas in the 1920s, *Dimetrodon* was the single most common genus in most locales. That went against all the laws of bioenergetics, unless the predator had a very low metabolism *and* its interaction with the ecosystem permitted it to reach its maximum theoretical abundance. I invested half a year in measuring every specimen Romer possessed, and extended the census to all the other samples of finbacks housed here in the United States and abroad. In nearly every quarry, in every formation, in every habitat, the result was the same: The finback predators were the first or second most common genera among all the fauna. The theory of predator-to-prey ratios had proved stunningly reliable with relation to the cold-blooded finbacks. If metabolic level was the primary agent determining the abundance of predators, then cold-blooded *Dimetrodon* should have been highly abundant in all kinds of predator-prey systems, and over species of prey in many different habitats. The hundreds of specimens I catalogued showed that was precisely the case. *Dimetrodon* was extremely catholic in its choices of prey. In sediments laid down in quiet Permian lakes, the commonest large prey was a fish-eating reptile, *Ophiacodon*. At other sites deposited in swampy floodplains, *Ophiacodon* was rare but another semi-aquatic species, the big-headed amphibian *Eryops,* assumed the role of supplying food to the finback.

Dimetrodon had no choice but semi-aquatic, fish-eating prey, because at this stage of evolutionary history the large plant-eaters hadn't developed very far and were only rarely of good size. But could *Dimetrodon* have maintained its extraordinary abundance in normal ecosystems, where it fed on land-living vegetarians? This was a key question, because ultimately *Dimetrodon*'s predator-to-prey ratio had to be compared to that for meat-eating dinosaurs which fed almost exclusively on plant-eaters. Fortunately, *Dimetrodon*'s fossil record did include two dry, floodplain habitats where a big plant-eater—a buck-toothed reptile called *Diadectes*—was its most common prey. In these two habitats *Dimetrodon* proved to be as successful and abundant as it was in the systems where fish-eaters were its main fare.

Were there any ecological situations where *Dimetrodon* could not keep its populations at a very high level? I found only one. Al Romer had excavated a highly unusual quarry in Texas called the Geraldine Bone Bed. There the fossil skeletons were mingled with segments of fossil logs and dark, carbon-rich stains which permeated the entire mass of sediment. The Geraldine was the strangest of all *Dimetrodon*'s habitats. Animals usually common elsewhere were rare or absent entirely—*Eryops,* for example. And animals usually very rare in the rest of Early Permian beds were superabundant at Geraldine—the big finback herbivore *Edaphosaurus* and the eel-like amphibian *Archeria*. Romer concluded that the Geraldine Bone Bed was the remains of a stagnant backwater,

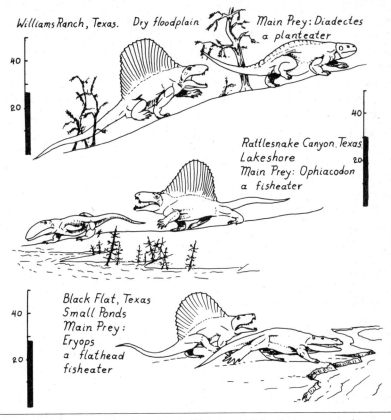

Williams Ranch, Texas. Dry floodplain — Main Prey: Diadectes a planteater

40

20

40

20

Rattlesnake Canyon, Texas
Lakeshore
Main Prey: Ophiacodon
a fisheater

40

20

Black Flat, Texas
Small Ponds
Main Prey:
Eryops
a flathead
fisheater

Finback meat-eaters of the Permian—efficient coldbloods in every habitat. Finbacks provide the perfect test case for predator-prey theory. Finback anatomy is so primitive that all scientists agree cold-bloodedness was the only possible adaptive level. Therefore, finback predators should have been superabundant in most habitats—and they were. Quarries dug in floodplain sediment show high abundance, and so do quarries dug in lake limestone and pond mudstone.

a fetid swamp where rotting vegetation had choked the river channel. Those conditions kept *Dimetrodon* out. Only a few tiny juvenile specimens were found. But this was the sole exception to the rule that *Dimetrodon* maintained extremely high abundances.

At this point it might be valuable to explain a bit more about precisely how predator-to-prey ratios are calculated. A bioenergetic census isn't made on raw numbers of specimens alone but on estimates of total body weight as well. Body size determines how much food energy a predator requires. And the size of the prey's carcass, of course, determines the amount of meat available to the predator. My technique for calculating predator-to-prey ratios involved two steps. First, I sculpted scale models from careful reconstructions of the animal's appearance in life. Top, side, and cross-sectional views of the skeleton served as the basis for restoring major muscle masses in clay. When a model was completed, I immersed it in water to measure its volume. Once this volume was measured, it was a simple step to calculate the volume of the full-sized animal. And since nearly all vertebrate bodies are almost as dense as water (the average carcass is 95 percent as heavy as the equivalent water volume), the live weight could be figured with precision.

After the live weights were determined for all the common species, I made a census of the total meat available, the total weight represented by all the fossils from the habitat under study. Sometimes these calculations profoundly changed the traditional picture of predator-and-prey relations. For example, in the floodplains haunted by *Dimetrodon,* the commonest prey were smallish, flat-bodied amphibians, including the boomerang-headed *Diplocaulus.* Paleoecologists had traditionally reconstructed these ecosystems with *Diplocaulus* in the role of chief meat-supplier to the fin-backed predator. However, most *Dimetrodons* were medium-sized and big animals, between twenty and a hundred pounds in weight. *Diplocaulus* weighed only a pound or two on average, at best a Permian hors d'oeuvre for the bigger animal. All the *Diplocaulus* carcasses together didn't amount to enough meat to keep even one mated pair of *Dimetrodons* alive and healthy. The two large species of prey, *Eryops* and the *Diadectes,* were only one tenth as common as *Diplocaulus,* but those animals were a hundred times heavier, on average. An adult *Eryops* or *Diadectes* weighed about two hundred pounds, heavier than most adult *Dimetrodons.* So most of the meat supply for *Dimetrodon* came from the rarer but bigger species. This illustrates an important general rule: Large predators obtain most of their food from large prey.

Armed with new confidence in this method of predator-prey

analysis, I returned to my primary goal—evaluating the dinosaurs' metabolism. Thanks to a census by a Canadian paleontologist, a complete count of the very rich dinosaur beds in the Judith Delta and later sediment was available. These formations yield Late Cretaceous fauna, duckbill and horned dinosaurs, and other plant-eaters, all of which were hunted by the tyrannosaurids. If orthodoxy were correct, tyrannosaurs should have been as common as *Dimetrodon* had been hundreds of millions of years before. But, if my hypothesis were right, dinosaurs would show the same low predator-to-prey ratio as fossil mammals. To compare these dinosaurs to warm-blooded mammals, I calculated predator-to-prey ratios from some recent publications that supplied counts for saber-toothed cats and hyenalike hunters found in South Dakota. When I calculated the body weight of each fossil saber-tooth and of its prey and added all the columns of data, the final ratio between the mammal predators and their total available prey proved nearly identical to the tally for dinosaurs—both the tyrannosaurs and the mammals added up to between 3 and 5 percent of the weight of their prey. The case for warm-blooded tyrannosaurs was beginning to look good. If only one dinosaur habitat had the same predator-to-prey ratio as one fossil mammal habitat, the case for warm-bloodedness would obviously have been weak. But, in fact, dozens of fossil dinosaurs and dozens of fossil mammals from the full variety of sediments exhibited the identical range. This consistent pattern had only one logical interpretation: Dinosaurs and mammals were fundamentally similar in their metabolic needs and both had a much higher metabolism than cold-bloods like the finbacks.

I published summaries of my findings in *Nature* and in *Scientific American*. Dale Russell from the Canadian National Museum was the first to notice a curious twist in the evidence for my case. Dinosaurs did indeed have a much lower ratio to their prey than did finback reptiles or spiders. But their ratios were still higher than those obtaining today in the Serengeti, in Indian game parks, or in most ecosystems today where large mammals are the top predators. Predatory dinosaurs average about 3.5 percent of their prey. In the best-studied modern game park, the Serengeti, the predators average only one tenth of 1 percent or less—in other words, their prey is nearly a thousand times greater in number than the predators. The average ratio to prey of all modern predatory mammals is 1 percent or less—three or four times less than the ratio of the predatory dinosaurs.

If dinosaurs had been as warm-blooded as modern lions, why were their predator-to-prey ratios so much higher? That was a question requiring an answer. Dale Russell concluded that the dinosaurs' metabolic rates must have been three or four times lower than the mammals'. But I believe his conclusion was incorrect. The predator-to-prey ratios for fossil mammals average about 3 or 4 percent, much higher than those calculated for today's mammals, and identical to those of the dinosaurs. Do these averages imply that the extinct mammals were cold-blooded? That is hardly likely. The extinct predators that established the percentages were perfectly normal mammals—saber-toothed cats, hyenalike carnivores, giant wolflike bears. Nothing in their anatomy has ever suggested they were cold-blooded. Paleontologists who had studied them have universally assumed—correctly, I think—that they were as warm-blooded as any modern mammals.

This evidence, however, did present a lovely paleontological paradox: Dinosaurs—supposedly cold-blooded—and fossil mammals—supposedly warm blooded—both exhibited the same predator-to-prey ratios, which were higher than those of any modern mammal habitat. Did such an apparent paradox have a solution? I suspect it will be found in a proper understanding of a basic geological axiom called "uniformitarianism." Usually defined as meaning that the present is the key to the past, the central assumption of uniformitarianism is the idea that the natural processes seen in operation today are the only forces that were at work in the past. In general, that is a reliable assumption. But taken to an extreme, the concept is used to argue that all ancient ecosystems were organized exactly like present-day habitats. Such an argument would insist that no extinct warm-blooded predator could reach a ratio of 4 percent of its prey because warm-blooded carnivores today rarely attain that level. Extreme uniformitarianism would also be forced to maintain that 3 or 4 percent ratios for fossil mammals were wrong, the result of unknown distortions in the processes of fossilization.

Such criticism of the argument from predator-to-prey ratios assumes that today's world is normal and typical of all of the earth's history. That is simply not the case. In many ways modern ecosys-

The Serengeti Today   0.3% predators

The Eocene Epoch   50 million years ago

4.4%
predators

Why the Serengeti lion is so inefficient. Five sets of forces team up to make a lion's life hard on the Serengeti Plains: 1) The most common plant-eater is the wildebeest, an antelope that prefers the wide-open, treeless plains where a lion has a hard time stalking prey unawares; 2) Severe summer droughts kill off thousands of wildebeest each year, and most of this meat is wasted as far as the lion is concerned; 3) Some plant-eaters are so big and aggressive that lions can't make kills; 4) Human herders and hunters harass the meat-eaters; and 5) Human tourists in minibuses add more aggravation.

But back in the Eocene Epoch, warm-blooded mammal predators had a much easier time. Dense forest gave lots of opportunities for ambush and most of the plant-eaters were small enough to be caught and killed easily.

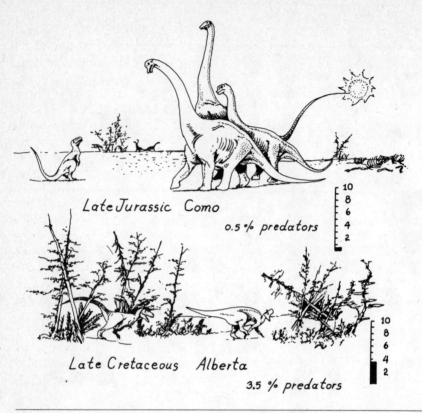

Late Jurassic Como

0.5 % predators

Late Cretaceous Alberta

3.5 % predators

Dinosaurian predators played by hot-blooded rules. Some dinohabitats were like the Serengeti—summer droughts killed masses of plant-eaters, woodlands were open and made ambush difficult, and the plant-eaters were huge. The Jurassic at Como was like this, and here the predators were rare and inefficient. But in the Late Cretaceous of Alberta, the plant-eaters were much smaller, the forest was denser, and the summers were far less dry. So the Alberta predators were more common and more efficient.

---

tems are abnormal, distorted by unusually dry climates and by the intrusions of human activity. The following figures are instructive. Under ideal conditions, like those found in a game park or a well-run zoo, lions require a minimum of ten times their own weight in meat per year to live healthily and reproduce. So 10,000 pounds of meat suffice as a full year's supply for a family of lions weighing a total of 1,000 pounds (one 350-pound male, two 250-pound females, and three 50-pound cubs). To supply this, a herd of deer or antelope weighing 20,000 pounds would be required. In this ideal situation, therefore, the predator-to-prey ratio can be found by dividing 1,000 pounds of lion by 20,000 pounds of prey—5 percent. Why, then, are the ratios in the Serengeti only one tenth of 1 percent?

The answer is that the grasslands and woodlands of the Serengeti are very far from ideal. The savannah covered by short grass is poor hunting ground because there isn't sufficient cover to allow the lions or hyenas to approach their prey. As a consequence, the predators are inefficient and do not catch enough prey to make an ecological difference, so the vegetarian herds grow bigger and bigger. And humans compound the situation. Herdsmen and ranchers kill off predators to protect their livestock. Poachers and pelt hunters kill for the skins. Hordes of tourists insist on harassing the predators during their hours of rest. In consequence, the Serengeti predators never build their populations to full potential. Is it any wonder why the predator-to-prey ratios are so far below the maximum possible with prey multiplying so abundantly and predators multiplying at such a minimum? Clearly, the predator-to-prey ratios of this modern game park are most unreliable guides for any understanding of the past.

Most of the habitats frequented by fossil mammals or dinosaurs were not nearly as hard on predators as the Serengeti. The ancient ecosystems were not generally as treeless and, of course, were free of any interference from humans. It would not be surprising therefore to find higher predator-to-prey ratios obtaining in the fossil samples. In addition, many ancient predators enjoyed the advantage of being well-enough armed to attack successfully even the largest plant-eater. Today's lions and hyenas are not big enough to kill healthy adult rhinoceroses, elephants, or water buffalo. But that is not typical of the situation during the entire Age of Mammals. Most of the time in the past, the carnivores were large and strong enough to assault the biggest prey; for example, the giant wolf-bear *Pliocyon* found in Nebraska eight million years ago averaged five to six hundred pounds. If it hunted in packs, *Pliocyon* could have killed elephant-sized prey with ease. In general, the typical mammalian fauna of the past included far more formidable top predators than do any of today's ecosystems. Since fewer plant-eaters were immune to attack, the top predators would have been more efficient at culling them and building up the predator-to-prey ratio. On balance, these considerations suggest a solution to the paradox presented by the difference between ancient and modern ratios. Warm-blooded predators could achieve ratios as high as 5 percent when climate was favorable, when they were strong enough, and when there were no humans to befoul the sample.

Would this general picture also have applied to the dinosaurs? In other words, were some of their habitats sufficiently favorable so that as warm-blooded predators they could attain ratios as high as 4 or 5 percent? The habitat least favorable to predatory dinosaurs would certainly have been the barren sand dunes of Outer Mongolia during the Cretaceous Period. This environment offered little cover, water was scarce, and droughts must have been severe. Plant-eaters are commonly found in these red sands, but meat-eaters are rare and mostly of small size. The total mass of predators to prey was far below 1 percent. But across the North Pacific, the Late Cretaceous habitats in Alberta were ideal for predators. Forested deltas and floodplains provided ample cover for attacking, and the predators were very large, powerful, and nimble—adult tyrannosaurs grew to two tons, enough to attack the rhinoceros-sized duckbills and horned dinosaurs. In Alberta, the ratio of predator to prey averaged 4 percent, much higher than in Mongolia. These percentages clearly mean that tyrannosaurs were probably as warm-blooded as the saber-toothed cats and wolf-bears that took over the top predator roles many millions of years later.

Quite interesting as a test case was the situation during the Late Jurassic in Wyoming. Here the conditions for predators were not as trying as in the barren dunes of Mongolia nor as easy as in the densely forested deltas of Alberta. The conditions were intermediate—Morrison Formation sediments show that these habitats contained more trees than the dune fields but suffered longer dry seasons than the Alberta deltas. *Allosaurus* was the most common predator of the Morrison, and it didn't enjoy the same advantages tyrannosaurs would later have. Allosaurs averaged one ton in adult weight, large by modern standards, but tiny compared to the twenty-

and thirty-ton brontosaurs. Logically, then, the rules of predator-to-prey relationships derived from warm-blooded animals predict that *Allosaurus* would not have been able to reach as high a level as tyrannosaurs could later because they wouldn't have been able to cull their prey as effectively. And this prediction is justified by the evidence. The predator-to-prey ratios of the allosaurs averaged 1.5 percent, lower than for tyrannosaurs.

Orthodox paleontologists greet these arguments from predator-to-prey ratios with incredulity. But they make ecological sense. Predatory dinosaurs exhibited very low ratios (1 percent or less) in the same sorts of difficult habitats where warm-blooded mammals obtained low ratios. And they achieved higher ratios in more favorable situations, exactly as did the extinct mammals. Most important point of all was that both dinosaur predator-to-prey ratios and fossil and living mammal predator-to-prey ratios averaged far, far lower than those of certifiably cold-blooded reptiles.

There is also important independent confirmation to be derived from the evidence from the microstructure of bones discussed in a previous chapter. All the extinct groups whose bone texture indicated fast growth—dinosaurs, protomammals, and mammals—also had low predator-to-prey ratios. Moreover, my arguments from predator-to-prey ratios caused such controversy that I sought additional supports as well. In 1982, it occurred to me that footprints could serve as further proof.

It seemed logical that warm-blooded animals would be forced to move around for food at much higher average speeds than their cold-blooded cousins because high metabolism demands a more or less continuous supply of calories, hence a more or less continuous search for them. If correct, this notion implies that fossil footprints would be good indicators of the number of required calories. Since they usually record unhurried, and not maximum, speeds, fossil footprints should be reliable indicators of the average intensity of foraging.

I tested this idea by calculating the speeds indicated by footprints made by living species. As expected, average walking speed today is much higher in mammals than in cold-blooded amphibians and reptiles. I then calculated walking speeds from the footprints of fossil mammals. They fell into the range of their modern, living relatives, confirming that extinct mammals had as high a metabolism as modern. Next, I turned to testing this idea against the footprints of the very primitive reptiles and the amphibians of the Coal Age—species more archaic in bone structure than living lizards. Once again, the hypothesis proved out. Average speed in the ancient Coal Age and Early Permian animals was exceptionally low, only one to two miles per hour. This was dramatic proof that a leisurely mode of foraging was in fact a concomitant of cold-bloodedness.

Having established the reliability of this approach, I finally turned to calculating the average walking speeds for the dinosaurs and for their ancestors of the Triassic, the thecodonts. There was absolutely no ambiguity in the figures: Dinosaurs and the thecodonts before them moved through their Mesozoic world at average speeds as high as those of warm-blooded mammals and much higher than those of the cold-bloods. Such speeds make evolutionary sense only if metabolic rates were constantly demanding ingestion of calories at a thoroughly warm-blooded rate.

There are, then, three ways of accomplishing the apparently impossible, of counting calories in species long extinct. And the conclusion to be drawn from all three methods coincides precisely. The fossil animals that have low predator-to-prey ratios are the ones that also have bone texture indicating fast growth, and also had high average walking speeds. The picture of dinosaurs painted by these three sources of information is one where the metabolic levels were set on high, where the drama of growing to adulthood was acted out speedily, and where the cast of characters was in constant motion about the Mesozoic landscape, engulfing prodigious quantities of indispensable food.

PART 5
DYNASTIC FRAILTY AND THE PULSES OF ANIMAL HISTORY

19

PUNCTUATED EQUILIBRIUM: THE EVOLUTIONARY TIMEKEEPER

Scientific thinking often benefits from the throwing of "bombs"—the publication of ideas so revolutionary that one half of the profession is scandalized, while the other half is captivated by the prospect of daring new solutions to old problems. Even when the revolutionary idea finally proves not entirely correct, the natural tendency to accept orthodoxy unchallenged is beneficially shaken. And certainly the vigorous reexamination of facts and conclusions provokes the creation both of new methodologies and new ideas. Darwin's *Origin of Species,* published in 1859, was such a "bomb." European scientists had been excavating fossils for sixty years, and were anchored to the conviction that the species they were finding were units of the Creator, fixed and unchanging through time. Such a conception of species corresponded perfectly with the prevailing theological conviction that man was a special creation entirely apart from all other organic beings. Darwin's "bomb" was the first well-researched argument in favor of the idea that one species could indeed evolve into another, and then another, and so on until the disorganized primordial slime had been transformed into something as complex and elegant as a Guernsey cow or an Anglican bishop.

The intellectual upheaval that followed upon Darwin's publication of his ideas is well known. Religion and religious thinking have never recovered in Western society from the vast defeat they

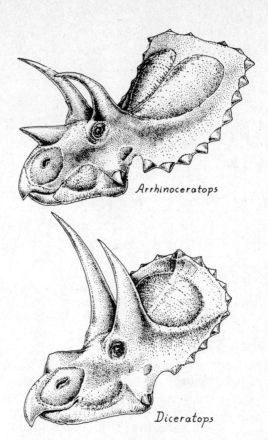

Arrhinoceratops

Diceratops

Fast evolutionary turnover. Most dinosaur genera were very short-lived, lasting only six million years or less. *Arrhinoceratops* and *Diceratops* are two examples—these horned dinosaurs had a geological life-span of only four million years.

suffered when they tried to resist the conclusions about nature that he stimulated. The contest was very like a war. And, as in most wars, the winners wrote the history. In the long run, Darwin won. The evidence for the evolutionary transmutation of one species into another became so strong that by 1900, nearly all scientists had converted to "Darwinism" of one form or another. Darwin became an heroic figure, the champion of rational thought. His opponents became stubborn obscurantists, superstitious defenders of organized religion despite massive amounts of scientific evidence.

Not all of Darwin's opponents, however, were in that camp. Many honest, soundly trained paleontologists, hardworking scholars who weren't religious bigots at all, saw in the rocks abundant evidence to suggest that species were fixed and largely unchanging. Darwin himself was not a professional paleontologist. As the naturalist on H.M.S. *Beagle,* he gathered the fossil bones of giant ground sloths he found on the Argentine pampas, and sent them on to Sir Richard Owen for study, his own anatomical expertise being too weak. So he was open to the charge of being a mere amateur when it came to fossils. Perhaps he did not realize how good the evidence for long-lasting and unchanging species was. By the 1850s, paleontology was already a mature science. English, French, German, and Russian scientists had thoroughly studied the rock strata, and had gathered many thousands of fossil specimens. These experts had no trouble recognizing the same species of elephant, for example, across vast expanses of geography and through enormous thicknesses of strata. The same was true for dinosaurs. The conclusion seemed inescapable: Species were fixed units that did not change much through time and across space.

In 1859, Hugh Falconer, a prominent English paleontologist with a wide reputation as an expert on fossil elephants, did support Darwin. But he urged Darwin to modify his belief that most of evolution was a slow and continuous process. Falconer was convinced that the evolution of one species into another was a sudden event, and that most of the time species remained unchanged during the long intervals between the sudden transmutations. Unfortunately, Falconer died in 1866, before his modified view of Darwinism could become widely known. When science jumped onto the Darwinian bandwagon around 1900, all these quite legitimate objections were lost sight of. Primitive reptiles, protomammals, dinosaurs, pterodactyls, mammoths, and saber-toothed cats were all supposed to have evolved gradually, one species imperceptibly into another, all through the History of Life. The fossil facts did not read that way, but everyone assumed the "missing" intermediates between species had existed. The fossil record was just too incomplete to preserve many of the transitional stages. Darwin had argued exactly this line to counter his critics.

Shortly before I entered graduate school at Harvard in 1972, the biggest paleontological "bomb" of the century had been dropped by Niles Eldredge of the American Museum in New York and Stephen Jay Gould of Harvard. They published an article entitled "Punctuated Equilibria." This rather short paper produced enormous repercussions. Eldredge and Gould had rediscovered the idea that species might not change continuously through time. They combined this idea with another developed by twentieth-century naturalists, that new species usually form in small breeding populations isolated from the main body of the old parent population. The "punctuated" part of their idea was the notion that most evolutionary change happens suddenly, like a punctuation mark in a sentence. The "equilibrium" part was that most of the time, most species are not evolving at all because conditions in large, stable populations do not permit changes to occur. Eldredge and Gould insisted that the fossil record was not nearly so poor as Darwin believed. The rarity of missing links between species, so bothersome to a century of Darwinists, was in fact a faithful representation of evolutionary history. If most new species form suddenly, in isolated populations, then only rarely would fossils preserve that fleeting evolutionary moment. And if typical, widespread species endure millions of generations without changing, then the fossil record will consist mostly of long-lived invariant species. It can be said that, in their own terms, Eldredge and Gould had rediscovered Hugh Falconer's position.

Two things convinced me that punctuated equilibrium was an idea worth considering. First of all was Professor Bernie Kummel's endorsement. His own work on fossil armored squid made him enormously well informed about how fossil species changed through strata. And his habit of mind was extremely rigorous, very little given to unsound flights of speculative fancy. I expected only the most qualified discussion of the theory from him. I was therefore all the more impressed by his unequivocal endorsement. I questioned Bernie one evening at a party for us students, "Is Steve right?" "Yup," said Bernie. He pointed out that it had been known all along that really good fossil strata always showed species lasting through millions of years. Only misguided geneticists insisted that species evolved all the time.

The second thing to convince me was my own studies the following summer. With Bernie Kummel's endorsement of punctuated equilibrium deeply impressed upon me, I went off to Como Bluff to test the idea in the field among the brontosaur habitats.

Como Bluff and Sheep Creek nearby were nearly perfect for testing punctuated equilibrium with fossil dinosaurs. Good skeletons could be found from the bottom to the top of the Morrison Formation, strata 250 feet thick representing approximately a million years. And the habitats represented in these strata clearly changed dramatically throughout this time—from shallow, seasonally dry lakes and swamps, to well-drained floodplains crossed by large, shifting streams, to poorly drained floodplains dotted by carbon-rich ponds. If orthodox genetic theory were correct, then the Como dinosaurs should have evolved continuously throughout this time to match the changing demands of their environment. If punctuated equilibrium were correct, the fossils should reveal species of dinosaurs that remained static, unresponsive to local conditions until a new species suddenly appeared.

Most of the anatomical fieldwork for my study had already been done. American paleontologists have published superbly illustrated studies of the Como dinosaurs, comparing specimens from the different quarries with loving attention to every detail of the bones and muscle processes that covered them. My task was chiefly to fill a few gaps in the data—about where the old quarries had been dug, in what strata of the rock, and in what sort of sediment.

The Sheep Creek *Brontosaurus* stands in the museum at the University of Wyoming. It is a splendid skeleton from the lake limestones close to the very bottom of the Morrison Beds. I surveyed every square inch of it for my notes so as to compare it with Yale's *Brontosaurus* from a quarry high up on Como Bluff, and with the New York Museum of Natural History's *Brontosaurus* from Nine-Mile-Crossing, a quarry in an in-between layer. My final notes contained a record of *Brontosaurus* through hundreds of thousands of breeding generations, spanning many major environmental shifts and climatic changes. Therein was contained absolutely no evidence for continuous evolutionary change. *Brontosaurus* had remained fixed in its adaptation through a million years.

Not only did *Brontosaurus* remain static in form for a very long time, but when it did change, it seemed to jump forward with a quick evolutionary spurt. Adult brontosaurs found at Como had thigh bones of about six feet (1,750 millimeters) maximum length, indicating a total length of body of just under seventy feet. The Sheep Creek *Brontosaurus* was that size, as were the one at Yale,

How new species of brontosaur appear suddenly in rock layers (measurements in feet and meters show the rock level in the Morrison Formation)

the one in New York, and many others. No change in size was indicated in the entire million-year span recorded at Como. But in Colorado, in beds laid down a bit later than those at the top of Como Bluff, much larger brontosaurs are found: gigantic specimens, with thigh bones about six and one-half feet (2,000 millimeters) long. If the story of the rocks is read literally, *Brontosaurus*'s development was punctuated. Long epochs had passed without change, followed by the sudden appearance of a new, larger species.

Brontosaurs were not the only dinosaurs from Como to support the concept of punctuated equilibrium. *Allosaurus,* the contemporary predator, remained fixed at one adult size—a thigh about three feet (850 millimeters) long—through the entire span of strata from the lower Morrison Beds right up to near the top of the formation on the Bluff. But in Colorado, in those same beds that yield giant brontosaurs, are also found giant allosaurs, with thighs nearly 1,200 millimeters long. Just as was the case with the brontosaurs, *Allosaurus* had stayed fixed in equilibrium throughout the million years recorded at Como, and then suddenly changed, producing a new, much larger species when the Colorado beds were laid down. Another brontosaur, *Camarasaurus,* seems to have followed the same pattern.

The discovery that punctuated equilibrium was a valid conception of how evolution works, and that it therefore applies to the history of dinosaurs, generates some intriguing new methods of measuring the pace of evolution. Since the original theory was suggested, Eldredge and Gould, its proponents, and Steve Stanley at Johns Hopkins, have concentrated on finding out how different families of animals display their own particular rate of evolutionary change. As a general rule, most species change very little from the time they first appear until their final extinction. But the total length of time each species exists varies a great deal. Stanley points out that some clams are stubborn evolutionary sluggards that seem to abhor alteration excepting on rare occasions. Most types of clams go on for many millions of years with hardly any adaptive shifts. Mammals are quite the reverse. Their species appear almost frenetic in their haste to evolve into something new. The average mammalian species therefore lasts nowhere near as long as the average species of clam.

A very well-preserved segment of fossil history, like the one at Como, permits a computation of how long the average species and genus of dinosaur lasted. And that can be compared with the rates of change computed for warm-blooded animals on the one hand and cold-blooded on the other. Rates of evolutionary change must somehow be linked to the metabolic rate of any given organism. When metabolism is very high, the high physical costs of its living must drive the animal to be an aggressive competitor or predator. And the rapid reproduction typical of warm-blooded animals tends to fill habitats to overcrowded levels much more quickly than the more leisurely breeding schedules characteristic of cold-bloods. An ecological community full of warm-blooded species is therefore a tough environment where the resident species jostle one another for food and water, breeding sites and burrows, all year round. In this sort of environment, the average species will not last long in terms of geological time before it is driven to extinction either by a new species, or by a combination of old species, or by an adverse change of climate.

Clams, by contrast, with their low metabolism, move around very little to accomplish the tasks of their adult lives. It is consequently not surprising that species of clams last much longer than warm-blooded species. Crocodiles and turtles are far more active than clams, but are still sluggish compared to the fully warm-blooded mammals. If dinosaurs resembled cold-blooded reptiles metabolically, their rate of evolutionary change would very nearly match that found in crocodiles or large turtles. But if the dinosaurs' metabolism was heated, then the average life span of one of their species or genera would have to be short, like a mammal's.

One of the best places to study the rate of evolutionary changes in dinosaurs is in the Late Cretaceous deltas of Wyoming, Montana, and Alberta. There the changes through the last ten million years of the dinosaurs' history can be followed. As might be expected, the turtles and crocodiles show very little change in these strata. The genera representing these cold-bloods hold on through formation after formation. But what about the dinosaurs here? Their evolutionary pace stands out as quite different. New species and genera kept appearing and eliminating older ones at quite a brisk tempo, geologically speaking. The average genus of dinosaur lasted for only a fraction of the time of that of the average crocodile.

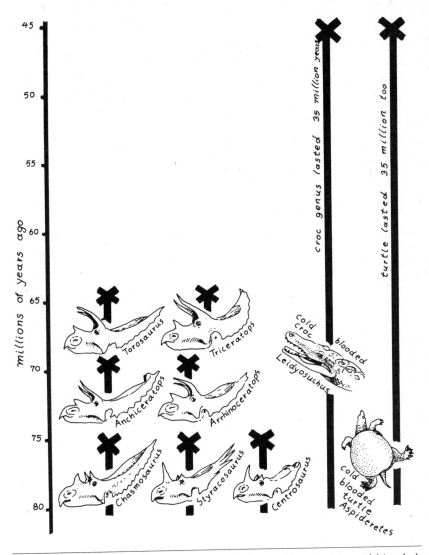

Fast evolutionary turnover—A warm-blooded trait. Thoroughly cold-blooded creatures—crocodiles and turtles—have a spaced-out evolutionary style, and genera last for thirty million years or more without much change. But dinosaur genera had a much brisker replacement rate. The horned dinosaurs evolved so fast that the average genus lasted only five or six million years.

A faster rate of evolutionary change manifests itself in the higher levels of classification as well. On average, three or four species of dinosaur composed a genus, and six to twelve genera a family. Families of cold-blooded genera are almost indestructible because their slow rate of change implies that only very rarely will all the genera in one family become extinct. For example, the family that includes all modern crocodilians is the Crocodylidae, whose debut occurred way back in the Mid Cretaceous, nearly a hundred million years ago. The snapping turtle's family is as old, and so is the soft-shelled turtle's. For all the cold-blooded reptiles, the average endurance of a family is fifty-five million years. But families of dinosaurs change much more quickly. Their average life span is only twenty-five million years, almost exactly the same as for mammals. In sum, the pace of the dinosaurs' evolution suggested by punctuated equilibrium is both fast and furious—much too high to be consistent with the concept of cold-blooded dinosaurs.

Another aspect of the evolutionary pace also places the dinosaurs in the warm-blooded category: diversification of species. Truly warm-blooded families of genera diversified themselves quickly, constantly splitting into new genera and new species that fill unoccupied niches of the ecology and bump older species from those they fill. The great American paleontologist Henry Fairfield Osborn called this proliferation "adaptive radiation." By contrast, cold-blooded animals are slow to increase their share of the ecosystem, and their "adaptive radiations" (with a few exceptions) are rather lethargic affairs.

Giant tortoises are a good illustration. They represent a family of terrestrial plant-eaters that first evolved in Eocene times, fifty million years ago. But tortoises were almost unbelievably slow to diversify, and even now some zoologists would place all the species of the last five million years into one single genus, *Geochelone*. Compare such a conservative pattern with that of the duckbill dinosaurs. The first of them appeared fifteen million years before the end of the Cretaceous. Within the next ten million years they had expanded so quickly that seven distinct genera can be found in one small outcrop of the Judith River Formation. Horned dinosaurs also exhibited such aggressive expansion during the same period and produced five or six genera in the Formation. These are rates of expansion every bit as fast as those clocked by the big mammalian families during the Age of Mammals.

One part of the orthodox story does appear to be unassailable, an ineradicable fact safe from even the wildest heretic: Dinosaurs are indeed all extinct. The fact of their extinction is the cornerstone underlying the orthodox belief that dinosaurs were maladapted failures. Recently, after a lengthy and intense disputation with an Old Guard paleontologist, he summarized his argument with what for him was a rhetorical question, "If your dinosaurs were so hot, how come they're all dead?"

Dinosaurs are incontrovertibly dead. But that does not prove what orthodoxy believes about them. Paradoxically, the extinction of the dinosaurs is strong evidence that their biology was heated to levels far above those of typical reptiles. The basic principle is simple: The higher the metabolic needs of a group of species, the more vulnerable it is to sudden and catastrophic extinction. What is the best natural design for avoiding extinction? The answer is animals with a lethargic metabolism, like the alligator or the large turtle. Each individual is greatly resistant to drought or famine. And corporately, the species is therefore resistant to extinction. An entire family of lethargic species would be most difficult to kill off all at once, worldwide. Conversely, the most effective way to design species that are almost guaranteed to die off completely is to endow them with the highest, most compulsive need for calories and protein. Any major perturbation of the environment might render each and every species extinct at one blow.

The Cretaceous extinctions were the most massive in the history of the terrestrial ecosystem. But some families of species sailed through the crisis without suffering any noticeable effects. The survivors included reptiles and amphibians, families with low metabolic needs and very sluggardly evolutionary rates—gill-breathing salamanders, monitor lizards, alligators, crocodiles, soft-shelled turtles, snapping turtles, and the long-snouted champsosaurs. But dinosaur after dinosaur became extinct until none was left. If dinosaurs were so hot, how come they're all dead? This question answers itself. Dinosaurs went extinct *because* they were so hot.

20

THE KAZANIAN REVOLUTION: SETTING THE STAGE FOR THE DINOSAURIA

Dinosaurs should never be taken out of their historical context! To appreciate the adaptive dexterity of the Dinosauria, they must be viewed in their place within the succession of evolutionary dynasties. One of the greatest flaws in the orthodox conception of dinosaurs is that it ignores the evolutionary patterns prior to their appearance. The evolutionary development and destiny of the beasts that preceded the dinosaurs are essential to understanding their real place in the history of life. Such a context renders the argument for their warm-bloodedness logical, and nearly irrefutable.

The land ecosystem has hosted four great megadynasties throughout the entire history of life on earth. Megadynasty I consisted of the very primitive reptiles and amphibians of the Coal Age and the Early Permian. All the dinosaurs filled Megadynasty III, and mammals fill Megadynasty IV. Orthodoxy maintains, remember, that dinosaurs were cold-blooded sluggards, so the progression from Megadynasty III to IV is made to appear like a great advance in physiological sophistication. But there's an enormous problem with this view of the sequence. Megadynasty II, the one preceding the dinosaurs, belonged to the protomammals, generally known as the Order Therapsida. They definitely included the immediate ancestors of genuine mammals—and the advanced protomammals showed many signs of mammal-style adaptations in their

Tartarian saber-toothed protomammals of the Late Permian. These five-hundred-pound gorgonopsians had warm-blooded evolutionary style.

body and skull. Most paleontologists therefore are willing to believe these later protomammals had already developed some degree of warm-bloodedness. Now, if they were warm-blooded, it is very strange indeed that they subsequently lost their dominant position to the supposedly cold-blooded dinosaurs. To clarify this question, we shall have to investigate Megadynasty II further. How did the Therapsida rule their world? How had they replaced the archaic creatures of Megadynasty I? And why did they yield to the dinosaurs?

Before the emergence of Megadynasty II, the world was ruled by primitive reptiles and their amphibian neighbors. They had such an archaic structure that nearly every scientist who has studied this

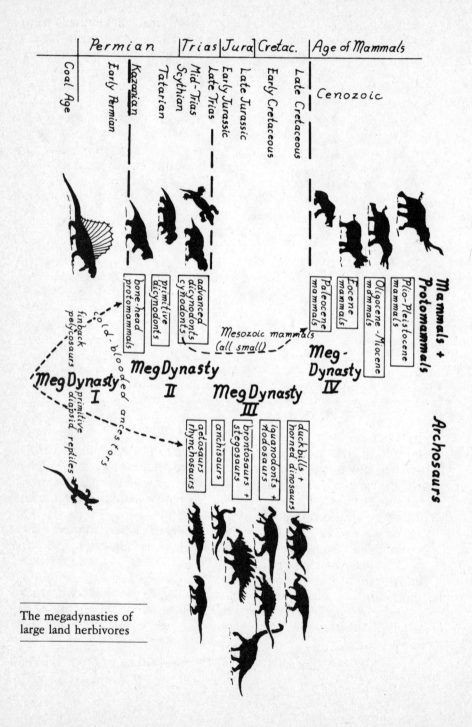

The megadynasties of
large land herbivores

time has come away convinced that all the large land animals were entirely cold-blooded. The newer, heretical thinkers concur. Megadynasty I's top predator was *Dimetrodon*. Its predator-to-prey ratios and bone texture clearly demonstrate that it and all its contemporaries were slow-growing creatures equipped with low metabolism. Their fossil footprints record a very slow average pace. And the limb joints of *Dimetrodon* and its contemporaries were designed only for slow walking and crawling; the hip socket was too shallow to contain a strong thrust of the thigh from running, and the bony crest on the shin bone was too weak to provide much leverage for the knee-opening muscles. Taken altogether, this earliest Megadynasty must have created a world in slow motion, where the fastest gait was a lumbering waddle.

What sort of evolutionary punctuation might be expected in such an age? If the theory that metabolism controls the tempo of evolution contains any truth, it yields three predictions: (1) Genera would survive unchanged for immense periods of time; (2) The rate of proliferation of species would be extremely slow; (3) Sudden, mass extinction would not be possible. All three of these predictions prove out. *Dimetrodon,* its vegetarian cousin *Edaphosaurus,* and the other chief genera of the dynasty endured for about twenty million years—far longer than the genera of dinosaurs or mammals lasted in Megadynasty III and IV. Moreover, in Megadynasty I the ecological niches were undersaturated, because new types did not develop quickly. On the dry floodplains there was only one genus of large vegetarian, *Diadectes,* and only the one big predator, *Dimetrodon.* Such an impoverished system appears very underpopulated compared to most dinosaur habitats. And unlike the dinosaurs, which suffered several mass extinctions, nothing similar ever struck the ecological community ruled by *Dimetrodon.* Genera went extinct one by one, with new genera entering according to a rather leisurely schedule.

When the first protomammals appeared, *Dimetrodon* and the entire somnolent world of Megadynasty I passed away. This was an extraordinary upheaval, which I call "the Kazanian Revolution"—eventually it will set the stage for the appearance of the dinosaurs. The Kazanian Epoch is named from the old Russian province of Kazan, west of the Ural Mountains. There, the most ancient protomammals are found in the red-stained sediment and yield their bones to the careful spadework of Soviet paleontologists. The Kazanian families burst into the evolutionary drama with unprecedented energy. They started out as slender-limbed, wolf-sized predators, and rapidly expanded their empire into nearly every role in the ecology, sweeping away the *ancien* cold-blooded *régime*. Within a few million years Kazanian protomammals had taken over all the carnivorous roles—large, medium, and small—nearly all the herbivorous roles, and produced dozens of small, insect-eating species as well. Never before had the ecosystem witnessed such a spectacular proliferation of new species from a single family. The Kazanian Revolution was the first terrestrial example of explosive "adaptive radiation."

Right from the beginning, the Kazanian protomammals stuffed whole clusters of species into each role. The best preserved of these are found in the red beds of the South African Karoo and display a richness far greater than anything recorded before. Protomammals produced four different families of predators, with eight or ten different species, to prowl through the floodplains and forests of the Karoo. Biggest of them were the dome-headed anteosaurs, the size of polar bears, with thick, bony buttresses over their eyes for head-butting in the mating season. Anteosaurs were armed with a great row of long teeth that meshed together to clamp down on prey. Predators from other families, the size of wolves and jaguars, displayed a wide variety of lethal devices for dealing with their prey. Plant-eaters were numerous, too. Five families and a score of species munched their way through the greenery of the ancient Karoo.

Why did these Kazanian protomammals evolve so quickly? They produced new species at very heated rates, and adapted them very speedily, so that most lasted only a few million years before they were replaced. The evolutionary tempo of the Kazanian appears as fast as that of our own Class Mammalia during the Age of Mammals. Yet they suffer from the same bias maintained against the dinosaurs. They included some large species—up to one or two tons—but paleontologists dismiss those as behemoths of low metabolism that had to use their bulk to keep them warm. As a student at Harvard, I became interested in these Kazanian protomammals because they displayed the same evolutionary vigor I discerned in the dinosaurs. And I began to suspect that both they and the dinosaurs had been warm-blooded.

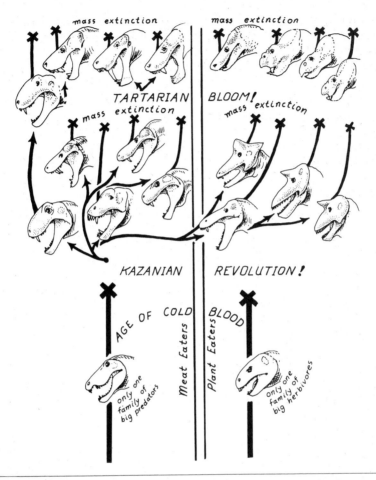

First explosion of the warm-bloods! Before the Kazanian Epoch, the evolutionary style was slow and spaced out. Species lasted for many millions of years and there was only one common large predator family and only one common large plant-eating family in most habitats on the land ecosystem. But suddenly the Kazanian protomammals burst upon the evolutionary stage and in a few million years branched out into five separate meat-eating families and four separate plant-eating families. Habitats were filled to overflowing with fast-evolving species. And soon after this great evolutionary boom, there was the first gigantic crash—a mass extinction that wiped out most of the Kazanians. A few protomammals survived, and a new evolutionary bloom followed—the Tartarian radiation. (Each head portrait represents one family in this chart.)

A Kazanian scene: *Trochosaurus,* a protomammal, attacks an herbivorous pareiasaur.

My desire to study the Kazanians firsthand took me to Cape Town, South Africa, where 90 percent of the specimens are housed. My hypothesis was that if the Kazanians had been the first warm-blooded animals, then they would have required a great deal of food per year, and therefore the predators would be rare. Large samples of fossils are needed to prove such an hypothesis. And fortunately for me, one man, Liewe Dirk Boonstra, had poured an entire lifetime into excavating the Karoo and carefully sorting the species. Almost singlehandedly, he had built a detailed picture of the Kazanian world. His storehouse of fossils left little room for doubt: Although the Kazanian predators had been diverse and fast-evolving, they had been rare compared to the plant-eaters, very rare. The total body weight of all the Kazanian predators represented only 7 percent of the total for the entire fauna. This was a predator-to-prey ratio as low as in some mammal faunas, strong support for the case for warm-bloodedness.

I did not wish to rely on that argument alone, so I sought corroboration from a totally separate line of evidence—the microtexture of bone. At the very same time I was examining the protomammals in Boonstra's collection, Armand de Ricqlès was cutting bone samples from them in Paris. Each of us was doing his work unknown to the other, but both of us suspected something special was to be learned from the Kazanians. De Ricqlès published his results in *Annales de Paléontologie;* I published mine in *Scientific American.* Our separate lines of detective work converged on the same conclusion: The Kazanians had been a new phenomenon in the history of life, the first vertebrates whose bone microtexture indicated fast growth, the first ecosystem whose predators were rare, the first with a warmed-up metabolism.

Consistent with a warm-blooded metabolism, the Kazanian therapsids would have developed a more sophisticated design in the mechanics of their limbs. The older style limbs of the Coal Age wouldn't have been adequate for warmed-up metabolic needs. If de Ricqlès and I were correct about Kazanian metabolism, major adaptive remodeling should have been manifest in limb joints, adaptions for faster speeds. On this point, Boonstra once again supplied most of the preliminary material; his excavations had recovered literally dozens of good skeletons, and he had published precise diagrams of every limb from shoulder to wrist, hip to ankle. It turned out that both the shoulder and hip sockets of the Kazanians were much deeper than any found in the Coal Age. They had obviously been built to withstand much more powerful pressures from the muscles of the limbs. The knee joints indicated the Kazanians were designed for fast, bouncy gaits—the crests for supporting the extensive muscles of the knee were massively developed. When those muscles contracted on a one-ton protomammal of the Kazanian, the great beast would surely have bounded forward into a lively run. Clearly, the shuffling age of the Carboniferous was over; the Age of Trots had begun.

All the pieces of the Kazanian puzzle seemed to fall into place with unusual ease. Warm-blooded metabolism and fast-growing rates were consistent with the overall picture of fast evolutionary rates. They were also corroborated by the remodeling of limbs for faster moving speeds. And finally, they were consistent with the new sexual vigor implied by the head-butting armament commonly found. All of these things implied the Kazanian protomammals had extra energy to burn.

In Kazanian times, the Karoo experienced cool winters because it was closer to the South Pole than it is now. To remain active all year round, the smaller animals might well have needed some kind of insulation. Orthodox paleontology insists that hair is a uniquely mammalian invention, the adaptive badge of our own Class Mammalia. Several dissenters have however suggested that perhaps hair evolved long before the first true mammal appeared in the Late Triassic. Perhaps hair did evolve at the very beginning of warm-bloodedness, back in the Kazanian, fully forty million years before the earliest true mammals. Maybe our picture of the Kazanians should include shaggy protomammals stalking their prey through a winter snowstorm, hot breath steaming from their nostrils.

A last, quite important, piece of the evolutionary puzzle also falls easily into place concerning the Kazanian: mass extinction. Warm-blooded protomammals would of course have been vulnerable to catastrophic die-offs. And exactly such a disaster did cut most of them down. After a reign of five to ten million years, nearly all of them went extinct, leaving only a few surviving groups. This disaster was the earliest truly mass extinction in the history of land ecosystems. The Kazanian protomammals paid the price of their warmed-up metabolism.

In the badlands of the Karoo Basin and the outcrops northwest of the Urals, the end of the Kazanian and the beginning of the Tartarian Epoch are clearly marked. At the dawn of the Tartarian, the surviving protomammals rebounded into ecological dominance, exploding into a riot of new species, genera, and families. Dicynodonts ("two-tuskers") replaced the extinct domeheads as the big plant-eaters. Saber-toothed gorgons replaced anteosaurs as giant meat-eaters. This second wave of protomammals produced a very rich array of species, just as the first had. And they evolved quickly.

Were the Tartarians also warm-blooded? Almost certainly. De

Ricqlès found a mammal-type bone texture in all of the Tartarian protomammals. And I found very low predator quantities. Samples from both the Karoo and from Russia revealed that predators made up only 5 to 12 percent of the total preserved fauna.

The high metabolism of the Tartarian protomammals is also attested by the sudden collapse of their system. Their dominance lasted only a few million years before it crashed into a mass die-off corresponding with the very end of the Permian Period. During the next epoch, the Scythian, yet another wave of proto-mammals evolved to replace the vanished Tartarians. This third wave also exhibited all the signs of warm-bloodedness. Apparently, once it had evolved, the genie was out of the bottle, never again to be imprisoned. After the first takeover by warm-blooded families in the Kazanian, each new wave of animals of Megadynasty II displayed all the marks of high metabolism—fast-growth bone structure, ecosystems where predators were rare, and vulnerability to mass extinction. And the most important thing to remember

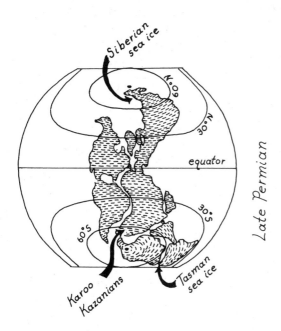

World geography in the Kazanian Epoch. All the continents were jammed together in one mass, and the climate in the far north and far south had cold winters. The Karoo protomammals must have had some sort of warm-blooded adaptation to cope with the weather.

about those three waves of protomammals is that they existed long before the appearance of the first dinosaur.

The disaster that wiped out most of the Tartarian groups at the end of the Permian Period was also an opportunity for new families to expand their ecological roles. One such group consisted of protomammals that survived from the Tartarian. Several dicynodonts survived to produce a new adaptive radiation in the Scythian and flourished tremendously right down through the rest of the Triassic Period. Two new groups entered the role of large predator and herbivore. One of these was the cynodonts, with their doglike faces. This group had been limited to a small body size in Tartarian times. But extinction of all the Tartarian top predators took the lid off their evolution. The Scythian cynodonts were able to evolve into wolf-sized predators. These cynodonts were the protomammals closest to the heart of Harvard's Al Romer because they included the direct ancestors of all true mammals, from platypuses and 'possums to monkeys, apes, and ourselves.

The other group that took advantage of the Tartarian extinctions to expand their roles during the Scythian were the Archosauria, the group that would evolve crocodiles, pterodactyls, and dinosaurs. The earliest archosaurs of the Scythian Epoch were those big predators the "crimson crocodiles" (Family Erythrosuchidae), named after the red stain on the bones of the first specimens discovered. All through the Triassic, from the Scythian till the close of the Period, a titanic ecological battle was waged between the advanced protomammals, led by the dicynodonts and cynodonts and the Erythrosuchidae and their descendants. On the outcome of this conflict balanced this history of the Mesozoic Era. Had the protomammals won, they and their descendants would have dominated the ecosystem during the Jurassic and the Cretaceous. If the crimson crocodiles and their descendants won, a totally new evolutionary line would gain control.

If the clash of mighty empires is your favorite historical fare, the Triassic is irresistible. Two mighty evolutionary dynasties collided in direct competition: the advanced protomammals against the early Archosauria. At first, the protomammals appeared to recapture most of their lost glory. The two-tuskers regained their dominant position as the big herbivore. And they evolved very advanced limb muscles, arranged nearly exactly like those of primitive mammals. Mammal loyalists can be proud of these Triassic

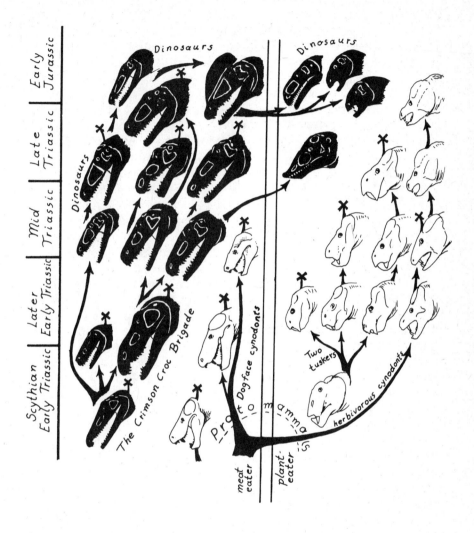

Archosaurs displaced the protomammal dynasties. As the Triassic wore on, more and more archosaur families—shown here in black—invaded the ecological roles of large predator. By Late Triassic times, all the large predator roles were taken over and the archosaurs had begun to invade the herbivore guilds. And finally, in Early Jurassic times, the dinosaurs secured complete control of both plant-eater and meat-eater components of the land vertebrate system.

two-tuskers because they demonstrate how mammalian design can win great ecological success. They can be even prouder of the cynodonts, for these animals ascended the evolutionary *Scala Naturae* even more rapidly. Unlike the rather tubby two-tuskers, the predatory cynodonts evolved sleek profiles, elongated bodies, and slender limbs designed for rapid movement. They also produced the most advanced, most mammal-like faces, teeth, and jaws. When a Triassic cynodont snarled, it bared teeth that strongly resembled a wolf's—the large canines were located far forward in the doglike snout, the mouth front was lined with short nipping teeth (incisors), and behind the canines ran a long row of teeth with multiple cusps for slicing and chewing. The muscles of their jaws were biomechanical marvels. The two muscles involved pulled across each other, an arrangement allowing the cynodonts to bite hard without placing excessive strain on the joint of the jaw bones. We humans today enjoy the advantages of such a jaw joint inherited from our cynodont ancestors of the Triassic. Our outer jaw muscles (the masseters) pull upward and forward while our upper set of muscles (the temporals) pull upward and backward, their combined crossing action preventing stress from building up in our jaw joint. Every time we tackle an especially tough steak, we should harbor a little thought of thanks to our cynodont ancestors.

No doubt that the Triassic protomammals were the best and the brightest ever produced by the protomammals. And, in sharp contrast to their pervasive bias against any notion of warm-blooded dinosaurs, orthodox paleontologists have been more than willing to accept the idea that advanced cynodonts were warm-blooded creatures. No one voiced surprise when Armand de Ricqlès announced he had found mammal-type bone texture in the Triassic dicynodonts and cynodonts. Orthodoxy had always maintained, after all, that advanced protomammals were physiologically far more sophisticated than the reptiles were.

All these assumptions might lead one to believe these dog-faced protomammals exercised unassailable hegemony over their ecosystem. But that was not the case. The biggest, strongest, most dangerous predators of the Early Triassic were not cynodonts or any sort of protomammal at all. That role belonged to those crimson crocodiles, the shock troops of the rival empire. They attained a weight of half a ton, and were armed with dinosaurlike heads three feet long, and saw-edged teeth.

Orthodox paleontologists and mammal loyalists cannot escape the meaning of the ecological takeover by these crimson crocodiles. Before they evolved, the honors for top predator had gone to protomammals—the anteosaurs of the Kazanian, the gorgons of the Tartarian. After the Tartarian mass extinction, the role of top predator lay open for the taking, for whichever group was best adapted to claim it. Filling that role is an ecological challenge like no other, because the competition is bloody and merciless. Top predators struggled for hunting territory and for scavenging rights over the biggest carcasses. Natural selection is therefore especially unforgiving when it comes to top predators—the survivors will be the fastest, meanest, and most efficient species every time.

Now, even though the cynodonts were among the fastest and most efficient predators available at the opening of the Triassic Period, there can be no doubt that the crimson crocodiles eventually seized undisputed possession of the top predator niches. Not only did the early archosaurians fill the top predator roles, they, their descendants, and close relatives (as a group, formally known as the Thecodontia) usurped more and more of the medium and small predatory niches as well as the Triassic Period went on. Nowhere else does the fossil history of life display such a clear-cut case of evolutionary imperialism; as the archosaurs' reign expanded, the cynodonts' contracted. By the Mid Triassic, the cynodonts had retreated to small and medium-small predators and to herbivores. Meanwhile, the thecodonts filled the ecosystem with fox-sized, wolf-sized, lion-sized, and polar bear-sized carnivorous species, from a few pounds to half a ton as adults.

By Late Triassic times, the erythrosuchians' descendants had branched out into two distinct river-and-lake groups of fish-eaters. The proterochampsids, with their flat heads and long snouts, propelled themselves through the waters by means of their large hind legs. The heavily armored phytosaurs, on the other hand, swam by virtue of their crocodile-like tails. The crimson crocodiles' descendants also developed into an herbivorous group, the aetosaurs. They developed body armor top and bottom, with extra protection in some species of curved, bony spikes over the shoulders. For their time, the aetosaurs were the best-protected land animals that had evolved anywhere.

Working on my senior thesis at Yale, I did a great deal of

Early Triassic class warfare: A pair of dog-faced cynodonts, *Cynognathus,* are threatened by two 1,000-pound *Erythrosuchus.*

meditating about the success of the erythrosuchians. Could it be that *Erythrosuchus* and all its relatives had had superior adaptive equipment? Had the thecodonts in fact been warm-blooded? If they had been, the evidence would have to come from the microtexture of their bones and their predator-to-prey relations. And both forms of evidence came tumbling into the laboratory during the 1970s. Armand de Ricqlès found mammal-type bone texture in *Erythrosuchus* itself and in its close kin. I counted *Erythrosuchus* specimens in museums from Cape Town to Berkeley. The case for high metabolism was every bit as conclusive as it was for the advanced protomammals.

There was therefore nothing at all paradoxical about the success of the crimson crocodiles. This vigorously evolving group had wrested control of the predatory roles from the protomammals because of the erythrosuchian anatomical equipment, which had been equal to, or better than, the best produced by the two-tuskers or the dog-faces. This was an heretical conclusion indeed because the erythrosuchians and all the Thecodontia were uncles of the dinosaurs, the evolutionary cousins of the direct ancestors of the true Dinosauria. There was the spark of dinosaurness about everything the thecodonts did, and the earliest true dinosaurs, of all sorts, shared many of the anatomical features of the Thecodontia. Most notable of these were the extra openings in the side of the snout and in the lower jaw that had been the trademarks of *Erythrosuchus* and other early thecodonts. All the early dinosaurs' skulls were characterized by the loose, open construction directly modeled on the thecodonts. Without any doubt, the dinosaurs had inherited fundamental adaptive equipment from thecodont ancestors, and they owed much of their success to the momentous developments among the first crimson crocodiles of the Earliest Triassic.

All the many thecodont families went extinct at the end of the Triassic. But their end corresponded with the beginning of the first great Age of Dinosaurs. As the Jurassic Period began, the terrestrial ecosystem was once again riddled with unfilled niches, and into these ecological opportunities streamed a horde of new species. In Late Triassic times the dinosaurs had been a minority group, but in the Jurassic every single large land predator and herbivore role was filled by their newly evolving species.

When we place the history of the dinosaurs into its proper context, therefore, the high metabolic adaptations of the Dinosauria are not at all surprising. They were the inevitable results of evolutionary processes that had begun long before the first true dinosaur appeared. Dinosaurs weren't the first dynasty of warm-blooded creatures. Neither were they the first to have a fast-paced evolutionary tempo, punctuated by sudden mass extinction. The world's ecosystems had been shaken out of their plodding, cold-blooded rhythm long before the dinosaurs made their entry into

Two lines of the crimson-croc takeover: the armored aeotosaur *Desmatosuchus* and a big-headed rauisuchid predator. Both from the Late Triassic.

the evolutionary race. The Kazanian protomammals were the pioneers of high metabolism, revolutionizing the rules of competition and predation in the epochs of the Late Permian. Once high metabolic adaptations had been introduced onto the ecological stage, no group of large land vertebrates could hope to achieve dominance without such physiological equipment. So when the first dinosaurs began elbowing their way into the roles of large predator and herbivore late in the Triassic Period, they were simply employing the same strategy for success that had been followed by their predecessors of the Early Triassic, and by the Tartarian protomammals before that, and by the Kazanians before that.

The untenable nature of orthodox views about cold-blooded dinosaurs stands revealed in the context of this progression from the Kazanian to the Late Triassic. If dinosaurs were 100 percent cold-blooded, with a metabolic system no more sophisticated than a lizard's, then the Age of Dinosaurs amounted to an inexplicable Age of Throwbacks, a monumental step backward in the progression of life on land, a return to the slow-motion conditions of the Coal Age. It makes no historical sense to believe the dinosaurs were cold-blooded. All the fabric of fossil evidence comes together to weave a coherent story of an unbroken succession of warm-bloods following one another down through the ages, from the Late Permian, straight through the entire Mesozoic Era—the Triassic, Jurassic, and Cretaceous—and finally, consistently, into our own Age of Mammals.

21

THE TWILIGHT OF THE DINOSAURS

The mass murder that marked the end of the Cretaceous Period seems to attract all manner of solutions. Perfectly respectable scientists, who pride themselves on their caution when dealing with their own specialty, indulge in the wildest flights of fancy when it comes to cracking the mystery of the Cretaceous killer. I keep a file of published "solutions." Among its contents, it is suggested the dinosaurs died out "because the weather got too hot"; "because the weather got too cold"; "because the weather got too dry"; "because the weather got too wet"; "because the weather became too hot in the summer and too cold in the winter"; "because the land became too hilly"; "because new kinds of plants evolved which poisoned all the dinosaurs"; "because new kinds of insects evolved which spread deadly diseases"; "because new kinds of mammals evolved which competed for food"; "because new kinds of mammals ate the dinosaurs' eggs"; "because a giant meteor smashed into the earth"; "because a supernova exploded near the earth"; "because cosmic rays bombarded the earth"; or, "because massive volcanoes exploded all around the earth."

It has always seemed a bit strange to me that otherwise sober scientists should leap to conclusions about the extinction of the dinosaurs. Perhaps, as Zorba the Greek told us, scientists and nonscientists alike are seduced into believing far-fetched solutions because we all need a little madness. The events in question had

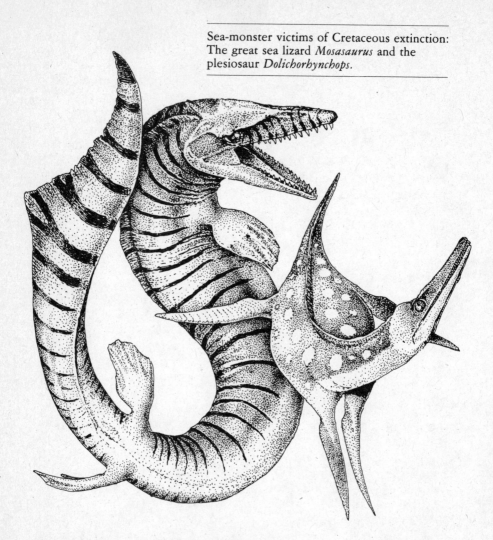

Sea-monster victims of Cretaceous extinction: The great sea lizard *Mosasaurus* and the plesiosaur *Dolichorhynchops*.

no eyewitnesses and were heroic in size, larger than life, unlike anything we see in our modern world. As a consequence, we are lured, attracted by the notion that mysteries of heroic scale require solutions of equally heroic scale, solutions totally different from the mundane, day-to-day events we experience all around us.

I am firmly convinced that all the great mysteries—the Mayan

Pyramids, Stonehenge, the stone heads on Easter Island, and the dinosaur extinctions—are solvable. In fact, I believe most have already been solved long ago. But the solution is usually so obvious, so nonfantastic, that its very mundaneness comes as a jolt. So what is the obvious, mundane solution to the great mass murder of the Cretaceous? Let us build up the evidence piece by piece, before we speak the name of the murderer.

The attempt to solve the crime of mass extinction begins by encountering one of the basic problems in criminology: the reliability of circumstantial evidence. Scientists have occasionally hoped dinosaur carcasses would yield direct evidence about the agent of death. In the 1920s, one paleontologist concluded the duckbills had all died from some ghastly poison because their skeletons were contorted into what looked like postures of agony. But it turned out those skeletons were contorted—the neck twisted upward and backward—because this was the normal posture for any corpse since the muscles and tendons of the neck contracted after death. And in fact, fossil skeletons seldom if ever yield a clue about the cause of death at the end of the Cretaceous or any other time. Thin sections from dinosaur bones usually show no obvious signs of pathology. But even this is not conclusive since most fatal diseases leave no clear mark on bone.

The only clues for finding the murderer, then, are those residing in the circumstantial evidence—the conditions at the scene of the crime. The problem with such evidence, however, is the great difficulty in separating the relevant facts from the mass of irrelevant details. A very standard procedure adopted by those attempting to solve the mystery is to pore over the many details we know about the circumstances surrounding the final death of the dinosaurs in North America, with scattered bits and pieces of information from elsewhere in the world. The hypothesis here is that some shift in the habitat must have doomed the dinosaurs. So investigation must concentrate on what changed at the end of the Cretaceous. The crippling flaw in this method of approach is that habitats in the world are never really stable, they are always changing through time. If we look for evidence of environmental change anywhere in geological history, therefore, we are certain to find it.

As an example of the problems inherent in this method of dealing with the circumstantial evidence, paleontologists interested in plants have found that winters became cooler at the end of the Cretaceous. As a result, many geologists concluded the dinosaurs died from climatic chill. But such a judgment was premature. Perhaps the chilling trend was only an innocent environmental bystander, a change in habitat that just happened to occur at the same time the real agent of death was killing off the Dinosauria. It cannot be considered guilty merely because it was present at the scene of the crime.

The single most important first step for judging the evidence about the mass extinction is unfortunately the one usually ignored: the search for a repeated pattern through time. All too often, the extinction of the dinosaurs is viewed as a single, isolated outbreak of evolutionary mayhem, an ecological St. Valentine Day's Massacre inflicted upon the denisons of the Cretaceous plains and forests. If the extinction were indeed a unique, never-repeated event, then it would be nearly impossible to sort out the irrelevant coincidences from the true trail of the killer. But if the true culprit was a repeat killer that struck the ecosystem again and again all through geological history, we would be presented with a far superior chance of sifting out the irrelevant evidence. Repeated attacks from the same agent under a variety of circumstances will eventually reveal a *modus operandi,* the characteristic pattern of the criminal.

The first step in solving the mystery of the great Cretaceous mass extinction, then, is to ask, Were there any others? The answer is a resounding affirmative. Mass extinction struck at the end of the Permian, when the Tartarian families of gorgons disappeared along with their dicynodont prey, and at the end of the Triassic, when the big two-tuskers on land and the long-bodied fish-lizards at sea died out. It also struck at the end of the Jurassic Period when many (but not all) lines of dinosaurs died out. Altogether the stratigraphic record indicates eight sudden mass extinctions among the dominant families of large, land-dwelling vertebrates. The most recent occurred only ten thousand years ago when most of the giant species of mammal—mammoths, mastodons, super-large camels, saber-toothed cats, and others—perished.

The next step is to ask whether these mass extinctions follow any coherent pattern. Again, the answer is affirmative—Baron Cu-

| Triassic | Jurassic | Cretaceous | Eocene |

Giant Eelwhale *Basilosaurus*

Sealizard *Mosasaurus*

Meerkrokodil *Metriorhynchus*

Long Bodied Fishlizard *Cymbospondylus*

Fast Whale *Zygorhiza*

Double Penguin *Dolichorhynchops*

Fast Fishlizard *Baptanodon*

Fast Fishlizard *Delphinosaurus*

Pulses of extinction hit long-bodied and compact-bodied sea monsters simultaneously. The long-bodied eel-shaped whale of the Eocene was just the last wave of evolutionary replacement that followed a mass extinction. And the compact Eocene whales—like *Zygorhiza*—were the last wave of replacement of the fast-swimming guild.

Cretaceous sea monsters in the deep and the shallows. Marine habitats came in two types: 1) the very shallow, weed-choked seas that spread north to south across the middle of North America; and 2) the clear, deep water off the continental coasts, like the Pacific Ocean along California. Long-bodied sea lizards dominated the shallows, but the deep waters hosted much larger populations of long-necked swan lizards. Both habitats were struck by mass extinction at the end of the Cretaceous.

vier's law of land-sea simultaneity applies. Each time the land eco-system suffers mass extinctions, the oceanic system suffers as well. As dinosaurs were snuffed out at the end of the Cretaceous, the great sea lizards, and the snake-necked plesiosaurs were also dying out, as were a host of large and small invertebrates, from coral-like oysters to shelled squid and microscopic plankton. This same land-sea simultaneity marked the Tartarian disaster, the Jurassic extinctions, and all the other times of Great Dying, including the one that struck our class Mammalia at the end of the Eocene Epoch during the Age of Mammals. The dual land-and-sea nature of these extinctions automatically eliminates a long list of potential agents. The Cretaceous die-off cannot be explained by the evolution of poisonous plants, for example, because the sea creatures would not have been affected by their toxicity.

The next step is to observe that the extinctions hit the land and the sea at the same time but in different ways. The entire saltwater marine system suffered, the extinction being as complete among the tiny, planktonic animals as among the giant sea serpents. The extinctions eliminated most large animals but left freshwater swimmers nearly untouched. Crocodiles, alligators, and freshwater turtles changed little from the Cretaceous to the post-Cretaceous. Another difference between the land and the sea was that small land animals did not suffer as much as the large ones. The big dinosaurs of the Cretaceous disappeared totally, but many families of lizards and mammals passed right through the disaster without losing their evolutionary stride. Finally—a point largely ignored by most scientists—the land extinctions struck hardest at the most dynamic, rapidly evolving groups of large creatures, the families that showed the highest rate of producing species and the most vigorous rates of adaptation. These are the very groups for which there is the strongest evidence of warm-blooded metabolism—the protomammals, the crimson crocodiles, the pterodactyls, the dinosaurs, the large mammals of the Eocene Epoch.

Extinctions on land possess another peculiarity. On average, plants do not suffer as much as the plant-eaters. The land plants of the Late Cretaceous did suffer some extinctions but nothing compared to the wholesale devastation experienced by the plant-eating dinosaurs. The relative immunity of plants holds true for the other previous and subsequent mass extinctions.

At this point, the *modus operandi* of the agent for the mass extinctions is revealed in some detail. The suspect: (1) kills on land and sea at the same time; (2) strikes hardest at large, fast-evolving families on land; (3) hits small land animals less hard; (4) leaves large cold-blooded animals untouched; (5) does not strike at freshwater swimmers—most of these creatures are cold-blooded, so criterion (4) applies; (6) strikes plant-eaters more severely than plants.

Can any known agent of extinction operate according to this pattern? At present many scientists do believe so. As of this writing, the scientific press is full of discussion about the newest solution to the mass extinction—one that is literally unearthly. The 26 Million Year Death Star is supposedly a giant heavenly body that strikes down the global ecosystems as it brushes past the earth in a repeated series of near collisions. Such extraterrestrial theories are hardly new.

The impact of meteorites was proposed as the agent of extinction decades ago. No direct evidence for these existed until Walter Alvarez, a chemist specializing in microanalysis of rare elements, discovered the now famous iridium layer. Iridium is a "noble metal," similar to platinum but denser, and like platinum or gold it does not easily form compounds with other elements. According to classic astrochemical theory, iridium is extremely rare on the earth's surface, but much more abundant in celestial bodies such as meteors, asteroids, and dead stars.

Alvarez did not originally start out to investigate for iridium. His initial concern with Cretaceous sediment was quite routine; he was trying to find ways of identifying the last layer of sediment formed immediately before the Cretaceous Period ended. If that layer could be identified, geologists could use that horizon as a standard for comparing the sequence of geological events all over the world. This type of marker has valuable applications, because it is usually very hard to date strata in Europe relative to layers in America. Occasionally layers of sediment carry distinctive chemical trademarks across millions of square miles because volcanic activity can spread clouds of fine-particle dust over the world's oceans, depositing unique concentrations of minerals in the mud at the sea bottom.

Alvarez used detectors of ultra-high sensitivity on ocean sediment at Gubbio, Italy, where an unusually good sequence of strata was laid down at the very end of the Cretaceous. An abrupt change in plankton fossils marked the layer where paleontologists would place the end of the Cretaceous. Right exactly at that point Alvarez stumbled upon a striking geochemical marker, with totally unexpected implications—a thin zone rich in iridium.

Alvarez and his co-workers announced their discovery and stunned the paleontological community with their conclusion: A giant meteor had struck down the world of dinosaurs. The central idea was that such a huge meteor (or asteroid), smashing into the earth at the very end of the Cretaceous, would blanket an immense area of the earth with its extraterrestrial cargo of iridium because the explosion of the celestial mass would send up vast clouds, full of iridium-rich dust. Such dust clouds would subse-

Victims of mass extinction: Long-bodied whales like *Basilosaurus* and primitive porpoiselike whales like *Zygorhiza* died out at the end of the Eocene Epoch.

quently settle down onto the earth's surface and meld onto the face of the globe.

Walter Alvarez's theory gained converts essentially because a giant celestial collision could explain part of the peculiar selectivity of the great extinctions. When the meteoric mass struck the earth, the resulting dust clouds would blacken the sky, obstructing sunlight. Plants would die, as would the plant-eaters and finally the carnivores as temperatures dropped under the deadly umbrella of dust. Cold-blooded creatures could hide in their burrows and wait until the dust settled because their low metabolism would permit long fasts with no ill effects. The lack of extinction among crocodiles and turtles would therefore be easily explained. Small animals would suffer some extinctions (some mammal families did die out), but they generally have burrows to hide in to avoid the consequences of the dust-induced chill. And since many small animals are omnivores, they could have survived by feeding on the carcasses of the dinosaurs that had succumbed. Many species of plant would survive in dormant states—seeds, spores, underground tubers, and bulbs—so the mild effects on the plant world would also be explicable. For those of us who are convinced the dinosaurs were warm-blooded, the great dust cloud could explain why all of them were wiped out. Their high metabolism combined with large size made them especially vulnerable because they could not wait out the disaster. Even the repeat nature of mass extinctions could be explained. A comet could follow a regular cycle of crashes with the earth, a trajectory of collisions repeated every time the comet's and the earth's orbits coincided. A mathematical analysis published in 1983 claimed that such extinctions struck regularly, every 26 million years, so the agent has even been dubbed with a name, the 26 Million Year Death Star.

I do believe that extinctions come in cycles. I do not believe the theory of a bolt from the cosmos. An astronomer friend of mine from Boulder challenged me about this. I advocate a wide variety of heresies about the dinosaurs, so why could I not accept the theory of their extinction based on the striking meteor and the resulting iridium layer? My defense is simple. I champion heresies only if they fit the facts better than orthodoxy.

The theory of the great meteoric explosion fails to fit the facts in one major area. It insists that the extinctions were sudden, catastrophic. All the dinosaurs supposedly died out in a few dozen years, or approximately that. But for quite a while now, orthodox paleobotany has maintained the extinctions were spread over tens of thousands of years or more. And no question, this time orthodoxy has got it right. Paleontologists working in Montana claim they observe a gradual extinction of dinosaurs and Cretaceous mammals and a gradual build-up of new groups of Mammalia, destined for world domination in the next era. In fact, with few exceptions (Dale Russell of Canada's National Museum being chief among them), paleontologists are in rare agreement. The last extinctions were not a single weekend of colossal slaughter but a drawn-out process requiring thousands and even millions of years.

What we require here is a careful, bed-by-bed analysis of the fossil faunas. From them a precise schedule of the extinctions must be established. If the last dynasty of dinosaurs proves to have been dwindling for millions of years before the iridium layer was formed, the theory of the cosmic collision loses all its validity. Can such a detailed timetable be worked out? Unfortunately, the easiest and most popular way to do it is also the most misleading: Counting the number of genera in the fossil sample we possess. It is impossible to know all the genera of dinosaur that lived at one time. It is only possible to identify a fraction of the total because many were so rare in life they had little chance of becoming fossilized. Thus the more skeletons we discover, the more genera we are likely to find. The only reliable way to compare the quantity of genera is to juxtapose formations that contain the same number of identifiable specimens. The earliest of the three Late Cretaceous formations in Alberta, the Judith River, has produced several hundred skeletons; yet the next layer up, the Scollard, has produced only forty. Obviously there is an enormous drop in the number of genera between the two formations, but perhaps this is merely the result of the smaller sample taken in the Scollard.

Dale Russell has used a mathematical procedure called "rarefication analysis" to correct the numbers here. Rarefication analysis determines how many genera of dinosaur would be found in each formation if the number of skeletons were the same for each. He concluded that the big drop between the Judith and the Scollard would not appear as large if the same number of specimens were available from both formations. Russell therefore decided there was no crisis present among the dinosaurs in Scollard times. He insisted that no significant decrease in genera of dinosaurs occurred until the very end of the last formation, the Edmonton–Hell Creek, when they all died off at once.

Can a correct answer be found here? Did the extinctions begin millions of years before the iridium layer was laid down, or did they happen suddenly, precisely at the time of the alleged cosmic collision? To answer this question it is important to remember that both genera and species of dinosaur had been dying out all through the Cretaceous—all through the Mesozoic, in fact. What made the final Cretaceous extinctions special is that no new wave of species appeared to replace those that had died out. In one sense, that is the essential point of all mass extinctions—the rates of extinction outpace the production of new species, so whole groups simply run out of species entirely. But Russell is correct in arguing that there must have been many more dinosaurs near the very end than are known. New species and genera of dinosaurs undoubtedly kept evolving until near the final days of the Edmonton–Hell Creek. But he is wrong in insisting that the world of the dinosaurs was suffering no ills before those last days. A very important ingredient of the ecosystem was already falling to dangerous levels back in Scollard times.

The dangerously declining parameter of the ecology was evenness, what ecologists call equability. When ecosystems are healthy and well insulated from extinction, no single genus dominates. There will be several nearly equally abundant genera in each ecological category—several large plant-eaters, several big meat-eaters, and so on. Judith River times were precisely like that. No single genus of dinosaur was dominant; the chief roles were shared. Three large duckbills were common: *Corythosaurus, Lambeosaurus, Prosaurolophus,* and the horned dinosaurs were represented by three fairly common genera: *Centrosaurus, Chasmosaurus,* and *Styracosaurus.* Such evenness is expressed in mathematical terms by what is called "Simpson's index." The fauna of the Judith scores a 3.2 on the Simpson scale.

The Simpson index is formulated to respond to an implicit question: What is the probability of meeting two individuals of the same genus in a row if animals were being met at random? If the fauna in a given area is very even, with many equally common genera, two individuals of the same genus would not be met in a row very often. Hence a high Simpson score indicates a low probability of two-in-a-row, and that means high evenness. Simpson's index is easily computed: (1) Take the commonest species. (2) Take its share of the total population, say one third of the total. (3) Square this fraction and the result is the probability of meeting that species twice in a row. Thus if a species represents one third of the whole fauna, two-in-a-row probability is $(\frac{1}{3})^2$, which equals $\frac{1}{9}$. (4) Now repeat this calculation for all species. (5) Add all the two-in-a-row probabilities together, and divide into one to yield the index. For example, assume four genera of dinosaur made up $\frac{1}{3}$, $\frac{1}{4}$, $\frac{1}{4}$, and $\frac{1}{6}$ of the total. The two-in-a-row probabilities are $\frac{1}{9}$, $\frac{1}{16}$, $\frac{1}{16}$, and $\frac{1}{36}$. Converting fractions to decimals, the probabilities are 0.11, 0.06, 0.06, and 0.03. Adding together, they equal 0.26. $1 \div 0.26$ is 3.8 units. And 3.8 is the Simpson index. By contrast, if one species represented 99.99 percent of the fauna, then the two-in-a-row probability would be 0.999, and Simpson's index would be about 1.0.

The dinosaurs of the Judith River enjoyed a rich, even ecosystem, one of the most even ever evolved. But the next layer up, the Scollard, is very uneven. One genus of duckbill, *Saurolophus*, made up 75 percent of all the big specimens, and others were quite rare. Simpson's index for this time falls to 1.4, a low score. Something was happening—new species were not evolving adaptations fast enough to permit them to take a more even share of the ecosystem. And similar low evenness continued through the next formation, the Edmonton–Hell Creek, for which Simpson's index is only 1.3. One genus, *Triceratops*, made up 70 to 80 percent of the finds of large dinosaurs. This unbalanced situation endured for two million years before the final extinction.

I first became aware of the pattern of evenness in the evolution of the dinosaurs when I was a graduate student. It was clear that, even if a cosmic collision had killed off the very last dinosaurs, all the dynasties had already been badly weakened from some other cause. I turned to checking the other great extinctions to determine whether there had been a disturbance of evenness before the final collapse on those occasions as well. I invested three years collecting evidence from the museums of Africa, Europe, and the United States. After counting and measuring thousands of skulls

evenness index = 3.8   = 1.38

Judith-style	Lance-style
1/6	1/7
1/4	
1/4	5/6
1/3	

Judith River dinosaurs enjoyed high evenness—no one species dominated. But Lance faunas were unhealthily uneven—*Triceratops* made up 80 percent or so of all the big dinosaur sample.

and skeletons, the killing agents' mode of operating came into sharp focus. Every well-recorded mass extinction fit the pattern of the Late Cretaceous. Long before each final extinction, a decay in evenness had occurred. The saber-toothed gorgons of the Tartarian Epoch are a perfect case in point. Their ecosystem precisely displayed the typical three stages: a time of faunal richness; a subsequent decay of evenness; and the final collapse, when the gorgons disappeared entirely.

The historical pattern followed by mass extinctions simply does not support the theory of a Death Star's killing off faunas suddenly, within a few years' time. What, then, does the iridium layer mean? I am not certain. Sediment-depositing processes had slowed down at the very end of the Cretaceous when the iridium-rich layer was laid down. Some geologists have therefore suggested that terrestrial volcanoes might have produced the iridium, which became concentrated because it was mixed with unusually small amounts of sediment. On the other hand, Erle Kauffman, a Boulder paleontologist specializing in the Cretaceous, believes a celestial body did strike the earth. But he views the crash as the coup de grâce to a dying ecosystem already suffering from massive problems. According to this hypothesis, the dinosaurs were diminishing long before the great collision, but the final celestial blow put the finishing touch to the moribund system.

In any event, history proves celestial collisions cannot be the chief culprits in the collapse of ecosystems. At best they are accessories. But that leaves the more important question unanswered. What is it that attacks the evenness of an ecosystem? For an answer, I am surprised at how little attention is paid to the old, well-thought-out theory of the shallow seas. The best answer for the extinction of the great sea animals is that their favorite haunts disappeared when the warm, shallow seas drained off the continents. And the best answer for the extinction of the open-water, deep-sea creatures is that surface water becomes colder and more thoroughly mixed with deep water when the shallow seas drain off. These changes in the ocean are well documented. There is absolutely no need for an extraterrestrial hypothesis for those extinctions when there is a perfectly good explanation on earth.

The well-established drain-mix-and-cool theory of extinction for the ocean, however, leaves it hard to determine how those disturbances could affect large, active land animals like the Tartarian gorgons or the Cretaceous dinosaurs. Climates do cool a bit on land when shallow seas drain off, because shallow bodies of water act as thermal buffers. But large, active animals are usually more resistant to cold than smaller ones; yet the mass extinctions struck at them harder. There must therefore be something more than a cooling trend of the temperature contributing to killing off the vigorous giants.

The probable culprit was a natural agent so ordinary and earthbound it seems totally devoid of glamour compared to the hypotheses of death-dealing cosmic collisions. And that culprit was clearly described as long ago as 1925 by the great paleontologist Henry Fairfield Osborn.

Let us observe the historical sequence that unfolds on the land during the mass extinctions. Shallow seas drain off, so that land

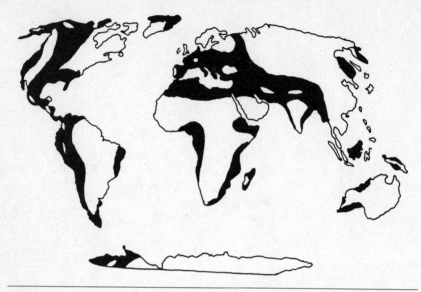

Evolutionary good times—when the seas spread over the continents. Shallow water from the oceans covered huge areas of the continents during the middle of the Cretaceous Period (inundated area shown here in black). All this water made the climate warm and humid on the lowland landscapes along the shores, but when the seas drained off the continents, the winters got cooler and the summers got drier.

areas once underwater become dry and regions that had been separated from each other become connected by land bridges or island chains. At the same time, mountain-building forces weaken so that there are fewer barriers dividing the terrestrial regions. Such changes of course require millions of years. But the net result is a more homogenized ecosystem where species can pass more easily from one end of a continent to another, and from one continent to another. Such easy intercontinental exchange can be found precisely at the end of the Cretaceous. Until late in the Cretaceous, Mongolia had supported a quite different fauna from that of North America. There were many advanced mammals and protoceratopsid dinosaurs in the Central Asiatic Highlands not found in Alberta, Montana, and Wyoming. But very late in the last epoch of

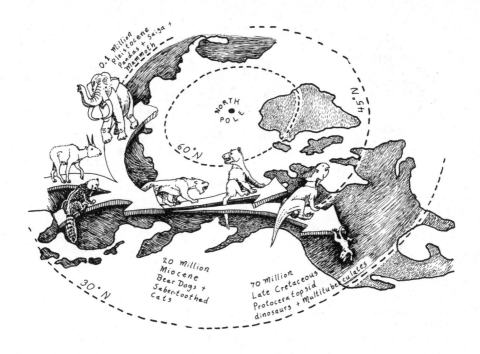

How Asia exports its species to the U.S. All through the history of life, Eurasia has exported hordes of species across the Bering Strait whenever sea level was low enough and Arctic climate mild enough. Seventy million years ago, protoceratopsid dinosaurs and some early mammals (multituberculates) came over; twenty million years ago came big bear dogs and saber-toothed cats; and a hundred thousand years ago came mammoths, tundra antelope (saiga), and pandas. And American species passed eastward in the opposite direction at the same time.

the Cretaceous Period, the Asiatic mammals and dinosaurs began appearing in North America. These immigrants could only have passed over the Bering Land Bridge where the northeastern tip of Asia met Alaska.

Other well-studied times of crisis show the same symptom, a stepped-up exchange of species across continents. Large land mammals were extremely hard hit by extinctions at the beginning of the Eocene Epoch, a time of extraordinarily brisk exchange of species all across Europe, Asia, and North America. Could such interchanges over the continents cause extinction?

There can be no question about the answer here. It certainly could. One of the unshakable tenets of animal geography is that the most extreme consequences are possible when foreign species move into a new region. Every species of reptile, bird, and mammal carries its own unique load of parasites and disease organisms. And many foreign organisms will find no native enemy to hold them in check, so they will run amok. All the worst outbreaks of disease that have swept through mankind or its domestic stock have ultimately come from the introduction of foreign species. The Black Plague came from somewhere in Asia before it swept through Europe. Rinderpest, an Asian cattle disease, got into Africa when Lord Kitchener's army used Indian cattle to haul cannon up the Nile to fight the Mahdi in the late 1800s. Released among the native African hooved stock, Rinderpest became the Black Death of the antelopes, massacring millions of ruminants from the Sudan to the Cape. Even now, after a century of attempts to control it, game wardens worry more about Rinderpest than any other threat to the continent's wildlife.

Osborn was aware of the Rinderpest's history, and he made a special study of the international exchange of species. It was clear to him that if just one disease, from just one foreign species, could wreak such unprecedented havoc across a whole continent, then the most appalling catastrophes could occur when entire faunas— scores of species from previously separated regions—mixed together. Disaster would be inevitable. As long as species remain in one biological region, they adapt to their predators, competitors, and parasites. This is often referred to as the law of co-evolution. On a large time scale, co-evolution over millions of years will usually allow an entire ecosystem to adjust itself in literally innumerable ways. But when two continents mix their faunas, each group will be challenged by enemies for which there has been no co-evolutionary preparation.

Germs and bacteria are not the only death-dealing tourists. Larger animals can function in the same fashion. A well-intentioned New Yorker imported European starlings into Central Park a while ago to brighten Manhattan with the birds of English literature. Starlings are not pests in their native Old World habitats. But here in North America they are spreading like feathered locusts; no native species can stop them. Rabbits are minor nuisances in England. But released into Australia, which had evolved its fauna separately from Europe throughout the Age of Mammals, they exploded unchecked across the land. During most of the Age of Mammals, South America was an island continent. South American mammals and birds evolved into all manner of species found nowhere else (giant ground sloths twenty feet tall, saber-toothed pouched mammals, flightless killer birds larger than a lion). North American mammals crossed into South America only two million years ago when the isthmus formed at Panama. Among the immigrants were representatives of the elephants, jaguars, deer, tapirs, and wolves, to name only a few. These North American immigrants devastated the native fauna. Most of the big South American species went extinct, victims of predation and competition from the northerners, as well as of their diseases.

The Late Cretaceous world contained all the prerequisites for this kind of disaster. The shallow oceans drained off and a series of extinctions ran through the saltwater world. A monumental immigration of Asian dinosaurs streamed into North America, while an equally grand migration of North American fauna moved into Asia. In every region touched by this global intermixture, disasters large and small would occur. A foreign predator might suddenly thrive unchecked, slaughtering virtually defenseless prey as its populations multiplied beyond anything possible in its home habitat. But then the predator might suddenly disappear, victim of a disease for which it had no immunity. As species intermixed from all corners of the globe, the result could only have been global biogeographical chaos.

Such a scenario is hardly hypothetical, and it hardly requires an extraterrestrial hypothesis. Global disaster was simply the inevitable result of unleashing pests and pestilence on natives and foreigners alike. The worst effects would fall on the most widely traveled. Large land animals crossed geographical barriers easily, so they spread more havoc and suffered more. Small species cannot migrate as easily, because even a small river can block their progress. Therefore extinction caused by faunal mixing would always be hardest on the biggest, most active animal—exactly fitting the picture for all the great extinctions in geological history.

How would warm-bloodedness fit into this explanation? Would dinosaurs have been more at risk if they possessed high metabolism than if they had been merely good reptiles? Again the answer is certainly yes. Cold-blooded creatures with their very low metabolism do not travel well. The relatively small amount of their energy confines them to small home territories and very slow rates of geographical expansion. Only big energy spenders require large territories and constantly push at their geographical limits. The fastest-spreading land vertebrates should also be the largest, most metabolically active. And warm-bloodedness adds a further vulnerability as well. The most effective way to nurture a large crop of germs is to keep them constantly warm. A rattlesnake in the desert discourages such incubation because its temperature fluctuates from near freezing to 90°F within the space of a single day. But tissue kept warm by high metabolism would be ideal from a parasite's point of view. Today, animals with high, constant body temperatures (mammals and birds) have a much longer list of diseases carried than do reptiles and amphibians. Dinosaurs with high metabolism would have been at much greater risk of mass extinction during intercontinental exchange than would the giant, low-metabolism reptiles.

Such a scenario explains all the details of the mass extinction without resorting to extraterrestrial agents of any kind. the mundaneness of Osborn's theory is stunning to those fascinated by the hypothesis of a celestial collision. Perhaps a meteor or a large asteroid did strike the last populations of the dinosaurs. Maybe there is a place for an occasional bolt out of the heavens to kill off the remnants of a weakened ecosystem. But the overwhelming share of the credit (or blame) for the grand rhythm of extinction and reflowering of species on land and in the sea must surely go to the earth's own pulse and its natural biogeographical consequences.

22

DINOSAURS HAVE CLASS

A public lecture delivered on March 10, 1969, at Yale's Peabody Museum stated authoritatively the advantages we enjoy today over old-fashioned paleontologists. They believed the Dinosauria constituted a natural group, but we knew they were a miscellany of unrelated reptiles. As such, they did not deserve the recognition of a formal zoological label. Only natural groups—species descended from one common ancestor—merited such a label. The term "Dinosauria" ought therefore to be expunged from the lexicon, banished from our speech.

I feel especially bad about that lecture because it was part of a program given to teachers of high school science—but most of all, because I wrote it. At the time, I was serving as Docent in charge of Special Programs. It was my responsibility to ensure that each lecture contained only the most up-to-date material. Since how you defined a dinosaur was one of the most frequently asked questions, I made sure all the lectures reported the most modern theory: Dinosaurs were an unnatural group.

Ever since Darwin, most zoologists have insisted that "real" groups must have phylogenetic integrity, a unity of descent. To qualify as real and natural, a group would have to prove that all of its species traced their evolution back to one common ancestor. If that could be done, the group of species was entitled to a formal zoological label.

The fanged beaked dinosaur, *Heterodontosaurus,* about three feet long, from the Early Jurassic of South Africa. Heterodontosaurs had a twist-thumb claw, built like that of anchisaurs.

As science improved at reconstructing evolutionary trees during the 1880s and 1890s, under those criteria many zoological groups were stripped of their labels. The Pachydermata are an excellent example. Early nineteenth-century anatomists believed thick-skinned mammals were all somehow related. Consequently, horses, tapirs, rhinoceroses, elephants, and hippos were all formally identified as the Order Pachydermata. But in the 1880s, an avalanche of fossil data poured in demolishing the notion that "pachyderms" constituted a natural evolutionary unit. Rhinos, tapirs, and horses are indeed closely related; they stem from an ancestor much like the little Dawn Horse, *Eohippus*. But that group is not at all close to hippos, which trace back to a pig ancestor, and neither the horses nor the hippos are close to elephants. The case against the Pachydermata was overwhelming. And the name was stricken from the list of acceptable terms. The "pachyderms" were an unnatural mixture of three distinct pedigrees, sharing only a few unimportant resemblances in their skin and digestive systems, resemblances clearly resulting from independent evolutionary events.

When I was in college in the 1960s, it was commonly accepted that the case against the Dinosauria was equally strong— the dinosaurs were in fact two entirely separate groups, each tracing its ancestry to a different Triassic reptile (both ancestors were pseudosuchian reptiles of the Triassic). The implication was that each group of dinosaurs was no more closely related to the others than they all were to crocodiles or to birds. The public might have great affection for the Dinosauria, but the label was without scientific foundation. This was the accepted wisdom, but the evidence for it was not much discussed. It was, in fact, yet another piece of orthodoxy, rarely debated, never seriously challenged. And the notion that the term "Dinosauria" should not even be used matched nicely with the generally dismissive attitude at large toward the Mesozoic monsters.

Into the mid-nineteenth century, "Dinosauria" was accepted as a respectable scientific term. Paleontologists believed there were three major subgroups: the carnivorous Theropoda; the long-necked Sauropoda (brontosaurs); and the beaked herbivores, now called Ornithischia ("bird hips"). Out of these three groups, the Ornithischia had the most unusual skull and hip. It therefore became fashionable to divide the Dinosauria into two groups: the Order Saurischia for the carnivores and brontosaurs combined, and the Order Ornithischia. At first, this division did not undermine the belief that all dinosaurs were one natural group. Paleontologists wrote as though the Saurischia and the Ornithischia had evolved from a single, very primitive ancestor, the Ur-dinosaur.

By the 1920s, however, the view had become quite different. Paleontologists began to embrace the idea that the ancestry of the Ornithischia had been totally different from that of the Saurischia. Formal classification listed the Order Saurischia and the Order Ornithischia as distant cousins, alongside the Order Crocodilia, and the Order Pterosauria (flying reptiles). All these independent or-

WRONG!

THE UNNATURAL ORDER PACHYDERMATA

unknown pachyderm ancestor

CORRECT!

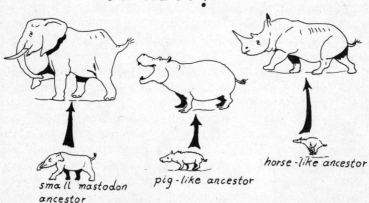

small mastodon ancestor

pig-like ancestor

horse-like ancestor

Example of an unnatural taxonomic act—the "Order Pachydermata." To qualify as a natural group, a clan of species must trace its ancestry back to one single common ancestor. In the early nineteenth century, naturalists lumped all the big, thick-skinned mammals into the Order Pachydermata. If pachyderms were truly a natural group, then rhinos, hippos, and elephants must have had a common ancestor. But fossil discoveries in the last half of the nineteenth century proved that the Pachydermata was hopelessly unnatural. In fact, rhinos, hippos, and elephants belong to three separate orders, and each of the three "pachyderms" evolved their big size and thick skin from separate small, thin-skinned ancestors.

---

ders were grouped together into the reptile subclass, Archosauria. Obviously, the Dinosauria was an unnatural group.

The evidence for this change in classification was in fact extremely scanty, and the idea was never thoroughly thought out or debated. An altogether rather half-baked idea established itself as the orthodox view of the relationships among the dinosaurs. Even at its inception, this belief was logically flawed: Just because the Order Ornithischia was more specialized in its skull and hip than the Order Saurischia did not prove that the two orders did not share a common ancestor. If the defining characteristics of the Ornithischia and Saurischia, all well known in 1920, were compared, more than a dozen were shared by both groups—and none of them were present in any other "reptile" group. But all of this was ignored in the press to adopt the view that ornithischians and saurischians were two separate groups issuing from two hypothetical ancestors. The change in conception was part of a general conceptual crisis that afflicted the whole field of evolutionary biology at that time. Paleontologists had somehow arrived at a view of evolution which I call the hub-and-spoke syndrome. Each major evolutionary line supposedly originated in one primitive, unspecialized stock, an evolutionary hub. Subsequently, all the advanced lines evolved outward—like separate spokes of a wheel—from that hub. Under the influence of this conception, paleontologists tended to invent wholly imaginary groups to serve as the ancestors for their grand theories. Poorly known fossils, represented by fragmentary skeletons, were often elevated to the status of "common ancestral stocks." The crimson crocodiles were treated this way. Some of those little-known Triassic creatures were lumped together as a hypothetical archosaurian ancestral stock, called the Pseudosuchia. As all this theory crystallized into textbook form, the Pseudosuchia became firmly established as the ancestral hub from which radiated all the separate archosaurian lines to become saurischians, ornithischians, crocodiles, pterodactyls, and birds.

Nearly everything written about the Pseudosuchia as the central hub for Archosaurian evolution is hypothetical. And owing to recent discoveries in South America and China, many pseudosuchian families are now far better known. Almost without exception, each family possessed strong distinctive specializations that disqualified it as a "generalized" ancestor for any more advanced group. A good illustration here is provided by the ornithosuchids,

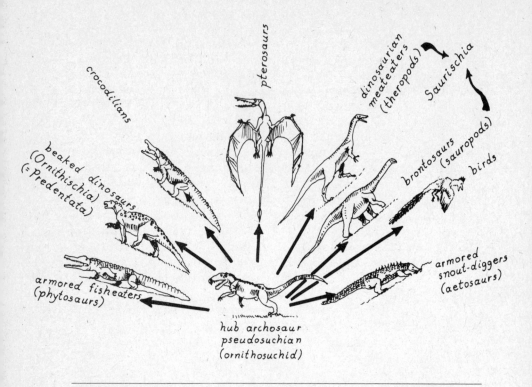

crocodilians

pterosaurs

dinosaurian
meat eaters
(theropods)

Saurischia

beaked dinosaurs
(Ornithischia)
(= Predentata)

brontosaurs
(sauropods)

birds

armored fish eaters
(phytosaurs)

armored
snout-diggers
(aetosaurs)

hub archosaur
pseudosuchian
(ornithosuchid)

Myth of the hub-and-spokes. Since the 1920s, most textbooks show the pseudosuchians as the central archosaur hub, with two or three separate spokes representing the different dinosaur clans. If this view is correct, the Dinosauria is an unnatural group.

---

the "bird-crocodiles." Both technical and popular books define them as the essential hub for all the Archosauria, the perfect common ancestor. As far as they were known in the 1930s, from poorly defined skeletons, the ornithosuchids did fit preconceptions about what the ancestor of the dinosaurs should look like. Their hind legs were longer than their forelegs and there was the vague suggestion of the bipedal locomotion common in primitive dinosaurs. By the 1970s, however, good, clear skeletons from Argentina revealed that the ornithosuchids possessed bizarre specializations—

for example, a huge, drooping snout, pinched in the side-to-side dimension—a development totally unexpected in an ancestor of the dinosaurs. Ornithosuchids were not dinosaur ancestors or anyone else's. They spent their entire evolutionary history evolving into more and more specialized ornithosuchids. They did not evolve into true crocodiles, pterodactyls, birds, or anything else among the ranks of the more advanced archosaurian groups.

The tale of the ornithosuchids also illustrates the underlying theoretical weakness of the hub-and-spoke conception of evolution. Common ancestral stocks may be posited at the base of an hypothetical family tree. But the real-life ecosystem is a mean, cruel place. In order to survive, any group must evolve to keep pace with the threats from new predators and competitors. It is simply not possible for a family of species to wait around for millions of years, twiddling its evolutionary thumbs as it were, until it receives the signal to evolve into something more important.

The system of nomenclature employed by biological scientists represents their understanding, their organization of the life they study. The system is clear and hierarchical. In zoology, the Kingdom embraces the widest group. Next beneath that comes the Phylum—and the term "Phylum Chordata" embraces all vertebrate animals. The Phylum is then divided into Classes and Subclasses, and each Class or Subclass is divided into Orders. Orders are then divided into Families, Families into Genera, and finally Genera into Species. It must be emphasized these are man-made classifications. And sometimes this system of nomenclature can warp our perception of the real evolutionary events. For example, the Archosauria is traditionally ranked as a subclass which is divided into the Order Saurischia, the Order Ornithischia, the Order Crocodilia, the Order Pterosauria, and as a primitive hub, the Order Thecodontia—reserved for the pseudosuchians and other crimson-crocodile groups. It makes for a tidy arrangement, everything in its proper place, especially if we plot each advanced order arising independently, as a spoke, from the hub order, the Thecodontia. But this orderliness produces a seductive bias toward reconstructing the history of evolution in oversimplified ways.

How can the pitfalls of such overly tidy thinking be avoided? Over the last several decades, scholars of evolution have developed methods for tracing the probable pattern of branching in the families they investigate. All their data indicates that evolution rarely follows the hub-and-spoke pattern. Evolution rather follows a pattern whereby a family emerges by early branchings, and each early branch branches again, so that the family tree finally resembles a tangled blueberry bush, a maze of ever smaller branchlets, bifurcating and ramifying in many directions. Is it then possible to make any sense of such a complicated pattern from the scattered fossils available for studying the history of the dinosaurs?

The most reliable method employed today consists of the search for what are called sister groups. Sister groups are quite easily defined. Suppose, for the moment, that the Ornithischia and Saurischia were sister groups. That would imply that they had descended from a common ancestor that was already specialized in certain distinctive ways. It would then be proper to expect an evolutionary indicator for this relationship, some feature that had evolved in the common ancestor and was then passed down to all the subsequent branches. Now, if orthodoxy were correct and the two orders of dinosaurs really did evolve independently from a primitive archosaurian hub, then no common heritages could be expected. And the only course of action for clarifying the relationships of the dinosaurs is to examine them thoroughly for shared adaptations that might mark each branching point.

Until 1971, I was myself a firm believer in the hub-and-spoke theory and the orthodox view of the dinosaurs as a group. The Dinosauria really consisted of three separate spokes—the Sauropoda, Theropoda, and the Ornithischia—since, in my opinion, even the Order Saurischia was artificial and the brontosaurs (Sauropoda) were a totally separate line from the carnivores (Theropoda). Each was ranked as an Order, and each supposedly evolved separately from the pseudosuchians. But my convictions were about to change.

In 1971, Peter Galton, then a postdoctoral fellow, was engaged in work on the very primitive anchisaurid dinosaurs. In the process, he made some really important discoveries on how the dinosaurs' claws worked—he was the first to discover the peculiar thumb-twists. Primitive meat-eating dinosaurs had huge, curved claws on their thumbs whose tips pointed inward when the fingers were flexed upward, but downward when the fingers were flexed downward—a most unusual arrangement. Galton also found exactly the same sort of thumb-twist in one group of primitive plant-eating dinosaurs, the anchisaurs, which were evolutionary uncles of the giant brontosaurs.

This twist-thumb was a blow to my belief in the artificiality of dinosaurs as a group. I had even been lecturing that the meat-eating dinosaurs had evolved quite independently from the anchi-

---

Kinship revealed in jaw, chest, and hand. Primitive beaked dinosaurs like *Heterodontosaurus* had a double breastbone, twist-thumb claw, and a low jaw joint just like that of anchisaurs. So anchisaurs must be very close relatives of the ancestors of all the Ornithischia.

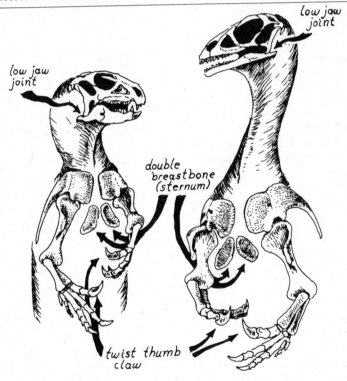

low jaw joint

low jaw joint

double breastbone (sternum)

twist thumb claw

Heterodontosaurus primitive beaked dinosaur

Anchisaurus

saurs-brontosaurs. But that twist-thumb was simply not to be found in any potential ancestor from the Triassic. The twist-thumb is a good example of what paleontologists call a "high-weight novelty," an adaptation so complex and unusual that it is highly unlikely it evolved twice in totally separate lines. The twist-thumb was therefore a very strong argument that the anchisaurs-brontosaurs were a very closely related sister group of the carnivores, and thus, that both groups had descended from one common ancestor—an original species that had first evolved the thumb.

It went against the grain in 1971 to be convinced that the theropods and sauropods were in fact a single evolutionary group. But I was about to embrace a far worse heresy. Orthodoxy maintained, as a kind of holy tenet, that the beaked dinosaurs, the Order Ornithischia, were in no way, shape, or form closely related to the other dinosaurs, the Order Saurischia. But in 1972, a little dinosaur from the Connecticut Valley was to blast this belief as well. The dinosaur in question was *Anchisaurus,* a five-foot herbivore originally discovered by Othniel Charles Marsh in the Brownstown quarries near Portland, Massachusetts. Peter Galton was at work on this skeleton, cleaning off areas still covered with rock, though the specimen had first been put on exhibition in 1880. As he removed the very hard sandstone, some quite unexpected bones came to light. The *Anchisaurus*'s upper hip bone (ilium) displayed unmistakable characteristics of the Ornithischia. Unlike other primitive saurischians, whose ilium was very short from front to back, *Anchisaurus* had an extra-long prong that stuck forward, which in life would have increased the size of the major muscle running from the ilium to the kneecap. But primitive ornithischians did have a very similar iliac prong. Was this dinosaur the ancestor of the ornithischians, or at lease a close cousin of the real ancestor? This was a very unorthodox question since it would imply that *Anchisaurus* was a missing link, an evolutionary bridge between the supposedly unlinkable Order Saurischia and the Order Ornithischia. If so, then the orthodoxy of 1970 had it wrong and the view of 1880 had been right: the entire Dinosauria constituted a single natural group.

Galton and I scrutinized every inch of the anchisaur skeleton. We constructed a list of the characteristics which might link anchisaurs to the Ornithischia, and in a brief time it grew to a formidable length. Dozens of adaptations linked these dinosaurs and distinguished them from all other archosaurs:

1. A slender, graceful neck shaped into an easy S-curve.
2. A shoulder structured with no collarbone or at best a very reduced one.
3. An upper arm bone (humerus) with a long shelf for the chest muscles.
4. A wrist structured without any prominent prong of bone on the rear inside (this is called the pisiform bone and is very evident in most reptiles and mammals, including man).
5. A long, forward-jutting prong on the ilium.
6. A hip socket formed as a wide hole between the three hip bones.
7. A birdlike hinge in the ankle, where two small ankle bones were firmly fused onto the lower ends of the shank (the joint here was formed between the two ankle bones and the lower part of the ankle).
8. A breast bone (sternum) divided into two parts that lay side by side.

All these observations constituted a fair case for *Anchisaurus* as a missing link. Peter Galton and I were feeling proud of our fledgling heresy until we did some reading in the old monographs that dated from before 1920. Most of the characteristics we had recognized had already been identified by the earlier scholars. And Thomas Henry Huxley had made nearly as good an argument for the naturalness of the Dinosauria as a single group in 1880. Somehow all of this was simply forgotten after 1930.

There remained one nagging difficulty that obstructed our argument for the dinosaurs as a single group: All of the anchisaurs and all of the primitive meat-eating dinosaurs possessed that very distinctive twist-thumb, but not one of the ornithischians had it. Ornithischian thumbs were usually short and terminated in a blunt hoof, not a curved claw. To clinch our theory, an ornithischian with a twist-thumb that ended in a claw was indispensable. And what happened next was nearly a miracle of serendipity.

The miraculous find took the form of Fuzz Crompton's fanged ornithischian. Fuzz—the nickname derived from his days in South Africa, when his bushy hair made him a standout in that socially

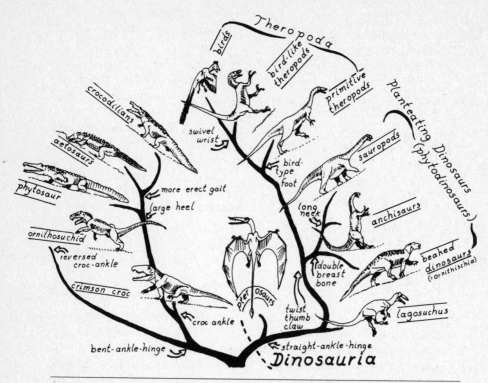

Theropoda
birds
bird-like theropods
primitive theropods
Plant-eating Dinosaurs (phytodinosaurs)
crocodilians
swivel wrist
sauropods
aetosaurs
bird-type foot
more erect gait
large heel
long neck
anchisaurs
phytosaur
ornithosuchid
reversed croc-ankle
double breast bone
beaked dinosaurs (=ornithischia)
crimson croc
pterosaurs
twist thumb claw
lagosuchus
croc ankle
bent-ankle-hinge
straight-ankle-hinge
Dinosauria

Recent discoveries suggest that the archosaur family tree was complex, with many basal branches. On this chart, some key evolutionary developments are shown marking the major splitting points. Straight-ankle hinge lines seem to mark the true Dinosauria, and double breastbones (sterna) mark the plant-eating dinosaurs.

conservative atmosphere—Crompton was director of the museum at Harvard. He had excavated a fine three-foot-long ornithischian from the Early Jurassic beds of Lesotho. This fossil caused quite a stir—it was the finest specimen from such an early geological age. The animal was primitive in many ways, and—most striking of all its features—it had fangs, large, sharp teeth arranged in pairs in the front of its mouth. Most ornithischians were herbivores, so possessed no dangerous biting teeth. Crompton's fanged ornithischian therefore presented a nice puzzle. What precisely had it used those fangs for?

What was absolutely riveting about this ornithischian, however, was its hand: the fingers were all long, not stubby like those of other primitive ornithischians, and the thumb possessed a wicked claw mounted on a twisted bone. There could be no doubt that here was the missing link Peter Galton and I had concluded we needed. This animal proved that the Order Ornithischia was the sister group of the Order Saurischia. *Anchisaurus* and Crompton's fanged ornithischian taken together made the argument. *Anchisaurus* was a saurischian leaning forward toward the Ornithischia, while Crompton's beast was an ornithischian leaning backward toward the Saurischia. The two dinosaurs were very close cousins, and they proved that the Dinosauria were a natural group.

Excited by our conclusion, Peter Galton and I rushed off a paper to the British journal *Nature,* announcing the resurrection of the Dinosauria as a legitimate scientific term. This marked the first time in half a century that anyone had made a serious case for their naturalness. All hell broke loose. We expected debate, discussion, dissent—and we certainly got it. But as good fortune would have it, shortly after Galton and I had published our piece, the brilliant Argentine paleontologist José Bonaparte published his work on the lineage of the dinosaurs—work that Galton and I had no knowledge of until that moment. Bonaparte was similarly unaware of our work, but he had arrived at the very same conclusions we had: The dinosaurs were a natural group and the Ornithischia had evolved from an anchisaurlike ancestor. Bonaparte argued that the earliest ancestor of all dinosaurs had been something like *Lagosuchus,* the tiny bunny-croc of the Mid Triassic. In fact, except for a few details, José Bonaparte's description of the dinosaurs' family tree was nearly identical to ours. And the fine nature of his work was a powerful support for this conception of the dinosaur's evolution.

While Peter Galton and I were at it, we also went one step further in our resurrection of the Dinosauria. We made them unextinct. We accomplished this by a simple rearrangement of the formal scientific nomenclature. We placed the birds into the Dinosauria. And if birds are members of the Dinosauria, then the dinosaurs are not extinct.

John Ostrom had proved that birds were direct descendants of small, advanced carnivorous dinosaurs. Traditional classification placed the birds in their own Class Aves and the dinosaurs in the Class Reptilia, because birds were feathered, warm-blooded fliers with advanced hearts and lungs, whereas dinosaurs were scaly-skinned, cold-blooded beasts with only limited capacity for vigor. But we were convinced the birds had inherited their heart-lung system and their warm-bloodedness from dinosaurs. Of course, dinosaurs hadn't flown. But the small, predatory dinosaurs had all the necessary adaptive prerequisites for evolving into flight. And feathers of some sort may well have insulated the body of some theropod dinosaurs. It might even appear that birds owed most of their distinctive adaptations to their dinosaur ancestors. Birds might never have evolved flight if their dinosaur forebears had not undergone a long history of evolutionary transformation into ever more active, fast-moving, warm-blooded predators. It was neither fair nor accurate to deny the dinosaurs credit for evolving into birds. It was therefore only proper to demote the Class Aves to a sub-division of the Class Dinosauria (or Class Archosauria with dinosaurs as a subclass).

The notion of birds as dinosaurs gave conservative zoologists yet another issue over which to protest. And after a lecture on the topic I delivered in Philadelphia, a woman arose to ask whether this meant her parakeet was dangerous! Some large dinosaurs obviously were most unbirdlike, *Diplodocus* or *Triceratops,* for example. But the bipedal predators were very avian in structure. And the small, advanced predators like *Deinonychus* were so close to *Archaeopteryx* in nearly every detail that *Archaeopteryx* might be called a flying *Deinonychus,* and *Deinonychus* a flightless *Archaeopteryx.* There simply was no great anatomical gulf separating birds from dinosaurs. And that implies dinosaurs are not extinct. One great, advanced clan of them still survives in today's ecosystem and the more than eight thousand species of modern bird are an eloquent testimony to the success in aerial form of the dinosaurs' heritage.

Finally, I suggest the standard terminology applied to dinosaurs stands in need of radical reorganization. Most popular books about dinosaurs today employ the traditional classification and divide them into Saurischia and Ornithischia. But the distinction implied by this nomenclature is misleading, if not obfuscatory.

Traditionally, herbivorous dinosaurs are not placed into one natural group, they are separated into two "orders"—the anchisaurs and brontosaurs are put into the Saurischia, and all the beaked dinosaurs into the Ornithischia. This separation is damaging because it obscures the fact that beaked dinosaurs are close relatives of anchisaurs. The ornithischians descended from anchisaurlike saurischians, just as the brontosaurs trace from a close relative of *Anchisaurus*. Therefore all the plant-eating dinosaurs of every sort really constitute one, single natural group branching out from one ancestor, a primitive anchisaurlike dinosaur. And a new name is required for this grand family of vegetarians. So I hereby christen them the Phytodinosauria, the "plant dinosaurs."

Of course, all the carnivores are also descended from a common ancestor that first evolved that birdlike hind foot—three toes to the front and one turned backward and inward. These meat-eaters already enjoy a good name, the theropods. Now, birds should be placed in as a subdivision of the Theropoda.

There are some very primitive, very early carnivorous dinosaurs from the Triassic that are presently hard to define. They had not yet evolved the birdlike foot or the expanded hip bone (iliac blade) found in all other predatory dinosaurs. Until these archaic creatures are better known, they can informally be left as a group of ancient uncles of the theropods.

At the very base of this system for classifying dinosaurs must be placed *Lagosuchus,* the bunny-croc, and its kind. And this raises another interesting wrinkle. Pterodactyls were most probably the evolutionary products of *Lagosuchus* or a very similar animal. They too are traditionally assigned their own order, the Order Pterosauria, but this arrangement obscures the very close relationship between early pterodactyls and early dinosaurs. It would be far clearer to make the Pterosauria a subdivision of the Dinosauria as well.

At the broadest level, then, how would this resurrected Class Dinosauria fit into the overall classification of land vertebrates? This is an important question and care must be taken. If the Dinosauria were to be located in the Class Reptilia, irretrievable damage would be done; once again, the dinosaurs would be subjected to more guilt by association—arguments that dinosaurs were cold-blooded because reptiles are, and so on. No, definitely not, the Dinosauria are not Reptilia Vera. And while we are at it, those uncles of the

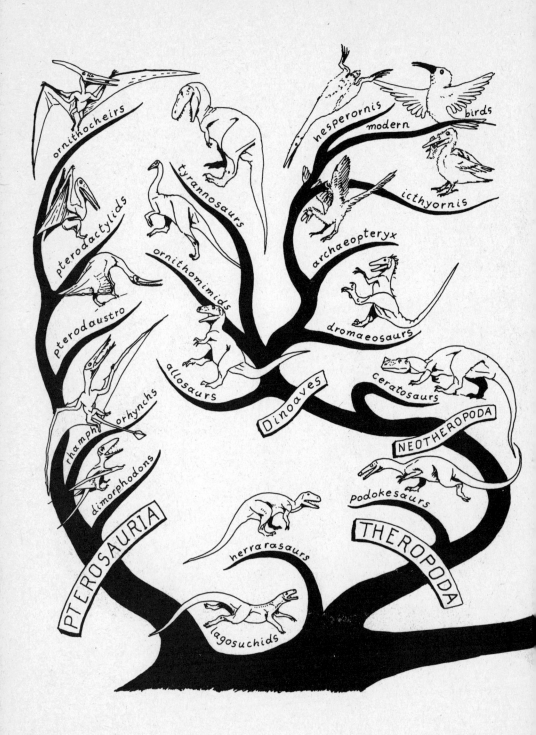

The Dinosaurian Family Tree:
Each figure represents a family.

dinosaurs, the crimson crocodiles, should also be taken out of the Class Reptilia. Most of them had all the basic adaptations of warm-bloodedness—fast growth, fast evolution, low predator-to-prey ratios (though not as low as the dinosaurs'). What I am proposing, then, is that we should remove the entire Archosauria from the Reptilia. (The same ought to be done for our own ancestors, the protomammals of the Late Permian and the Triassic. These fellows are usually left in the Order Therapsida in the Class Reptilia. They don't belong there. Even the earliest Kazanian therapsids displayed the telltale signs of warm-bloodedness in their bone structure and predator ratios.)

I proposed this sort of classification in 1975 in an article I published in *Scientific American*. Most taxonomists, however, have viewed such new terminology as dangerously destabilizing to the traditional and well-known scheme that has been with us since the time of Baron Cuvier. I cannot see any benefit to be gained by refusing to remove the dinosaurs (and the therapsids) from the confines of the Reptilia. Classification is a type of scientific definition, and definitions should help express our perceptions of nature, not hinder them. As long as textbooks and museum labels unreflectively repeat the message "Dinosaurs are reptiles," it will be difficult to establish an intelligent debate about the true nature of the dinosaurs' adaptations. Some of the orthodox paleontologists act as though the dinosaurs must be assumed cold-blooded until their warm-bloodedness is proved beyond any reasonable doubt. That is at least highly unscientific. And it certainly represents "argument by definition"—dinosaurs are reptiles, reptiles are cold-blooded, therefore dinosaurs were cold-blooded.

A truly scientific skeptic would start by assuming neither cold-bloodedness nor warm-bloodedness, and then reevaluate the evidence without prior terminological bias. So long as the Dinosauria remain stuck in the Class Reptilia, this type of analysis is nearly impossible. Let dinosaurs be dinosaurs. Let the Dinosauria stand proudly alone, a Class by itself. They merit it. And let us squarely face the dinosaurness of birds and the birdness of the Dinosauria. When the Canada geese honk their way northward, we can say: "The dinosaurs are migrating, it must be spring!"

NOTES AND REFERENCES

The literature on dinosaurs and other fossil vertebrates is a sprawling mass of short and long contributions, with many of the short technical papers being excellent but written in inaccessible jargon and many of the popular summaries being dreary repetitions of the "cold-blooded musheater in the swamps" myths. So I have listed here the best overall summaries that have good bibliographies, plus some of the old gems that have been forgotten, plus some key papers on important aspects of physiology and ecology.

GENERAL REFERENCES

The two milestone volumes are: 1) the AAAS Select Symposium 28, Westview Press, 1980 (ALMOST out of print—call the publisher so they will add another printing); and 2) the Los Angeles County Museum of Natural History Special Colloquium *Dinosaurs Past and Present,* LACM Press (1986). Nearly every important paper about warm-bloodedness, pro and con, are cited in these two volumes. The difference in tone between the two is remarkable. The 1980 AAAS book was unapologetically skeptical—even the title *Cold Look at the Warm-Blooded Dinosaurs* suggested that belief in warm-blooded Dinosauria was rash and beyond the boundaries of level-headed science. But the LACM volume contains articles by those who reconstruct dinosaurs and their world, and, with few exceptions, the artists, anatomists, and paleontologists accord the dinosaurs a much, much higher level of locomotor energetics than was widely believed six years ago. Sylvia Czerkas, the editor and organizer of the LACM colloquium, said to me after the conference, "You must be feeling pretty good, seeing your ideas vindicated more and more." Maybe so. At least the general attitude is shifting away from the view that dinosaurs must be assumed to be cold-blooded in all points and any contrary evidence dismissed with a "harrumph."

Czerkas, Sylvia, ed., *Dinosaurs Past and Present,* Los Angeles County Museum Special Symposium (Los Angeles: LACM Press, 1986).
Thomas, Roger D. K. and Everett C. Olson, eds., *A Cold Look at the Warm-Blooded Dinosaurs,* AAAS Selected Symposium 28 (Boulder, Colo.: Westview Press, 1980).
Wilford, John Noble, *The Riddle of the Dinosaur* (New York: Alfred A. Knopf, 1985).

1. BRONTOSAURUS IN THE GREAT HALL AT YALE

Notes:

I use *Brontosaurus* not *Apatosaurus* even though, according to the International Code of Zoological Nomenclature, the latter is the legal name. Al Romer used to complain that "rules of nomenclature should serve the cause of science, not the other way round." The same man—Yale's Professor Marsh—coined both *Apatosaurus* and *Brontosaurus;* the former name is just a bit older, but the latter is much, much better known by the public at large. Science should take every opportunity to divest itself of unnecessary obscurantism, and so I will use *Brontosaurus.* The type specimen (the specimen used first to define the genus) of *Brontosaurus* is the wonderfully complete skeleton mounted at Yale, and I'm sure that Marsh's ghost won't mind a bit when I use *Brontosaurus* in preference to *Apatosaurus.* The nomenclatural Law of Priority—the rule that says the legal name is the oldest name based on an adequate type specimen—was originally developed to honor the first discoverer of a species or genus and to stabilize the system of names. Using *Brontosaurus* honors Marsh, who discovered the genus, and certainly reduces confusion and instability when scientists communicate to the public.

Speaking of genera . . . it's common practice to talk about dinosaurs and other extinct vertebrates in the generic sense, not in the specific. Most popular and technical articles speak of *Triceratops* and *Allosaurus* and do not identify the species—for example, *Triceratops horridus* or *Allosaurus fragilis.* That's like talking about all the dog species together in one lump— the dog genus *Canis* contains the coyote species, *Canis latrans,* the wolf *Canis lupus,* the red wolf, *Canis niger,* and quite a few others. All the *Canis* species are very similar in anatomy, so it's useful shorthand to use *Canis* in general discussions of mammal evolution where we want to compare cats with dogs and weasels. When we say *Brontosaurus,* we invoke a closely knit group of species, including the most common *Brontosaurus excelsus,* the very rare *Brontosaurus louisae* (found only at Dinosaur National Monument), and the super-huge *Brontosaurus ajax.* These three differ in details of bony anatomy about to the same degree that closely related modern species of bird, crocodile, and mammal differ among themselves today. And therefore we can conclude that the three *Brontosaurus* species were "good" species—they each represent a distinct breeding unit that did not exchange genes very often with the other species. One reason paleontologists are reluctant to talk in species, rather than in generic units, is that it's often very hard to tell closely related species apart from fragmentary fossil skeletons. But things are changing. Recent intensive study of dinosaurs allows much greater precision in sorting out clusters of sibling species. A curious exception to tradition is that nearly everyone talks of *Tyrannosaurus rex* when the genus *Tyrannosaurus* is discussed. There are other species in the genus, but *T. rex* is the best known and the generic-plus-specific names sound so good together that *Tyrannosaurus rex* is just irresistible to the tongue.

Andrews, Roy Chapman, *All About Dinosaurs* (New York: Random House, 1953).
Bowler, Peter J., *Fossils and Progress: Paleontology and the Idea of Progressive Evolution in the Nineteenth Century* (New York: Science History Publications, 1976). Now unfortunately out of print, this is one of the best books on the subject.
———, *Evolution: The History of an Idea* (Berkeley: University of California Press, 1984).
Clemens, W. W., "Fossil Mammals of the Type Lance Formation, Wyoming, Part I, Introduction; Multituberculata," *Univ. Calif. Publ. Geol. Sci.* 48 (1963): 1–105.
Colbert, Edwin H., *Dinosaurs: Their Discovery and Their World* (New York, E. P. Dutton, 1961).
———, *Men and Dinosaurs: The Search in Field and Laboratory* (New York: E. P. Dutton, 1968).
———, *A Fossil Hunter's Notebook: My Life with Dinosaurs and Other Friends* (New York: E. P. Dutton, 1980).
Eisenberg, John F., *The Mammalian Radiations: An Analysis of Trends in Evolution, Adaptation, and Behavior* (Chicago and London: University of Chicago Press, 1981).
Fenton, Carroll Lane, *Tales Told by Fossils* (Garden City, New York: Doubleday, 1966).
Goodrich, S. G. and Alexander Winchell, *The Animal Kingdom: Wonders and Curiosities* (New York: A. J. Johnson and Company, 1867).

Gregory, William King, *Evolution Emerging: A Survey of Changing Patterns from Primeval Life to Man* (New York: Macmillan, 1951).

Howard, Robert West, *The Dawnseekers: The First History of American Paleontology* (New York and London: Harcourt Brace Jovanovich, 1975).

Kurtén, Bjorn, *The Age of Mammals* (New York: Columbia University Press, 1972).

Lanham, Url, *The Bone Hunters* (New York and London: Columbia University Press, 1973).

Lovejoy, Arthur O., *The Great Chain of Being: A Study of the History of an Idea* (Cambridge, Mass. and London, England: Harvard University Press, 1936 and 1964).

Lull, Richard Swann, *Organic Evolution* (New York: Macmillan, 1929).

Mantell, Gideon Algernon, *The Medals of Creation, or, First Lessons in Geology, and the Study of Organic Remains*, 2nd ed. (London: Henry G. Bohn, 1853).

Reed, W. Maxwell, *The Earth for Sam: The Story of Mountains, Rivers, Dinosaurs and Men* (New York: Harcourt, Brace and Co., 1930).

Rudwick, Martin J. S., *The Meaning of Fossils: Episodes in the History of Paleontology* (New York: Neale Watson Academic Publications, Inc., 1972). Recently reissued by University of Chicago Press.

———, *The Great Devonian Controversy: The Shaping of Scientific Knowledge Among Gentlemanly Specialists* (Chicago and London: University of Chicago Press, 1985). A thorough and perceptive analysis of the geology of the 1830s in England (a decade when many dinosaurs were discovered), this book also includes a comprehensive bibliography.

Scott, William Berryman, *A History of Land Mammals in the Western Hemisphere* (New York: Macmillan, 1913).

Shelton, John S., *Geology Illustrated* (San Francisco and London: W. H. Freeman and Co., 1966).

Swinnerton, H. H., *Fossils* (London and Glasgow: Collins Clear-Type Press, 1960).

Thorpe, Malcom Rutherford, ed., *Organic Adaptation to Environment* (New Haven: Yale University Press, 1924).

Zim, Herbert S., *Alligators and Crocodiles* (New York: William Morrow and Co., 1952).

———, *Dinosaurs* (New York: William Morrow and Co., 1954).

3. MESOZOIC CLASS WARFARE: COLD-BLOODS VS. THE FABULOUS FURBALLS

Carr, Archie, and editors of *Life, The Reptiles* (New York: Time, Inc., 1963).

Ditmars, Raymond L., *The Reptiles of North America* (Garden City, N.Y.: Doubleday, 1953).

Grubb, D., "The growth, ecology and population structure of giant tortoises on Aldabra," *Phil. Trans. Royal Soc. London* (B) 260 (1971): 327–372.

Herzog, H. A., "An observation of nest opening by an American Alligator, *Alligator mississippiensis*," *Herpetolog.* 31 (1975): 446–447.

Hunt, R. H., "Maternal behavior in the Morelet's Crocodile, *Crocodylus moreleti*," *Copeia* (1975): 763–764.

McIlhenny, E. A., *The Alligator's Life History* (Boston: Christopher's Publishing House, 1935).

Minton, Sherman A., Jr., and Madge Rutherford Minton, *Giant Reptiles* (New York: Charles Scribner's Sons, 1973).

Noble, G. K., "Contributions to the Herpetology of the Belgian Congo Based on the Collection of the American Museum Congo Expeditions," *Bull. American Museum of Natural History* 49 (1924): 147–348.

Owen, R., "Description of remains of *Megalania prisca*, Part II," *Phil. Trans. Royal Soc.* 171 (1880): 1037–1050.

Pope, Clifford H., *The Reptile World* (New York: Alfred A. Knopf, 1955).

Porter, Kenneth R., *Herpetology* (Philadelphia, London, and Toronto: W. B. Saunders Co., 1972).

Romer, Alfred S., *Osteology of the Reptiles* (Chicago: University of Chicago Press, 1956).

Romer, Alfred Sherwood and Thomas S. Parsons, *The Vertebrate Body*, 5th ed. (Philadelphia, London, and Toronto: W. B. Saunders Co., 1977).

Schmidt, K. P., "Contributions to the Herpetology of the Belgian Congo Based on the Collection of the American Museum Congo Expeditions, I. Turtles, Crocodiles, Lizards and Chameleons," *Bull. Amer. Mus. Nat. His.* 39 (1919): 385–624.

———, "Contributions to the Herpetology of the Belgian Congo Based on the Collection of the American Museum Congo Expeditions," *Bull. Amer. Mus. Nat. His.* 49 (1923): 1–147.

Young, J. Z., *The Life of Vertebrates*, 3rd ed. (Oxford: Clarendon Press, 1981).

4. DINOSAURS SCORE WHERE KOMODO DRAGONS FAIL
Notes:

There is a confusing plexus of terms for bioenergetics and heat regulation. *Homeotherm* means "constant temperature"; *poikilotherm* means "varying temperature"; *heliotherm* means "using the sun to heat up"; *ectotherm* means "relying on outside heat sources"; *endotherm* means "relying on internal body heat"; *automatic endotherm* means "having a very high and constant basal metabolism that supplies sufficient heat for maintaining a high, constant body temperature"; *exercise endothermy* means "using the heat of exercise to warm up the body"; *non-shivering thermogenesis* (NSTG) means "extra heat produced, without shivering, to keep body temperature constant in an endothermic homeotherm." When I use the term "warm-blooded," I'm using it in the nineteenth-century sense—the temperature-regulation style of most birds and advanced mammals today, a style that employs automatic endothermy, and NSTG, and shivering, and usually some special cooling mechanism, like sweating or panting. No typical, living reptile is an automatic endotherm, nor does any living reptile species have much of a NSTG capacity. Some primitive mammals—such as hedgehogs and sloths—have much lower basal metabolism than typical mammals and birds, but even these primitive mammals can increase heat production enormously with NSTG, something no typical reptile can do. Since 1968, I have been persuaded that dinosaurs, all dinosaurs, were automatic endotherms with a high NSTG capacity—in other words, dinosaurs were equivalent to advanced birds and mammals.

Bakker, Robert T., "Dinosaur Physiology and the Origin of Mammals," *Evolution* 25 (1971): 636–658.

———, "Locomotor Energetics of Lizards and Mammals Compared," *Physiologist* 15 (1972): 278.

———, "Dinosaur Bio-energetics—a Reply to Bennett and Dalzell, and Feduccia," *Evolution* 28 (1974): 497–503.

———, "Experimental and Fossil Evidence for the Evolution of Tetrapod Bioenergetics," In D. M. Gates and R. B. Schmerl, eds., *Perspectives of Biophysical Ecology* (New York: Springer-Verlag, 1975), pp. 365–399.

———, "Dinosaur Renaissance," *Scientific American* 232 (1975): 58–78.

Bennett, A. F. and W. R. Dawson, "Metabolism," In C. Gans, ed., *Biology of the Reptilia* 5 (London: Academic Press, 1976).

Bonaparte, Jose F., "*Pisanosaurus mertii* Casamiquela and the origin of the Ornithischia," *J. Paleont.* 50 (1976): 808–820.

Calder, W. A. and J. R. King, "Thermal and Caloric Relations of Birds," in D. Farner and J. King, eds., *Avian Biology* Vol. IV (New York and London: Academic Press, 1974), pp. 260–293.

Charig, A. J., "Dinosaur Monophyly and a New Class of Vertebrates: A Critical Review," In A. d'A. Bellairs and C. B. Cox, eds., *Morphology and Biology of Reptiles* (London: Academic Press, 1976).

Cooper, M. R., "The Prosauropod ankle and dinosaur phylogeny," *S. Afr. J. Sci.* 76 (1980): 176–178.

Dawson, T. J., " 'Primitive' Mammals," in G. C. Whittow, ed., *Comparative Physiology of Thermoregulation*, Vol. III (New York: Academic Press, 1973), pp. 1–46.

Dawson, W. R. and J. W. Hudson, "Birds," in G. C. Whittow, ed., *Comparative Physiology of Thermoregulation*, Vol. 1 (London and New York: Academic Press, 1970), pp. 223–310.

Jansky, L., "Non-shivering Thermogenesis and Its Thermoregulatory Significance," *Biol. Rev.* 48 (1973): 85–132.

Langston, Wann., "Ziphodont Crocodiles," *Fieldiana, Geology* 33 (1975): 291–314.

Lull, R. S., "Dinosaurian Climatic Response," in M. R. Thorpe, ed., *Organic Adaptation to Environment* (New Haven: Yale University Press, 1924), pp. 225–279.

McNab, B. K., "The Energetics of Endotherms," *Ohio Journal of Science* 74 (1974): 370–380.

———, "The Evolution of Endothermy in the Phylogeny of Mammals," *American Naturalist* 112 (1978): 1–21.

Santa Luca, A. P., A. W. Crompton, and A. J. Charig, "A Complete Skeleton of the Late Triassic Ornithischian *Heterodontosaurus tucki*," *Nature* 264 (1976): 324–327.

Schmidt-Nielsen, Knut, *How Animals Work* (Cambridge, London, and New York: Cambridge University Press, 1972).

Spotila, J. R., P. W. Lommen, G. S. Bakken, and D. M. Gates, "A Mathematical Model for Body Temperatures of Large Reptiles: Implications for Dinosaur Ecology," *American Naturalist* 107 (1973): 391–404.

Thomas, Roger D. K. and Everett C. Olson, eds., *A Cold Look at the Warm-Blooded Dinosaurs*, AAAS Selected Symposium 28 (Boulder, Colo.: Westview Press, 1980).

5. **THE CASE OF THE BRONTOSAURUS: FINDING THE BODY**

Behrensmeyer, Anna K. and Andrew P. Hill, *Fossils in the Making: Vertebrate Taphonomy and Paleoecology* (Chicago and London: University of Chicago Press, 1980).

Dodson, Peter, A. K. Behrensmeyer, Robert T. Bakker, and John S. McIntosh, "Taphonomy and Paleoecology of the Dinosaur Beds of the Jurassic Morrison Formation," *Paleobiology* 6 (2) 1980: 208–232.

Efremov, E. A., "Taphonomy and the Geological Record," *Acad. Sci. USSR Publ. Paleont. Int.* 24 (1950): 3–176.

6. **GIZZARD STONES AND BRONTOSAUR MENUS**

Beebe, C. William, *The Bird: Its Form and Function* (New York: Henry Holt and Co., 1906).

Buick, T. L., *The Mystery of the Moa.* (New Plymouth, New Zealand: Thomas Avery and Sons, 1931).

7. **THE CASE OF THE DUCKBILL'S HAND**

Dodson, Peter, A. K. Behrensmeyer, Robert T. Bakker, and John S. McIntosh, "Taphonomy and Paleoecology of the Dinosaur Beds of the Jurassic Morrison Formation," *Paleobiology* 6 (2) 1980: 208–232.

Lull, Richard Swann and N. E. Wright, "Hadrosaurian Dinosaurs of North America," *Geological Society of America Special Paper* 40 (1942): 1–242.

Osborn, Henry Fairfield, "Fossil Wonders of the West," *Century Magazine* 68 (1904): 680–694.

Riggs, Elmer S., "Structure and Relationships of Opisthocoelian Dinosaurs Part II: The Brachiosauridae," *Field Columbian Museum Publication* 94, Geological Series II, 6 (1904): 229–247.

9. WHEN DINOSAURS INVENTED FLOWERS

Axelrod, D. I., "Mesozoic Paleogeography and Early Angiosperm History," *Bot. Rev.* 36 (1970): 277–319.

———, "Edaphic Aridity as a factor in Angiosperm Evolution," *American Naturalist* 106 (1972): 311–320.

Bakker, Robert T., "Dinosaur feeding behaviour and the origin of flowering plants," *Nature* London 274 (17 August 1978): 661–663.

Doyle, J. A. and L. J. Hickey, "Coordinated Evolution in Potomac Group Angiosperm Pollen and Leaves," *Am. J. Botany* 59 (6, pt. 2) 1972: 660.

Hickey, L. J. and Doyle, J. A., "Fossil Evidence on Evolution of Angiosperm Leaf Venation," *Am. J. Botany* 59 (6, pt. 2) 1972: 661.

Scott, Richard A., Elso S. Barghoorn, and Estella Leopold, "How Old are the Angiosperms," *American Journal of Science* (1960): 258-A, 284–299.

Seward, A. C., *Plant Life Through the Ages: A Geological and Botanical Retrospect* (New York: Macmillan; and Cambridge, England: Cambridge University Press, 1931).

Stebbins, G. Ledyard, "The Probable Growth Habit of the Earliest Flowering Plants," *Ann. Missouri Bot. Garden* 52 (1965): 457–468.

———, *Flowering Plants: Evolution Above the Species Level* (Cambridge: Harvard University Press, 1974).

10. THE TEUTONIC DIPLODOCUS: A LESSON IN GAIT AND CARRIAGE

Alexander, R. McN., "Estimates of speeds of Dinosaurs," *Nature* 261 (1976): 129–130.

Coombs, W. P., Jr., "Theoretical Aspects of Cursorial Adaptation in Dinosaurs," *Quarterly Review of Biology* 53 (1978): 393–418.

Farlow, J. O., "Estimates of Dinosaur Speeds from a New Trackway Site in Texas," *Nature* 294 (1981): 747–748.

Gambaryan, P. P., *How Mammals Run: Anatomical Adaptations* (Jerusalem and London: Israel Program for Scientific Translations; New York and Toronto: John Wiley and Sons, Halsted Press Division, 1974).

Guggisberg, C.A.W., *S.O.S. Rhino* (New York: October House, 1966).

Halstead, Murat, *Full Official History of the War with Spain* (New Haven: Butler and Alger, 1899).

Hanks, John, *The Struggle for Survival: The Elephant Problem* (New York: Mayflower Books, 1979).

Haubold, H., "Ichnia Amphibiorum et Reptilium fossilium," *Handbook of Palaeoherpetology* 18 (1971): 1–124.

———, *Die Fossilen Saurierfahrten* (Wittenburg: A. Ziensen Verlag, 1974).

Lull, Richard Swann, *A Revision of the Ceratopsia or Horned Dinosaurs* (New Haven, Conn.: Memoirs of the Peabody Museum of Natural History, Volume III, Part 3, 1933).

McGinnis, Helen J., *Carnegie's Dinosaurs* (Pittsburgh: Carnegie Institute, 1982).

Muybridge, Edweard, *Animals in Motion* (New York: Dover Books, 1957).

Romer, A. S., "The Pelvic Musculature of Saurischian Dinosaurs," *Bull. American Museum of Natural History* 48 (1923): 605–617.

Sternberg, C. M., "Dinosaur Tracks from Peace River, British Columbia," *National Museum of Canada Bulletin* 68 (1921): 59–85.

Thompson, D'Arcy Wentworth, *On Growth and Form* (Cambridge: Cambridge University Press, 1961).

11. MESOZOIC ARMS RACE

Notes:

Stegosaurus actually contains two quite different genera, as the term is used today. Marsh named *Stegosaurus* for species with very long legs, rather small, squat triangular plates, and four to eight tail spikes. Marsh named *Diracodon* for stegosaur species with relatively gigantic and very narrow-based triangular plates, four tail spikes, and limbs less elongated than in true *Stegosaurus*. The National Museum in Washington, D.C., has a mounted *Diracodon;* Yale has an eight-spike *Stegosaurus;* the American Museum in New York and the Carnegie Museum in Pittsburgh have four-spiked *Stegosaurus.*

The obturator prong traditionally is viewed as a new adaptation within certain beaked clans, but I disagree. Primitive archosaurs have very wide, platelike ischial bone shafts, with a perforation near the hip socket. I view the obturator prong as a remnant of the old ischiadic plate, cut away from in front by expansion of the perforation and from behind by a general slimming of the shaft. *Deinonychus* has an obturator prong and a beaked-dinosaur-type hip design, and in *Deinonychus* the prong is clearly a remnant of the ischiadic plate.

Lull, Richard Swann, "A Revision of the Ceratopsia, or Horned Dinosaurs," *Memoirs of the Peabody Museum National Institute* 3 (3) 1933: 1–135.

Nopsca, F. von, *Osteologia reptilium fossilium et recentium,* "Fossilium catalogus," Pars 27, Berlin (1926); also *Supplement,* "Fossilium catalogus," Pars 50 (1931). Annotated bibliography of important papers dealing with fossil reptiles and amphibians.

14. ARCHAEOPTERYX PATERNITY SUIT: THE DINOSAUR-BIRD CONNECTION

Beebe, C. William, *The Bird: Its Form and Function* (New York: Henry Holt and Co., 1906).

———, "Ecology of the Hoatzin," *Zoologica* I (2) 1909: 45–66.

———, G. Inness Hartley and Paul G. Howes, *Tropical Wild Life in British Guiana* (New York: The New York Zoological Society, 1917).

Feduccia, A. and H. B. Tordoff, "Feathers of Archaeopteryx: Assymetric Vanes Indicate Aerodynamic Function," *Science* 203 (1979): 1021–1022.

Heilmann, Gerhard, *The Origin of Birds* (New York: Appleton and Co., 1927).

Huxley, Thomas Henry, "On the Animals Which Are Most Nearly Intermediate Between Birds and Reptiles," *Annals and Magazine of Natural History* 4, February 1868.

———, "Remarks upon *Archaeopteryx lithographica,*" *Proceedings of the Royal Society* 16 (1868).

———, "Further Evidence of the Affinity Between the Dinosaurian Reptiles and Birds," *Quarterly Journal of the Geological Society of London* 26 (1870).

Marsh, Othniel Charles, "*Odontornithes, a monograph of the extinct toothed birds of North America,*" *Rept. Geol. Explor. 40th Parallel* 7 (1880): 1–201.

Ostrom, John H., "Osteology of *Deinonychus antirrhopus,* an unusual theropod from the Lower Cretaceous of Montana," *Bull. Peabody Mus. Natural Hist.,* 30, 165 S., 83 Abb., 13 Tab., (1969), New Haven.

———, "*Archaeopteryx* and the Origin of Birds," *Biological Journal of the Linnean Society* 8 (1976): 91–182.

Strahl, Stuart D., "A Bird Stranger than Fiction," *Animal Kingdom* 87 (5) 1984: 14–19.

Von Meyer, Hermann, "On the *Archaeopteryx lithographica,* from the Lithographic Slate of Solenhofen," *Annals and Magazine of Natural History* 9 (April 1862).

16. THE WARM-BLOODED TEMPO OF THE DINOSAURS' GROWTH

Bakker, Robert T., "Experimental and Fossil Evidence for the Evolution of Tetrapod Energetics," in D. Gates and R. Schmerl, eds., *Perspectives in Biophysical Ecology* (New York: Springer-Verlag, 1975), pp. 365–399.

———, "Dinosaur Heresy—Dinosaur Renaissance," in Roger D. K. Thomas and Everett C. Olson, eds., *A Cold Look at the Warm-Blooded Dinosaurs* (Boulder, Colo.: Westview Press, 1980).

Enlow, D. H. and S. O. Brown; "A Comparative Histological Study of Fossil and Recent Bone Tissues, Part II," *Texas J. Sci.* 9 (1957): 186–214.

——, "A Comparative Histological Study of Fossil and Recent Bone Tissues, Part III," *Texas J. Sci.* 10 (1958): 405–443.

Horner, John R., "Coming Home to Roost," *Montana Outdoors* 13 (1982).

——, "Evidence of Colonial Nesting and 'Site Fidelity' Among Ornithischian Dinosaurs," *Nature* 297 (1982).

——, "Cranial Osteology and Morphology of the Type Specimen of *Maiasaura peeblesorum* (Ornithischia Hadrosauridae) with discussion of its Phylogenetic Position," *J. Vert. Paleo.* 3 (1983).

——, "The Nesting Behavior of Dinosaurs," *Scientific American* 250 (1984).

Johnston, P. A., "Growth Rings in Dinosaur Teeth," *Nature* 278 (1979): 635–636.

Peabody, F. E., "Annual Growth Zones in Living and Fossil Vertebrates," *J. Morphology* 108 (1961): 11–62.

Pekelharing, C. J., "Cementum Deposition as an Age Indicator in the Brush-Tailed Possum, *Trichosurus vulpecula* Kerr (Marsupiala)," *Aust. J. Zool.*, 18 (1970): 51–80.

Ricqles, Armand de, "Evolution of Endothermy: Histological Evidence," *Evolutionary Theory* 1 (1974): 51–80.

——, "Tissue Structures of Dinosaur Bone: Functional Significance and Possible Relation to Dinosaur Physiology," in Roger D. K. Thomas and Everett C. Olson, eds., *A Cold Look at the Warm-Blooded Dinosaurs* (Boulder, Colo.: Westview Press, 1980), pp. 103–139. See also the list of papers by Ricqles in the Bibliography of this volume.

17. STRONG HEARTS, STOUT LUNGS, AND BIG BRAINS

Hopson, J., "Relative Brain Size in Dinosaurs," In Roger D. K. Thomas and Everett C. Olson, eds., *A Cold Look at the Warm-Blooded Dinosaurs* (Boulder, Colo.: Westview Press, 1980).

Jerison, H. J., *Evolution of the Brain and Intelligence* (New York: Academic Press, 1973).

Russell, Dale. "A New Specimen of *Stenonychosaurus* from the Oldman Formation (Cretaceous) of Alberta," *Canadian J. Earth Sci.* 6 (1969): 595–612.

18. EATERS AND EATEN AS THE TEST OF WARM-BLOODEDNESS

Auffenberg, W., "A Day with No. 19—Report on a study of the Komodo Monitor," *Animal Kingdom* 6 (1970): 18–23.

Bennett, A. F. and B. Dalzell, "Dinosaur Physiology: A Critique," *Evolution* 27 (1973): 170–174.

Edgar, W. D., "Aspects of the ecological energetics of the wolf spider *Pardosa* (Lycosa) *lugubris* (Walekenaer)," *Oecologia* (B) 7 (1971): 136–154.

Golley, F. B., "Secondary productivity in terrestrial communities," *Am. Zool.* 8 (1968): 53–59.

Moulder, B. C. and D. E. Reichle, "Significance of spider predation in the energy dynamics of forest-floor arthropod communities," *Ecol. Monogr.* 42 (1972): 473–498.

Olson, E. C., "Community evolution and the origin of mammals," *Ecology* 47 (1966): 291–308.

Sinclair, A.R.E. and M. Norton-Griffiths, eds., *Serengeti: Dynamics of an Ecosystem* (Chicago and London: University of Chicago Press, 1979).

Varley, G. C., "The Concept of energy flow applied to a woodland community," in A. Watson, ed., *Animal Populations in Relation to Their Food Resources* (Oxford: Blackwell, 1970), pp. 389–406.

19. PUNCTUATED EQUILIBRIUM: THE EVOLUTIONARY TIMEKEEPER

Eldredge, Niles and Stephen Jay Gould, "Punctuated Equilibria: an alternative to phyletic gradualism," in Schopf, T.J.M., ed., *Models in Paleobiology* (San Francisco: Freeman, Cooper, 1972), pp. 82–115.

Gould, Stephen Jay, *Ever Since Darwin* (New York: W. W. Norton, 1977).
———, *Hen's Teeth and Horse's Toes* (New York and London: W. W. Norton and Co., 1983).
———, *The Flamingo's Smile* (New York: W. W. Norton, 1985).
——— and Niles Eldredge, "Punctuated Equilibria: the Tempo and Mode of Evolution Reconsidered," *Paleobiology* 3 (1977) 115–151.
Stanley, Steven M., *Macroevolution: Pattern and Process* (San Francisco: W. H. Freeman and Co., 1979).

20. THE KAZANIAN REVOLUTION: SETTING THE STAGE FOR THE DINOSAURIA

Brink, A. S., "Speculations on some advanced Mammalian characteristics in the higher Mammal-like Reptiles," *Paleontologica Africana* IV (1956): 77–97.
Frakes, Lawrence, Elizabeth M. Kemp and John C. Crowell, "Late Paleozoic Glaciation: Part VI, Asia," *Bul. Geol. Soc. Amer.* 86 (1975): 454–464.
Romer, A. S., "The locomotor apparatus of certain primitive and mammal-like reptiles," *Bull. American Museum of Natural History* 46 (1922): 517–606.

21. THE TWILIGHT OF THE DINOSAURS

Notes:

It's absolutely crucial to get straight what is being debated when the term "dinosaur diversity" is used in arguments about the Cretaceous extinction. Since 1977 I have argued that the *evenness* of *large* dinosaur species suffered major declines long before the very end of the Period. Dale Russell argues that the *total* number of all dinosaur species, large and small, didn't change much until the very end. I believe that there is no question about an evenness decline for the one-ton-plus size category; *Triceratops* simply overwhelms all other genera for the last two million years or so in nearly every habitat. And arguments about total number of small dinosaur species are really premature—we don't have nearly enough skulls and skeletons and a census based on shed teeth will be biased toward predators, because, for some reason, meat-eaters are easier to tell apart from shed teeth than are herbivores.

Alvarez, Walter, Frank Asaro, Helen V. Michel and Luis Alvarez, "Evidence for a Major Meteorite Impact on the Earth 34 Million Years Ago: Implications for Eocene Extinctions," *Science* 216 (1982).
Bakker, R. T., "Tetrapod Mass Extinctions—A Model of the Regulation of Speciation Rates and Immigration by Cycles of Topographic Diversity," in A. Hallam, ed., *Patterns of Evolution* (Amsterdam: Elsevier Scientific Publishing Co., 1977), pp. 439–468.
Crosby, Alfred W., *The Columbian Exchange: Biological and Social Consequences of 1492* (Westport, Conn.: Greenwood Press, 1972).
Elton, Charles S., *The Ecology of Invasions by Animals and Plants* (London: Chapman and Hall, 1972). First published by Methuen and Co. Ltd., 1958.
Ganapathy, R., "A Major Meteorite Impact on the Earth 65 Million Years Ago: Evidence from the Cretaceous-Tertiary Boundary Clay," *Science* 209 (1980).
Gould, Stephen Jay, "The Belt of an Asteroid," *Natural History* 89 (1980).
Gottfried, Robert S., *The Black Death: Natural and Human Disaster in Medieval Europe* (New York: The Free Press, 1983).
Krassilov, V. A., "Climatic Changes in Eastern Asia as Indicated by Fossil Floras. I. Early Cretaceous," *Paleog. Paleocl. Paleoec.* 13 (1973): 261–274.
MacArthur, Robert H., *Geographical Ecology: Patterns in the Distribution of Species* (New York, Evanston, San Francisco and London: Harper and Row, Publishers, 1972).
McNeill, William H., *Plagues and Peoples* (Garden City: Anchor Books, Doubleday, 1976).
Osborn, Henry Fairfield, *The Age of Mammals in Europe, Asia and North America* (New York: Macmillan, 1910).
Russell, Dale A., "The Gradual Decline of the Dinosaurs—Fact or Fallacy?," *Nature* 307 (26) January 1984: 360–361.

Simpson, George Gaylord, *The Geography of Evolution* (Philadelphia and New York: Chilton Books, 1965).

Zinsser, Hans, *Rats, Lice and History* (Boston: Little, Brown and Co., 1934).

22. DINOSAURS HAVE CLASS
Notes:

There's a tale to tell about the famous clawed hand of the *Heterodontosaurus* found by Fuzz Crompton. Pete Galton and I published a diagram in *Nature* showing the sharp twist to the thumb claw, and, indeed, the specimen showed this twist clearly then. But the hand bones were broken by accident later, and glued back together incorrectly so the first finger joint lost its twist (you can see the mismatched glued ends in the specimen and in casts). Some scholars have been misled into thinking that there was no twist in this animal.

Most books about dinosaurs use the term "Ornithischia" for the beaked dinosaurs, but I prefer Marsh's "Predentata" because it's much more precise—it refers to the unique predentary bone that forms the beak-core in the lower jaw. "Ornithischia"—"Bird Hips"—is less precise because some nonbeaked dinosaurs evolved bird hips too, *Deinonychus* being an excellent example.

And most books give a rigid, formal hierarchy to dinosaur family trees; the Orders Ornithischia and Saurischia being broken down into suborders and infraorders. I don't have a complete alternative hierarchy yet, but some suggestions can be made. The Archosauria should be a Class; the Dinosauria would be a Subclass; the Phytodinosauria and Theropoda would be Infraclasses, with the birds a Superorder within the Theropoda, and the Sauropoda and Predentata Superorders within the Phytodinosauria. It's hard to place the anchisaurs (traditionally called "prosauropods"); they may be closer to either predentates or sauropods. The nondinosaur archosaurs are hard to subdivide cleanly right now; I would put the ugly beaked rhynchosaurs into the Archosauria, tentatively, even though they had "cold-blooded" bone microtexture. Archosaurs clearly are related to the other "diapsid" reptiles (those having two temporal fenestrae and another large fenestra in the palate, below the eyes); diapsids include lizards and snakes and other lesser groups (maybe turtles). The Diapsida should be a high-rank category—a Superclass?

Another Superclass is needed for the mammals, protomammals, and their uncles, the finback clan (pelycosaurs). The name "Theropsid" ("mammal-face") could be used. Protomammals and mammals, the warm-blooded theropsids, could be the Class Neotheropsida ("newer mammal-faces"). Thus the Neotheropsida and Archosauria are the two Classes defined by crossing the threshold into warm-bloodedness. What about the Class Reptilia? I advocate abandonment. Use "reptile" in the lower case only, as a loose term for non-warm-blooded, nonamphibian tetrapods.

Index

Illustrations are indicated by page numbers in *italics*.

A NOTE ABOUT THE AUTHOR

Robert T. Bakker started his heretical dinosaur studies as a Yale undergraduate in the 1960s and continued at Harvard Graduate School where he was elected to the elite Society of Fellows. He has taught every level from kindergarten to advanced graduate seminars and tells the story of the dinosaur revolution in a lively style without scientific jargon. A skilled artist, Professor Bakker illustrated the book with his own reconstruction of dinosaurs alive and in action. He is adjunct curator at the University Museum, the University of Colorado. His wife and co-excavator of bones and ideas, Constance Clark, edits books on African history and South America for the Westview Press in Boulder.